Behavioural Types:
from Theory to Tools

RIVER PUBLISHERS SERIES IN AUTOMATION, CONTROL AND ROBOTICS

Series Editors

SRIKANTA PATNAIK
SOA University
Bhubaneswar
India

ISHWAR K. SETHI
Oakland University
USA

QUAN MIN ZHU
University of the West of
England
UK

Indexing: All books published in this series are submitted to Thomson Reuters Book Citation Index (BkCI), CrossRef and to Google Scholar.

The "River Publishers Series in Automation, Control and Robotics" is a series of comprehensive academic and professional books which focus on the theory and applications of automation, control and robotics. The series focuses on topics ranging from the theory and use of control systems, automation engineering, robotics and intelligent machines.

Books published in the series include research monographs, edited volumes, handbooks and textbooks. The books provide professionals, researchers, educators, and advanced students in the field with an invaluable insight into the latest research and developments.

Topics covered in the series include, but are by no means restricted to the following:

- Robots and Intelligent Machines
- Robotics
- Control Systems
- Control Theory
- Automation Engineering

For a list of other books in this series, visit www.riverpublishers.com

Behavioural Types:
from Theory to Tools

Editors

Simon Gay
University of Glasgow, UK

António Ravara
Universidade Nova de Lisboa, Portugal

Published, sold and distributed by:
River Publishers
Alsbjergvej 10
9260 Gistrup
Denmark

River Publishers
Lange Geer 44
2611 PW Delft
The Netherlands

Tel.: +45369953197
www.riverpublishers.com

ISBN: 978-87-93519-82-4 (Hardback)
978-87-93519-81-7 (Ebook)

Contents

2 Contract-Oriented Programming with Timed Session Types 27

Nicola Atzei, Massimo Bartoletti, Tiziana Cimoli,
Stefano Lande, Maurizio Murgia, Alessandro Sebastian Podda
and Livio Pompianu

3 A Runtime Monitoring Tool for Actor-Based Systems 49

Duncan Paul Attard, Ian Cassar, Adrian Francalanza, Luca Aceto
and Anna Ingólfsdóttir

**5 The DCR Workbench: Declarative Choreographies
for Collaborative Processes   99**

Søren Debois and Thomas T. Hildebrandt

**6 A Tool for Choreography-Based Analysis
of Message-Passing Software   125**

Julien Lange, Emilio Tuosto and Nobuko Yoshida

15 Protocol-Driven MPI Program Generation   329

Nicholas Ng and Nobuko Yoshida

Preface

This book presents research produced by members of COST Action IC1201: Behavioural Types for Reliable Large-Scale Software Systems (BETTY), a European research network that was funded from October 2012 to October 2016. The technical theme of BETTY was the use of behavioural type systems in programming languages, to specify and verify properties of programs beyond the traditional use of type systems to describe data processing. A significant area within behavioural types is session types, which concerns the use of type-theoretic techniques to describe communication protocols so that static typechecking or dynamic monitoring can verify that protocols are implemented correctly. This is closely related to the topic of choreography, in which system design starts from a description of the overall communication flows. Another area is behavioural contracts, which describe the obligations of interacting agents in a way that enables blame to be attributed to the agent responsible for failed interaction. Type-theoretic techniques can also be used to analyse potential deadlocks due to cyclic dependencies between inter-process interactions.

BETTY was organised into four Working Groups: (1) Foundations; (2) Security; (3) Programming Languages; (4) Tools and Applications. Working Groups 1–3 produced "state-of-the-art reports", which originally intended to take snapshots of the field at the time the network started, but grew into substantial survey articles including much research carried out during the network [1–3]. The situation for Working Group 4 was different. When the network started, the community had produced relatively few implementations of programming languages or tools. One of the aims of the network was to encourage more implementation work, and this was a great success. The community as a whole has developed a greater interest in putting theoretical ideas into practice. The sixteen chapters in this book describe systems that were either completely developed, or substantially extended, during BETTY. The total of 41 co-authors represents a significant proportion of the active participants in the network (around 120 people who attended at least one meeting). The book is a report on the new state of the art created by BETTY in

xv

the area of Working Group 4, and the title "Behavioural Types: from Theory to Tools" summarises the trajectory of the community during the last four years.

The book begins with two tutorials by Atzei *et al.* on contract-oriented design of distributed systems. Chapter 1 introduces the CO_2 contract specification language and the Diogenes toolchain. Chapter 2 describes how timing constraints can be incorporated into the framework and checked with the CO_2 middleware.

Part of the CO_2 middleware is a monitoring system, and the theme of monitoring continues in the next two chapters. In Chapter 3, Attard *et al.* present detectEr, a runtime monitoring tool for Erlang programs that allows correctness properties to be expressed in Hennessy-Milner logic. In Chapter 4, which is the first chapter about session types, Neykova and Yoshida describe a runtime verification framework for Python programs. Communication protocols are specified in the Scribble language, which is based on multiparty session types.

The next three chapters deal with choreographic programming. In Chapter 5, Debois and Hildebrandt present a toolset for working with dynamic condition response (DCR) graphs, which are a graphical formalism for choreography. Chapter 6, by Lange *et al.*, continues the graphical theme with ChorGram, a tool for synthesising global graphical choreographies from collections of communicating finite-state automata. Giallorenzo *et al.*, in Chapter 7, consider runtime adaptation. They describe AIOCJ, a choreographic programming language in which runtime adaptation is supported with a guarantee that it doesn't introduce deadlocks or races.

Deadlock analysis is important in other settings too, and there are two more chapters about it. In Chapter 8, Padovani describes the Hypha tool, which uses a type-based approach to check deadlock-freedom and lock-freedom of systems modelled in a form of pi-calculus. In Chapter 9, Garcia and Laneve present a tool for analysing deadlocks in Java programs; this tool, called JaDA, is based on a behavioural type system.

The next three chapters report on projects that have added session types to functional programming languages in order to support typechecking of communication-based code. In Chapter 10, Orchard and Yoshida describe an implementation of session types in Haskell, and survey several approaches to typechecking the linearity conditions required for safe session implementation. In Chapter 11, Melgratti and Padovani describe an implementation of session types in OCaml. Their system uses runtime linearity checking. In Chapter 12, Lindley and Morris describe an extension of the web programming language Links with session types; their work contrasts with the previous two chapters in being less constrained by an existing language design.

Continuing the theme of session types in programming languages, the next two chapters describe two approaches based on Java. Hu's work, presented in Chapter 13, starts with the Scribble description of a multiparty session type and generates an API in the form of a collection of Java classes, each class containing the communication methods that are available in a particular state of the protocol. Dardha *et al.*, in Chapter 14, also start with a Scribble specification. Their StMungo tool generates an API as a single class with an associated typestate specification to constrain sequences of method calls. Code that uses the API can be checked for correctness with the Mungo typechecker.

Finally, there are two chapters about programming with the MPI libraries. Chapter 15, by Ng and Yoshida, uses an extension of Scribble, called Pabble, to describe protocols that parametric in the number of runtime roles. From a Pabble specification they generate C code that uses MPI for communication and is guaranteed correct by construction. Chapter 16, by Ng *et al.*, describes the ParTypes framework for analysing existing C+MPI programs with respect to protocols defined in an extension of Scribble.

We hope that the book will serve a useful purpose as a report on the activities of COST Action IC1201 and as a survey of programming languages and tools based on behavioural types.

Simon Gay
Chair, COST Action IC1201

António Ravara
Vice-Chair, COST Action IC1201

Acknowledgments

This publication is supported by COST.

COST (European Cooperation in Science and Technology) is a pan-European intergovernmental framework. Its mission is to enable break-through scientific and technological developments leading to new concepts and products and thereby contribute to strengthening Europe's research and innovation capacities. www.cost.eu

This publication is based upon work from COST Action IC1201, Behavioural Types for Reliable Large-Scale Software Systems, supported by COST (European Cooperation in Science and Technology).

EUROPEAN COOPERATION
IN SCIENCE AND TECHNOLOGY

COST is supported
by the EU Framework
Programme Horizon 2020

List of Contributors

Abel Garcia, *Department of Computer Science and Engineering, University of Bologna – INRIA FOCUS, Mura Anteo Zamboni 7, 40127, Bologna, Italy*

Adrian Francalanza, *Department of Computer Science, Faculty of ICT, University of Malta, Malta*

Alessandro Sebastian Podda, *University of Cagliari, Italy*

Anna Ingólfsdóttir, *School of Computer Science, Reykjavík University, Iceland*

Cosimo Laneve, *Department of Computer Science and Engineering, University of Bologna – INRIA FOCUS, Mura Anteo Zamboni 7, 40127, Bologna, Italy*

Dimitrios Kouzapas, *School of Computing Science, University of Glasgow, UK*

Dominic Orchard, *University of Kent, UK*

Duncan Paul Attard, *Department of Computer Science, Faculty of ICT, University of Malta, Malta*

Eduardo R. B. Marques, *CRACS/INESC-TEC, Faculty of Sciences, University of Porto, PT*

Emilio Tuosto, *University of Leicester, UK*

Florian Weber, *School of Computing Science, University of Glasgow, UK*

Francisco Martins, *LaSIGE, Faculty of Sciences, University of Lisbon, PT*

Garrett Morris, *University of Edinburgh, Edinburgh, UK*

Hernàn Melgratti, *Departamento de Computación, Universidad de Buenos Aires, Argentina* and *CONICET-Universidad de Buenos Aires, Instituto de Investigación en Ciencias de la Computación (ICC), Buenos Aires, Argentina*

Ian Cassar, *Department of Computer Science, Faculty of ICT, University of Malta, Malta*

Ivan Lanese, *Focus Team, University of Bologna/INRIA, Italy*

Jacopo Mauro, *Department of Informatics, University of Oslo, Norway*

Julien Lange, *Imperial College London, UK*

Laura Voinea, *School of Computing Science, University of Glasgow, UK*

Livio Pompianu, *University of Cagliari, Italy*

Luca Aceto, *School of Computer Science, Reykjavík University, Iceland*

Luca Padovani, *Dipartimento di Informatica, Università di Torino, Italy*

Massimo Bartoletti, *Università degli Studi di Cagliari, Italy*

Maurizio Gabbrielli, *Focus Team, University of Bologna/INRIA, Italy*

Maurizio Murgia, *Università degli Studi di Cagliari, Italy*

Nicholas Ng, *Imperial College London, UK*

Nicola Atzei, *Università degli Studi di Cagliari, Italy*

Nobuko Yoshida, *Imperial College London, UK*

Ornela Dardha, *School of Computing Science, University of Glasgow, UK*

Raymond Hu, *Imperial College London, UK*

Roberto Zunino, *Università degli Studi di Trento, Italy*

Roly Perera, *School of Computing Science, University of Glasgow, UK* and *School of Informatics, University of Edinburgh, UK*

Rumyana Neykova, *Imperial College London, UK*

Sam Lindley, *University of Edinburgh, Edinburgh, UK*

Saverio Giallorenzo, *Focus Team, University of Bologna/INRIA, Italy*

Simon J. Gay, *School of Computing Science, University of Glasgow, UK*

Søren Debois, *Department of Computer Science, IT University of Copenhagen, Rued Langgaards Vej 7, 2300 Copenhagen S, Denmark*

Stefano Lande, *University of Cagliari, Italy*

Thomas T. Hildebrandt, *Department of Computer Science, IT University of Copenhagen, Rued Langgaards Vej 7, 2300 Copenhagen S, Denmark*

Tiziana Cimoli, *University of Cagliari, Italy*

Vasco T. Vasconcelos, *LaSIGE, Faculty of Sciences, University of Lisbon, PT*

List of Figures

List of Tables

List of Abbreviations

AIOC	Adaptive Interaction Oriented Choreographies
AIOCJ	Adaptive Interaction Oriented Choreographies in Jolie
AMQP	Advanced Message Queuing Protocol
AOP	Aspect-Oriented Programming
API	Application Programming Interface
ASCII	American Standard Code for Information Interchange
ATM	Automated teller machine
BCT	Behavioral Class Table
BPMN	Business Process Model and Notation
CCP	Concurrent constraint programming
CCS	calculus of communicating systems
CFSM	Communicating Finite State Machines
cHML	Co-Safety HML
CML	Concurrent ML
CMMN	Case Management Model and Notation
CTA	Communicating timed automata
DCR	Dynamic Condition Response
EFSM	Endpoint Finite State Machine
EVM	Erlang Virtual Machine
FIFO	First In, First Out
FSM	Finite State Machine
GHC	Glasgow Haskell Compiler
GMC	Generalised multiparty compatibility
GSM	Guard Stage Milestone
HML	Hennessy-Milner Logic
HTTP	Hypertext Transfer Protocol
I/O	Input/Output
IDE	Integrated Development Environment
IT	Information Technology
ITU	IT University of Copenhagen
JPF	Java PathFinder

JSON	JavaScript Object Notation
JVM	Java Virtual Machine
JVML	Java Virtual Machine Language
lam	deadLock Analysis Model
LLC	Linear lambda-calculus
MC	Model Checking
MFA	Module, Function, Arguments
mHML	Monitorable HML
MOM	Message oriented middleware
MPI	Message-Passing Interface
MPST	Multiparty Session Types
OOI	Ocean Observatories Initiative
PCF	Programmable Computable Functions
POP3	Post Office Protocol 3
QoS	Quality of service
RAC	Resource Access Control
RE	Runtime Environment
RFC	Request for Comments
RMI	Remote Method Invocation
RPC	Remote Procedure Call
RV	Runtime Verification
sHML	Safety HML
SMS	Short Message Service
SMTP	Simple Mail Transfer Protocol
SOAP	Simple Object Access Protocol
SODEP	Simple Operation Data Exchange Protocol
TCP	Transmission Control Protocol
TCP/IP	Transmission Control Protocol/Internet Protocol
TST	Timed session type
UML	Unified modeling language
URI	Uniform Resource Identifier
VM	Virtual Machine
WS-CDL	Web Services Choreography Description Language
XML	eXtensible Markup Language
μHML	μ HML

1

Contract-Oriented Design of Distributed Applications: A Tutorial

Nicola Atzei[1], Massimo Bartoletti[1], Maurizio Murgia[1], Emilio Tuosto[2] and Roberto Zunino[3]

[1]Università degli Studi di Cagliari, Italy
[2]University of Leicester, UK
[3]Università degli Studi di Trento, Italy

Abstract

Modern distributed applications typically blend new code with legacy (and possibly untrusted) third-party services. Behavioural contracts can be used to discipline the interaction among these services. Contract-oriented design advocates that composition is possible only among services with compliant contracts, and execution is monitored to detect (and possibly sanction) contract breaches.

In this tutorial we illustrate a contract-oriented design methodology consisting of five phases: specification writing, specification analysis, code generation, code refinement, and code analysis. Specifications are written in CO_2, a process calculus whose primitives include contract advertisement, stipulation, and contractual actions. Our analysis verifies a property called honesty: intuitively, a process is honest if it always honors its contracts upon stipulation, so being guaranteed to never be sanctioned at run-time. We automatically translate a given honest specification into a skeletal Java program which renders the contract-oriented interactions, to be completed with the application logic. Then, programmers can refine this skeleton into the actual Java application: however, doing so they could accidentally break its honesty. The last phase is an automated code analysis to verify that honesty has not been compromised by the refinement.

1

All the phases of our methodology are supported by a toolchain, called Diogenes. We guide the reader through Diogenes to design small contract-oriented applications.

1.1 Introduction

Developing service-oriented applications is a challenging task: programmers have to reliably compose loosely-coupled services which can dynamically discover and invoke other services through open networks, and may be subject to failures and attacks. Usually, services live in a world of mutually distrusting providers, possibly competing among each other. Typically, these providers offer little guarantees about the services they control, and in particular they might arbitrarily change the service code (if not the Service Level Agreement *tout court*) at any time.

Therefore, to guarantee the reliability and security of service-oriented applications, one must use suitable analysis techniques. Remarkably, most existing techniques to guarantee deadlock-freedom of service-oriented applications (e.g., compositional verification based on choreographies [35, 21]) need to inspect the code of *all* its components. Instead, under the given assumptions of mutual distrust between services, one can only analyse those under their control.

1.1.1 From Service-Oriented to Contract-Oriented Computing

A possible countermeasure to these issues is to use *behavioural contracts* to regulate the interaction between services. In this setting, a service infrastructure acts as a trusted third party, which collects all the contracts advertised by services, and establishes sessions between services with compliant contracts. Unlike the usual service-oriented paradigm, here services are responsible for respecting their contracts. To incentivize such honest behaviour, the service infrastructure monitors all the messages exchanged among services, and sanctions those which do not respect their contracts.

Sanctions can be of different nature: e.g., pecuniary compensations, adaptations of the service binding [29], or reputation penalties which marginalize dishonest services in the selection phase [3]. Experimental evidence [3] shows that contract-orientation can mitigate the effort of handling potential misbehaviour of external services, at the cost of a tolerable loss in efficiency due to the contract-based service selection and monitoring.

1.1.2 Honesty Attacks

The sanctioning mechanism of contract-oriented infrastructures protects honest services against malicious behaviours of the other services: indeed, if a malevolent service attempts to break the protocol (e.g. by prematurely terminating the interaction), it is punished by the infrastructure. At the same time, a new kind of attack becomes possible: adversaries can try to exploit possible discrepancies between the promised and the actual behaviour of a service, in order to make it sanctioned. For instance, consider a naïve online store with the following process:

1. Advertise a contract to "receive a `request` from a buyer, and then either send the `price` of the ordered item, or notify that the item is `unavailable`";
2. Wait to receive a `request`;
3. Advertise a contract to "receive a `quote` from a package delivery service, and then either `confirm` or `abort`";
4. Wait to receive a quote from the delivery service;
5. If the quote is below a certain threshold, then `confirm` the delivery and send the `price` to the buyer; otherwise, send `abort` to the delivery service, and notify as `unavailable` to the buyer.

Now, assume an adversary which plays the role of a delivery service, and never sends the `quote`. This makes the store violate its contract with the buyer: indeed, the store should either send `price` or `unavailable` to the buyer, but these actions can only be performed after the delivery service has sent a `quote`. Therefore, the store can be sanctioned.

Since these *honesty attacks* may compromise the service and cause economic damage to its provider, it is important to detect the underlying vulnerabilities *before* deployment. Intuitively, a service is vulnerable if, in *some* execution context, it does *not* respect some of the contracts it advertises. Therefore, to avoid sanctions a service must be able to respect *all* the contracts it advertises, in *all* possible contexts — even in those populated by adversaries. We call this property *honesty*.

Some recent works have studied honesty at the specification level, using the process calculus CO_2 for modelling contract-oriented services [6–9], whose primitives include contract advertisement, stipulation, and contractual actions. Practical experience has shown that writing honest specifications is not an easy task, especially when a service has to juggle with multiple sessions. The reason of this difficulty lies in the fact that, to devise an honest specification, a designer has to anticipate the possible behaviour of the context, but at design time he does not yet know in which context his service will be run. Tools to automate the verification of honesty may be of great help.

1.1.3 Diogenes

In this paper we illustrate the Diogenes toolchain [1], which supports the correct design of contract-oriented services as follows:

Specification Designers can specify services in the process calculus CO_2. An Eclipse plugin supports writing such specifications, providing syntax highlighting, code auto-completion, syntactic and semantic checks, and basic static type checking.

Honesty checking of specifications Our tool can statically verify the honesty of specifications. When the specification is dishonest, the tool provides a counter example, in the form of a reachable abstract state of the service which violates some contract.

Translation into Java The tool automatically translates specifications into skeletal Java programs, implementing the required contract-oriented interactions (while leaving the actual application logic to be implemented in a subsequent step). The obtained skeleton is honest when the specification is such.

Honesty checking of refined Java code Programmers can refine the skeleton by implementing the actual application logic. This is a potentially dangerous operation, since honesty can be accidentally lost in the manual refinement. The tool supports this step, by providing an honesty checker for refined Java code.

1.2 Specifying Contract-Oriented Services in CO_2

A service in our modelling language consists of a CO_2 process. CO_2 is a process algebra inspired from CCS [28], and equipped with contract-oriented primitives: contract advertisement, stipulation, and contractual actions. Contracts are meant to model the promised behaviour of services, and they are expressed as session types ([34]).

We show the main features of our language with the help of a small case study, an online store which receives orders from customers.

1.2.1 Contracts

We first specify the contract C between the store and a customer, from the point of view of the store. The store declares that it will receive an order, and then send either the corresponding price, or declare that the item is unavailable. We formalise this contract as the following first-order binary session type [19]:

```
contract C { order? string . ( price! int (+) unavailable! ) }
```

Receive actions are marked with the symbol ?, while send actions are marked with !. The sort of a message (int, string, or unit) is specified next to the action label; the sort unit is used for pure synchronizations, and it can be omitted. The symbol _._ denotes prefixing. The symbol _(+)_ is used to group send actions, and it denotes an *internal* choice made by the store.

1.2.2 Processes

Note that contracts only formalise the interaction protocol between two services, while they do not specify *how* these services advertise and realise the contracts. This behaviour is formalised in CO_2 [6, 7], a specification language for contract-oriented services. For instance, a possible CO_2 specification of our store is the following:

```
1    specification Store {
2        tell x C .        // wait until session x is created
3        receive@x order?[v:string] . (
4            if *          // checks if the item is in stock
5            then send@x price![*:int]
6            else send@x unavailable! ) }
```

At line 2, the store *advertises* the contract C, waiting for the service infrastructure to find some other service with a *compliant* contract. Intuitively, two contracts are compliant if they fulfil each other expectations[1]. When the infrastructure finds a contract compliant with C, a new session is created between the respective services, and the variable x is bound to the session name.

At line 3 of the snippet above the store waits to receive an order, binding it to the variable v of sort string. At line 4, the store checks whether the

---

[1]More precisely, the notion of compliance we use here is *progress*, that relates two processes whenever their interaction never reaches a deadlock [4].

ordered item is in stock (the actual condition is not given in the specification). If the item is in stock, then the store sends the price to the customer; otherwise it notifies that the item is unavailable (lines 5-6). The sent price *:int is a placeholder, to be replaced with an actual price upon refinement of the specification into an actual implementation of the service.

1.2.3 An Execution Context

We now show a possible context wherein to execute our store. Although the context is not needed for verifying the store specification, we use it to complete the presentation of the primitives of our modelling language.

```
1    specification BuyerA {
2         tell y { order! string , price? int } .
3         send@y order![*:string] .
4         receive@y price?[n:int]
5    }
6
7    specification BuyerB {
8         tell y { order! string . ( price? int + unavailable?
                                       + availablefrom? string) } .
9         send@y order![*:string] .
10        receive {
11             @y price?[n:int]
12             @y unavailable?
13             @y availablefrom?[date:string]}
14   }
```

The contract advertised by BuyerA at line 2 is *not* compliant with the contract C advertised by the store: indeed, after sending the order, BuyerA only expects to receive the price — while the store can also choose to send unavailable. Therefore, any service implementing BuyerA will never be put in a session with the Store. Instead, the contract advertised at line 8 by BuyerB is compliant with C. Note that this is true also if the two contracts are not one dual of each other: indeed, BuyerB accepts all the messages that the store may send (i.e., price and unavailable), and it also allows for a further message (availablefrom), to be used e.g. to notify when the item will be available. Although this message will never be used by the Store, it could allow BuyerB to establish sessions with more advanced stores. The symbol + is used to group receive actions, and it denotes an *external choice*, one which is not made by the buyer. At lines 11-13, BuyerB waits to receive at session y one of the messages declared in the contract.

1.2.4 Adding Recursion

Note that our Store can only manage the order of a single item: if some buyer wants to order two or more items, she has to use distinct instances of the store. We now extend the store so that it can receive several orders in the same session, adding all the items to a cart.

We start by refining our contract as follows:

```
1  contract Crec {
2      addToCart? string . Crec
3      + checkout? . (
4              price! int . (accept? + reject?)
5              (+) unavailable!
6      )
7  }
```

The contract Crec requires the store to accept from buyers two kinds of messages: addToCart and checkout. When a buyer chooses addToCart, the store must allow the buyer to order more items. This is done by recursively calling Crec in the addToCart branch. When a buyer stops adding items to the cart (by choosing checkout), the store must either send a price or state that the items are unavailable. In the first case, the store allows the buyer to accept the quotation and finalise the order, or to reject it and abort.

A possible specification of the store using the contract Crec is as follows:

```
1  specification StoreRec { tell x Crec . Loop(x) }
2  specification Loop(x:session) {
3      receive {
4          @x addToCart?[item:string] -> Loop(x)
5          @x checkout? -> Checkout(x)
6      }
7  }
8  specification Checkout(x:session) {
9      if *       // checks whether the items are available
10     then
11         send@x price![*:int] .
12         receive {
13             @x accept?
14             @x reject?
15         }
16     else send@x unavailable!
17 }
```

The store StoreRec advertises the contract Crec, and then continues as the process Loop(x), where x is the handle to the new session. The process Loop(x) receives messages from buyers through session x. When it receives addToCart, it just calls itself recursively; instead, when it receives checkout, it calls the process Checkout. This process internally chooses whether to send the buyer a `price`, or to notify that the requested items are unavailable. In the first case, it receives from the client a confirmation, that can be either `accept` or `reject`.

A possible buyer interacting with `StoreRec` is the following:

```
1   specification BuyerC {
2       tell y { addToCart! string . addToCart! string . checkout!
            . (price? int . (accept! (+) reject!) + unavailable?)
                } .
3       send@y addToCart![*:string] .
4       send@y addToCart![*:string] .
5       send@y checkout! .
6       receive {
7           @y price?[n:int] ->
8               if * then send@y accept! else send@y reject!
9           @y unavailable? -> nil
10      }
11  }
```

Note that the buyer's contract is compliant with `Crec`, even though the store contract is recursive, while the buyer's one is not.

1.3 Honesty

In an ideal world, one would expect that services respect the contracts they advertise, in *all* execution contexts: we call *honest* such services. In this section we illustrate, through a series of examples, that writing honest services may be difficult and error-prone. Further, we show how our tools may help service designers in specifying and implementing honest services.

1.3.1 A Simple Dishonest Store

Our first example is a naïve CO_2 specification of the store advertising the contract C at page 5:

```
1   specification StoreDishonest1 {
2       tell x C .
3       receive@x order?[v:string] . (
4           if *
5           then send@x price![*:int]) }
```

The store above waits for an order of some item v. Then, it checks whether v is in stock (the actual test is abstracted by the `*:boolean` guard). If the item is in stock, the store sends a price quotation to the buyer (again, the price is abstracted in the specification).

Note that the store does nothing when the ordered item is not in stock. In this way, the store fails to respect its advertised contract C, which prescribes to always respond to the buyer by sending either price or unavailable. Therefore, we classify this specification of the store as *dishonest*.

In this paper we give an intuitive description of honesty, referring the reader to the literature [6, 7] for a formal definition. A specification A is honest when, in all possible executions, if a contract at some session requires A to do some action, then A actually performs it. Basically, this boils down to say that when A is required to send a message, then it does so. Likewise, when A is required to receive a message, then A is ready to accept any message that its partner may be willing to send. More in detail:

- if the contract is an *internal* choice a1!S1 (+) ... (+) an!Sn, then A must send a message having sort Si, and labelled ai, for some i;

- if the contract is an *external* choice a1?S1 + ... + an?Sn, then A must be able to receive messages labelled with *any* labels ai in the choice (with the corresponding sorts Si).

The honesty property discussed above can be automatically verified using the Diogenes honesty checker, which uses the verification technique described and implemented in [6]. This technique is built upon an abstract semantics of CO_2 which approximates both values (sent, received, and in conditional expressions) and the actual *context* wherein a specification is executed. Basically, the tool checks, through an exaustive exploration, that in every reachable state of the abstract semantics a participant is always able to perform some of the actions prescribed in each of her stipulated contracts. Since this is a branching-time property, a natural approach to verify it is by model checking. To this purpose we exploit a rewriting logic specification of the CO_2 abstract semantics and the Maude [12] search capabilities. This abstraction is a *sound* over-approximation of honesty: namely, if the abstraction of a specification is honest, then also the concrete one is honest. Further, the analysis is *complete* for specifications without conditional statements: i.e., if an abstracted specification is dishonest, then also its concrete counterpart is dishonest. If the abstractions are finite-state, we can verify their honesty by model checking a (finite) state space[2]. Our implementation first translates a

---

[2]Abstractions are finite-state in the fragment of CO_2 without delimitation/parallel under process definitions. For specifications outside this fragment the analysis is still correct, but it may diverge; indeed, a negative result [9] excludes the existence of algorithms for honesty that are at the same time sound, complete, and terminating in full CO_2.

CO_2 specification into a Maude term [12], and then uses the Maude model checker to decide the honesty of its abstract semantics.

The honesty checker outputs the message below, that reports that the specification StoreDishonest1 is *dishonest*. The reason for its dishonesty can be inferred from the following output:

```
result:   ($ 0)(
   StoreDishonest1[if exp then do $ 0 "price" ! int . 0 else 0]
   | $ 0["price" ! int . 0 (+) "unavailable" ! unit . 0]
)
honesty: false
```

This shows a reachable (abstract) state of the specification, where $ 0 denotes an open session between the store and a buyer.

The state consists of two parallel components: the state of the store

```
StoreDishonest1[if exp then do $ 0 "price" ! int . 0 else 0]
```

and the state of the contract at session $ 0, from the point of view of the store:

```
$ 0["price" ! int . 0 (+) "unavailable" ! unit . 0]
```

Such contract requires the store to send either price or unavailable to the buyer. However, if the guard exp of the conditional (within the state of the store) evaluates to false, the store will not send any message to the buyer, so violating the contract C. Therefore, the honesty checker correctly classifies StoreDishonest1 as dishonest.

1.3.2 A More Complex Dishonest Store

We now consider a more evolved specification of the store, which relies on external distributors to retrieve items. The contract D specifies the interaction between the store and distributors:

```
contract D { req! string . ( ok? + no? ) }
```

Namely, the store first sends a request to the distributor for some item, and then waits for an ok or no answer, according to whether the distributor is able to provide the requested item or not.

Our first attempt to specify a store interacting with customers and distributors is the following:

```
1    specification StoreDishonest2 {
2        tell x C .
3        receive@x order?[v:string] .
```

```
4        tell y D .
5        send@y req![v]  .
6        receive {
7            @y ok? -> send@x price![*:int]
8            @y no? -> send@x unavailable!
9        }
10   }
```

At line 2, the store advertises the contract C, and then waits until a session is established with some customer; when this happens, the variable x is bound to the session name. At line 3 the store waits to receive an order, binding it to the variable v. At line 4 the store advertises the contract D to establish a session y with a distributor; at line 5, it sends a request with the value v. Finally, the store waits to receive a response ok or no from the distributor, and accordingly responds price or unavailable to the customer (lines 6-9). The price *:int is a placeholder, to be replaced upon refinement.

The honesty checker classifies StoreDishonest2 as *dishonest*. The reason for its dishonesty can be inferred from the following output:

```
result: ("y",$ 0)(
    StoreDishonest2[tell "y" D. (...)]
    | $ 0["price" ! int . 0 (+) "unavailable" ! unit . 0])
honesty: false
```

This output shows a possible (abstract) state which could be reached by StoreDishonest2. There, $ 0 denotes an open session between the store and a buyer, while "y" indicates that no session between the store and a distributor is established, yet. The contract at session $ 0 requires the store to send either a price or an unavailability message. However, in the given state there is no guarantee to find a distributor, hence the store might be stuck in the tell, never performing the required actions at session $ 0. Because of this, the store does not fulfil the contract C, hence it is correctly classified as dishonest.

1.3.3 Handling Failures

We try to fix the specification StoreDishonest2 by adapting it so to consider the case where the distributor is not available. Let us refine the specification StoreDishonest2 as follows:

```
1    specification StoreDishonest3 {
2        tell x C .
3        receive@x order?[v:string] . (
4            tell y D .
```

```
5                    send@y req![v]  .
6                    receive {
7                        @y ok? -> send@x price![*:int]
8                        @y no? -> send@x unavailable!
9                    }
10         after * -> send@x unavailable!
11      )
12  }
```

Note that `StoreDishonest3` uses the construct `tell ··· after ···` at lines
4-10. This ensures that, if no session is established within a given deadline,
then the contract is *retracted* (i.e., removed from the registry of available
contracts), and the control passes to the `after` process. In particular, in our
`StoreDishonest3`, if no distributor is found, then D is retracted, and the store
performs its duties with the buyer by sending him `unavailable`. Since the
actual deadline is immaterial in this specification, it is abstracted here as `*`.

By running the honesty checker on the amended specification, we obtain:

```
result: ($ 0,$ 1)(
    StoreDishonest3
        [ retract $ 1 . ( ... )
        + ask $ 1 True . do $ 1 "req" ! string .
            ( do $ 1 "no" ? unit . do $ 0 "unavailable" ! unit .
                (...)
            + do $ 1 "ok" ? unit . do $ 0 "price" ! int . (...)
            )]
    | $ 0["price" ! int . 0 (+) "unavailable" ! unit . 0]
    | $ 1["req" ! string . ("no" ? unit . (0).Id + "ok" ? unit .
        (0).Id)]
    )
honesty: false
```

Note that `StoreDishonest3` is still dishonest. The output above shows a
reachable (abstract) state where the store has opened two sessions, $ 0 and
$ 1, with a buyer and a distributor, respectively. At session $ 0 the store
is expected to send either `price` or `unavailable` to the buyer. Now, the
store can perform do $ 0 "price" ! int only *after* receiving the input
from the distributor, i.e. after performing do $ 1 "ok" ? unit. Similarly,
the store can only perform the action do $ 0 "unavailable" ! unit after
the action do $ 1 "no" ? unit. Should the distributor fail to send either of
these messages, then the store would fail to honour its contract C with the
buyer. Therefore, the honesty checker correctly classifies `StoreDishonest3`
as dishonest. Note that, even if in this case the distributor would be dishonest
as well, (since it violates the contract D with the store), this does not excuse
the store from violating the contract C with the buyer.

1.3.4 An Honest Store, Finally

In order to address the dishonesty issues in the previous specification, we revise the store as follows:

```
1   specification StoreHonest {
2       tell x C .
3       receive@x order?[v:string] . (
4           tell y D .
5               send@y req![v] .
6               receive {
7                   @y ok?  -> send@x price![*:int]
8                   @y no?  -> send@x unavailable!
9                   after * -> (
10                      send@x unavailable!
11                      | receive {
12                              @y ok? -> nil
13                              @y no? -> nil
14                      }
15                  )
16              }
17          after * -> send@x unavailable!
18      )
19  }
```

The main difference between this specification and the previous one is related to the receive at session y. At line 9, after * represents the case in which no messages are received within a given timeout (immaterial in this specification). In such case, the store fulfils its contract at session x, by sending unavailable to the buyer. Further, the store also fulfils its contract at session y, by receiving any message that could still be sent from the distributor after the timeout.

Now the honesty checker correctly detects that the revised specification StoreHonest is honest.

1.3.5 A Recursive Honest Store

We reprise the specification of StoreRec in Section 1.2, by providing a recursive store which interacts with buyers (via contract Crec at page 7) and with distributors (via contract D).

```
1   specification StoreHonestRec {
2       tell x Crec . Loop(x)
3   }
4
5   specification Loop(x:session) {
6       receive {
```

```
7              @x addToCart?[item:string] -> Loop(x)
8              @x checkout? -> Checkout(x)
9          }
10    }
11
12    specification Checkout(x:session) {
13        tell y D .
14            send@y req![*:string] .
15            receive {
16                @y ok?   -> send@x price![*:int] .
17                    receive {
18                        @x accept?
19                        @x reject?
20                    }
21                @y no?   -> send@x unavailable!
22                after * -> (
23                    send@x unavailable! |
24                    receive {
25                        @y ok?
26                        @y no?
27                    }
28                )
29            }
30        after * -> send@x unavailable!
31    }
```

The specification StoreHonestRec handles the checkout of buyers in the process Checkout, which is identical to lines 4–14 in StoreHonest. The main difference with respect to StoreHonest is that StoreHonestRec can receive multiple requests from a buyer, via the recursive process Loop(x). Despite this complication, the specification is still verified as honest by Diogenes.

1.4 Refining CO_2 Specifications in Java Programs

Diogenes translates CO_2 specifications into Java skeletons, using the APIs of the contract-oriented middleware in [3]. This middleware collects the contracts advertised by services, establishes sessions between those with compliant contracts, and it allows services to send/receive messages through sessions, while monitoring this activity to detect and punish violations. More specifically, upon detection of a contract violation the middleware punishes the culprit, by suitably decreasing its *reputation*. This is a measure of the trustworthiness of a participant in its past interactions: the lower is the reputation, the lower is the probability of being able to establish new sessions with it.

1.4.1 Compilation of CO_2 Specifications into Java Skeletons

We illustrate the translation of CO_2 specifications into Java through an example, the StoreHonest given in the previous section. From it, we obtain the following Java skeleton[3]:

```
1   public class StoreHonest extends Participant {
2     public void run() {
3       Session x = tellAndWait(C);
4
5       Message msg = x.waitForReceive("order");
6       String v = msg.getStringValue();
7
8       try {
9         Session y = tellAndWait(D, timeoutP);
10        y.sendIfAllowed("req", v);
11
12        try {
13          Message msg_1 = y.waitForReceive(timeoutP,"ok","no");
14          switch (msg_1.getLabel()) {
15          case "ok": x.sendIfAllowed("price", intP); break;
16          case "no": x.sendIfAllowed("unavailable"); break;
17          }
18        }
19        catch (TimeExpiredException e) {
20          parallel(()->{x.sendIfAllowed("unavailable");});
21          parallel(()->{y.waitForReceive("ok","no");});
22        }
23      }
24      catch(ContractExpiredException e) {
25        //contract D retracted
26        x.sendIfAllowed("unavailable");
27      }
28    }
29  }
```

We comment below how the specification of StoreHonest at page 13 is rendered in Java.

- tell x C (at line 2) is translated into the assignment

  ```
  3   Session x = tellAndWait(C)
  ```

  The API method tellAndWait advertises the contract C to the middleware, and blocks until a compliant buyer contract is found. Then, it returns a new object, representing the newly established session between the store and the buyer.

---

[3]Minor cosmetic changes are applied to improve readability.

- `receive @x order?[v:string]` (at line 3) is translated into

```
5    Message msg = x.waitForReceive("order");
6    String v = msg.getStringValue();
```

where the call to `waitForReceive` blocks until the store receives a message labelled `order` on session x.

- The block `tell y D ... after * ...` (at lines 4–17) is translated in Java as the try-catch statement:

```
try {
    Session y = tellAndWait(D, timeoutP);
    ...
}
catch(ContractExpiredException e) {
    ...
}
```

The call `tellAndWait(D, timeoutP)` advertises the contract D; the second parameter specifies a timeout (in milliseconds) to find a compliant contract. If the timeout expires, the contract D is retracted, and an exception is thrown. Then, the exception handler performs the recovery action specified in the `after` clause by sending `unavailable` to the client.

- `send @y req![*:string]` (at line 5) is translated as

```
y.sendIfAllowed("req", stringP)
```

This method call sends a message labelled `req` at session y, blocking until this action is permitted by the contract.

- The `receive` block at lines 6–16 is translated into a try-catch statement

```
try {
    Message msg_1 = y.waitForReceive(timeoutP,"ok","no");
    ...
}
catch (TimeExpiredException e) {
        parallel(()->{x.sendIfAllowed("unavailable");});
        parallel(()->{y.waitForReceive("ok","no");});
}
```

The `waitForReceive` waits (until the given timeout) to receive on session y a message labelled either `yes` or `no`, throwing an exception in case the timeout expires. In such case, the `catch` block performs the recovery actions in the `after` clause of the specification. Namely, the service spawns two parallel processes, which send `unavailable` to the buyer, and receives late replies from the distributor.

Note that the timeout values `timeoutP`, as well as the order price `intP`, are just placeholders. Further, in an actual implementation of the store service, we may want e.g. to read the order price from a file or a database. This can be done by refining the skeleton, introducing the needed code to make the service actually implement the desired functionality.

1.4.2 Checking Honesty of Refined Java Programs

Note that when refining the skeleton into the actual Java application, programmers could accidentally break its honesty. In general, this happens when the refinement alters the interaction behaviour of the service. For instance, in an actual implementation of our store service, we may want to delegate the computation of `price` to a separated method, as follows:

```
public int getOrderPrice(String order) throws MyException {...}
```

and change the placeholder `intP` at line 15 of the generated code with an invocation `getOrderPrice(v)`. The method could read the order price from a file or a database, and suppose that, in that method, each possible exception is either handled or re-thrown as `MyException`. If `getOrderPrice` throws an exception, then the `sendIfAllowed()` at line 15 is not performed. Unless the store performs it while handling `MyException`, the store violates the contract with the buyer, and so it becomes dishonest.

To address this issue, the Diogenes toolchain includes an *honesty checker* for Java programs, to be used after refinement. This honesty checker is built on top of *Java PathFinder* (JPF [27, 37]). We define suitable *listeners* for JPF, to intercept the requests to the contract-oriented middleware, and to simulate *all* the possible responses that the application can receive from it. Through JPF we symbolically execute the program, in order to infer a CO$_2$ specification that abstracts its behaviour, preserving dishonesty. Once a specification is constructed in this way, we apply the CO$_2$ honesty checker discussed in Section 1.3 to establish the honesty of the Java program.

We can check the honesty of a Java program through the static method `HonestyChecker.isHonest(StoreHonest.class)`, which returns one of the following values:

- `HONEST`: the tool has inferred a CO$_2$ specification and verified its honesty;
- `UNKNOWN`: the tool has been unable to infer a CO$_2$ specification, e.g. because of unhandled exceptions within the class under test.

In our example, we just provide the following stub implementation of the method getOrderPrice:

```
@SkipMethod
public int getOrderAmount(String order) throws MyException {
    return 42; }
```

where the annotation @SkipMethod is interpreted by the honesty checker as follows: assume that the method terminates (possibly throwing one of the declared exceptions), and it does not interact with the contract-oriented middleware. For our refined store, the honesty checker returns UNKNOWN, outputting:

```
error details: MyException:
  This exception is thrown by the honesty checker.
      Please catch it!
  at i.u.c.store.StoreHonest.getOrderPrice(Store.java:30)
  at i.u.c.store.StoreHonest.run(Store.java:15)
  at i.u.c.honesty.HonestyChecker.runProcess(HonestyChecker.java
      :182)
```

As anticipated above, this output remarks that if getOrderAmount throws an exception, then the store is dishonest.

As a first (naïve) attempt to recover honesty, we further refine the store by catching MyException, and just logging the error in the exception handler:

```
try {
    ...
    case "ok": x.sendIfAllowed("price",getOrderPrice(v)); break;
    ...
}
catch (TimeExpiredException e) { ... }
catch (MyException e) { System.out.println("failed"); }
```

In this case, the honesty checker correctly classifies the store as DISHONEST, producing the following output:

```
result ($ 0,$ 1)(
    StoreHonest[0] |
    $ 0["price" ! unit . 0 (+) "unavailable" ! unit . 0] |
    $ 1[0])
honesty: DISHONEST
```

This output highlights the reason for dishonesty: StoreHonest[0] means that the store does nothing, while at session $ 0, it should send either price or unavailable to the buyer.

To recover honesty, rather than just logging the error, we also perform `x.sendIfAllowed("unavailable")` in the exception handler, in order to fulfil the contract with the buyer:

```
catch (MyException e) {
    System.out.println("failed");
    x.sendIfAllowed("unavailable");
}
```

With this modification, the Java honesty checker correctly outputs HONEST.

1.5 Conclusions

We have presented Diogenes, a toolchain for the specification and verification of contract-oriented services. Diogenes fills a gap between foundational research on honesty [6–9] and more practical research on contract-oriented programming [3]. Our tools can help service designers to write specifications, check their adherence to contracts (i.e., their honesty), generate Java skeletons, and refine them while preserving honesty. We have experimented Diogenes with a set of case studies (more complex than the ones presented in this tutorial); our case studies are available at `co2.unica.it/diogenes`.

The effectiveness of our tools could be improved in several ways, ranging from the precision of the analysis, to the informative quality of output messages provided by the honesty checkers.

The precision of the honesty analysis could be improved e.g., by implementing the type checking technique of [7], which extends the class of infinite-state processes for which honesty can be verified. More specifically, the type system in [7] can also handle some processes with delimitation and parallel composition under recursion.

Another form of improvement would be to extend the formalism and the analysis to deal with timing constraints. This could be done e.g. by exploiting the timed version of CO_2 [3] and timed session types [2]. Although the current analysis for honesty does not consider timing constraints (and therefore is unsound in such scenario), it can still give useful feedback when applied to timed specifications. For instance, it could detect that some prescribed actions cannot be performed because the actions they depend on may be blocked by an unresponsive context.

When a specification/program is found dishonest, it would be helpful for programmers to know which parts of it is responsible for contract violations. The error reporting facilities of Diogenes could be improved to this purpose:

this would require e.g., to signal what are the contract obligations that are not fulfilled, and in what session, and in particular which part of the specification/program should be fixed. Further, it would be useful to suggest possible corrections to the designer.

Another direction for future work is to formally establish relations between the original CO_2 specification and the refined Java code. In fact, our tools can only check that the user-refined Java code obtained from an honest CO_2 specification is honest, but this does not imply that the refined Java code still "adheres" to the specification. Indeed, improper refinements could drastically modify the interaction behaviour of a service, e.g. by removing some contract advertisements — while preserving honesty. An additional static analysis could establish that the CO_2 process inferred from the user-refined Java code is behaviourally related to the original specification. An alternative way to cope with this issue would be to enhance the generation of the skeletal Java program, by providing a more structured class hierarchy. More precisely, we could avoid accidental breaches of honesty by separating, in the generated skeleton, the part that handles the interactions from the parts to be refined. This could be done e.g. by inserting entry points to invoke classes/interfaces whose behaviour is defined apart, so separating the application logic and simplifying possible updates in the specifications.

1.5.1 Related Work

In recent years many works have addressed the safe design of service-oriented applications. A notable approach is to specify the overall communication behaviour of an application through a *choreography*, which validates some global properties of the application (e.g. safety, deadlock-freedom, *etc.*). To ensure that the application enjoys such properties, all the components forming the application have to be verified; this can be done e.g. by projecting the choreography to end-point views, against which these components are verified [35, 21]. Examples of how to embody such approach in existing programming languages and models are presented for C [33], for Python [30], and for the actor model [31]. All those approaches are based on Scribble [38], a protocol description language featuring multiparty session types [21]. The strict relations between multiparty session types and actor-based models such as communicating machines [15] has been used to develop a framework to monitor Erlang applications [18].

This top-down approach assumes that designers control the whole application, e.g., they develop all the needed components. However, in many real-world scenarios several components are developed independently, without knowing at design time which other components they will be integrated with. In these scenarios, the compositional verification pursued by the top-down approach is not immediately applicable, because the choreography is usually unknown, and even if it were known, only a subset of the needed components is available for verification. However, this issue can be mitigated when the communication pattern of each component is available. In fact, in such case if the set of components is compatible, it is possible to synthesise a faithful choreography [26] with a suitable tool [24]. Such choreography can then be used to distil monitors for the components that are not trusted (if any). The ideas pursued in this paper depart from the top-down approach, because designers can advertise contracts to discover the needed components (and so ours can be considered a *bottom-up* approach). Coherently, the main property we are interested in is *honesty*, which is a property of components, and not of global applications. Some works mixing top-down and bottom-up composition have been proposed in the past few years [5, 16, 25]. Recent works [32] have explored how to integrate the bottom-up approach with inference of multiparty session types from GO programs.

The problem of ensuring safe interactions in session-based systems has been addressed by many authors [10, 11, 13, 14, 17, 19, 21–23, 36]. When processes have a single session, our notion of honesty is close (yet different) to session typeability. A technical difference is that we admit processes to attempt interactions which are not mandated by the contract. E.g., the process:

```
1    specification P {
2       tell x { a! . b! } . (send @x a! | send @x b!)
3    }
```

is honest, while it would *not* be typeable according to most works on session types, because the action b is not immediately mandated by the contract.

Other, more substantial, differences between honesty and session typing arise when processes have more than one session. More specifically, we consider a process to be honest when it enjoys progress in *all* possible contexts, while most works on session typing guarantee progress in a given context. For instance, consider the process:

```
1  specification Q {
2    tell x { a! } . tell y { b? } . receive @y b? . send @x a!
3  }
```

We have that Q is *not* honest, because the action at session x is not possible if the participant at the other endpoint of session y does not send b. Note instead that Q would be well-typed in [20], even if some contexts R can lead Q to a deadlock. The interaction type system in [14] would allow to check the progress of Q|R, given a context R.

Acknowledgments This work has been partially supported by Aut. Reg. of Sardinia P.I.A. 2013 "NOMAD", and by EU COST Action IC1201 "Behavioural Types for Reliable Large-Scale Software Systems" (BETTY).

References

[1] Nicola Atzei and Massimo Bartoletti. Developing honest Java programs with Diogenes. In *Formal Techniques for Distributed Objects, Components, and Systems (FORTE)*, volume 9688 of *LNCS*, pages 52–61. Springer, 2016.

[2] Massimo Bartoletti, Tiziana Cimoli, Maurizio Murgia, Alessandro Sebastian Podda, and Livio Pompianu. Compliance and subtyping in timed session types. In *Formal Techniques for Distributed Objects, Components, and Systems (FORTE)*, volume 9039 of *LNCS*, pages 161–177. Springer, 2015.

[3] Massimo Bartoletti, Tiziana Cimoli, Maurizio Murgia, Alessandro Sebastian Podda, and Livio Pompianu. A contract-oriented middleware. In *Formal Aspects of Component Software (FACS)*, volume 9539 of *LNCS*, pages 86–104. Springer, 2015. http://co2.unica.itco2.unica.it

[4] Massimo Bartoletti, Tiziana Cimoli, and Roberto Zunino. Compliance in behavioural contracts: a brief survey. In *Programming Languages with Applications to Biology and Security*, volume 9465 of *LNCS*, pages 103–121. Springer, 2015.

[5] Massimo Bartoletti, Julien Lange, Alceste Scalas, and Roberto Zunino. Choreographies in the wild. *Science of Computer Programming*, 109:36–60, 2015.

[6] Massimo Bartoletti, Maurizio Murgia, Alceste Scalas, and Roberto Zunino. Verifiable abstractions for contract-oriented systems. *Journal of Logical and Algebraic Methods in Programming (JLAMP)*, 86:159–207, 2017.

[7] Massimo Bartoletti, Alceste Scalas, Emilio Tuosto, and Roberto Zunino. Honesty by typing. *Logical Methods in Computer Science*, 12(4), 2016. Pre-print available as: https://arxiv.org/abs/1211.2609

[8] Massimo Bartoletti, Emilio Tuosto, and Roberto Zunino. Contract-oriented computing in CO_2. *Sci. Ann. Comp. Sci.*, 22(1):5–60, 2012.

[9] Massimo Bartoletti and Roberto Zunino. On the decidability of honesty and of its variants. In *Web Services, Formal Methods, and Behavioral Types*, volume 9421 of *LNCS*, pages 143–166. Springer, 2015.

[10] Lorenzo Bettini, Mario Coppo, Loris D'Antoni, Marco De Luca, Mariangiola Dezani-Ciancaglini, and Nobuko Yoshida. Global progress in dynamically interleaved multiparty sessions. In *CONCUR*, volume 5201 of *LNCS*, pages 418–433. Springer, 2008.

[11] Giuseppe Castagna, Mariangiola Dezani-Ciancaglini, Elena Giachino, and Luca Padovani. Foundations of session types. In *ACM SIGPLAN Conference on Principles and Practice of Declarative Programming (PPDP)*, pages 219–230. ACM, 2009.

[12] Manuel Clavel, Francisco Durán, Steven Eker, Patrick Lincoln, Narciso Martí-Oliet, José Meseguer, and José F. Quesada. Maude: Specification and programming in rewriting logic. *TCS*, 2001.

[13] Mario Coppo, Mariangiola Dezani-Ciancaglini, Luca Padovani, and Nobuko Yoshida. Inference of global progress properties for dynamically interleaved multiparty sessions. In *COORDINATION*, volume 7890 of *LNCS*, pages 45–59. Springer, 2013.

[14] Mario Coppo, Mariangiola Dezani-Ciancaglini, Nobuko Yoshida, and Luca Padovani. Global progress for dynamically interleaved multiparty sessions. *Mathematical Structures in Computer Science*, 26(2):238–302, 2016.

[15] Pierre-Malo Deniélou and Nobuko Yoshida. Multiparty session types meet communicating automata. In *European Symposium on Programming (ESOP)*, volume 7211 of *LNCS*, pages 194–213. Springer, 2012.

[16] Pierre-Malo Deniélou and Nobuko Yoshida. Multiparty compatibility in communicating automata: Characterisation and synthesis of global session types. In *International Colloquium on Automata, Languages, and Programming (ICALP)*, volume 7966 of *LNCS*, pages 174–186. Springer, 2013.

[17] Mariangiola Dezani-Ciancaglini, Ugo de'Liguoro, and Nobuko Yoshida. On progress for structured communications. In *Trustworthy*

Global Computing (TGC), volume 4912 of *LNCS*, pages 257–275. Springer, 2007.

[18] Simon Fowler. An Erlang implementation of multiparty session actors. In *Interaction and Concurrency Experience*, volume 223 of *EPTCS*, pages 36–50, 2016.

[19] Kohei Honda, Vasco T. Vasconcelos, and Makoto Kubo. Language primitives and type disciplines for structured communication-based programming. In *European Symposium on Programming (ESOP)*, volume 1381 of *LNCS*, pages 22–138. Springer, 1998.

[20] Kohei Honda, Nobuko Yoshida, and Marco Carbone. Multiparty asynchronous session types. In *ACM SIGPLAN-SIGACT Symposium on Principles of Programming Languages (POPL)*, pages 273–284. ACM, 2008.

[21] Kohei Honda, Nobuko Yoshida, and Marco Carbone. Multiparty asynchronous session types. *J. ACM*, 63(1):9:1–9:67, 2016.

[22] Hans Hüttel, Ivan Lanese, Vasco T. Vasconcelos, Luís Caires, Marco Carbone, Pierre-Malo Deniélou, Dimitris Mostrous, Luca Padovani, António Ravara, Emilio Tuosto, Hugo Torres Vieira, and Gianluigi Zavattaro. Foundations of session types and behavioural contracts. *ACM Comput. Surv.*, 49(1):3:1–3:36, 2016.

[23] Naoki Kobayashi. A new type system for deadlock-free processes. In *Proc. CONCUR*, volume 4137 of *LNCS*, pages 233–247. Springer, 2006.

[24] Julien Lange and Emilio Tuosto. A toolchain for choreography-based analysis of application level protocols. Available at `https://bitbucket.org/emlio_tuosto/gmc-synthesis-v0.2`

[25] Julien Lange and Emilio Tuosto. Synthesising choreographies from local session types. In *CONCUR*, volume 7454 of *LNCS*, pages 225–239. Springer, 2012.

[26] Julien Lange, Emilio Tuosto, and Nobuko Yoshida. From communicating machines to graphical choreographies. In *ACM SIGPLAN-SIGACT Symposium on Principles of Programming Languages (POPL)*, pages 221–232, 2015.

[27] Flavio Lerda and Willem Visser. Addressing dynamic issues of program model checking. In *SPIN workshop on Model checking of software*, pages 80–102, 2001.

[28] Robin Milner. *Communication and concurrency*. Prentice-Hall, Inc., 1989.

[29] A. Mukhija, Andrew Dingwall-Smith, and D.S. Rosenblum. QoS-aware service composition in Dino. In *ECOWS*, volume 5900 of *LNCS*, pages 3–12. Springer, 2007.

[30] Rumyana Neykova. Session types go dynamic or how to verify your Python conversations. In *Workshop on Programming Language Approaches to Concurrency and Communication-cEntric Software (PLACES)*, volume 137 of *EPTCS*, pages 95–102, 2013.

[31] Rumyana Neykova and Nobuko Yoshida. Multiparty session actors. In *COORDINATION*, volume 8459 of *LNCS*, pages 131–146. Springer, 2014.

[32] Nicholas Ng and Nobuko Yoshida. Static deadlock detection for concurrent go by global session graph synthesis. In *International Conference on Compiler Construction (CC)*, pages 174–184. ACM, 2016.

[33] Nicholas Ng, Nobuko Yoshida, and Kohei Honda. Multiparty session C: safe parallel programming with message optimisation. In *Objects, Models, Components, Patterns (TOOLS)*, pages 202–218, 2012.

[34] Kaku Takeuchi, Kohei Honda, and Makoto Kubo. An interaction-based language and its typing system. In *PARLE*, pages 398–413, 1994.

[35] Wil M. P. van der Aalst, Niels Lohmann, Peter Massuthe, Christian Stahl, and Karsten Wolf. Multiparty contracts: Agreeing and implementing interorganizational processes. *Comput. J.*, 53(1):90–106, 2010.

[36] V. T. Vasconcelos. Fundamentals of Session Types. *Information and Computation*, 217:52–70, 2012.

[37] Willem Visser, Klaus Havelund, Guillaume Brat, SeungJoon Park, and Flavio Lerda. Model checking programs. *Automated Software Engineering*, 10(2):203–232, 2003.

[38] Nobuko Yoshida, Raymond Hu, Rumyana Neykova, and Nicholas Ng. The Scribble protocol language. In *Trustworthy Global Computing (TGC)*, volume 8358 of *LNCS*, pages 22–41. Springer, 2013.

2

Contract-Oriented Programming with Timed Session Types

**Nicola Atzei, Massimo Bartoletti, Tiziana Cimoli, Stefano Lande,
Maurizio Murgia, Alessandro Sebastian Podda and Livio Pompianu**

University of Cagliari, Italy

Abstract

Contract-oriented programming is a software engineering paradigm which proposes the use of behavioural contracts to discipline the interaction among software components. In a distributed setting, the various components of an application may be developed and run by untrustworthy parties, which could opportunistically diverge from the expected behaviour when they find it convenient. The use of contracts in this setting is essential: by binding the behaviour of each component to a contract, and by sanctioning contract violations, components are incentivized to behave in a correct and cooperative manner.

This chapter is a step-by-step tutorial on programming contract-oriented distributed applications. The glue between components is a middleware which establishes sessions between services with compliant contracts, and monitors sessions to detect and punish violations. Contracts are formalised as timed session types, which describe timed communication protocols between two components at the endpoints of a session. We illustrate some basic primitives of contract-oriented programming: advertising contracts, performing contractual actions, and dealing with violations. We then show how to exploit these primitives to develop some small distributed applications.

2.1 Introduction

Developing trustworthy distributed applications can be a challenging task. A key issue is that the services that compose a distributed application may be under the governance of different providers, which may compete against each other. Furthermore, services interact through open networks, where competitors and adversaries can try to exploit their vulnerabilities.

A possible countermeasure to these issues is to use *behavioural contracts* to discipline the interaction among services. These are formal descriptions of service behaviour, which can be used at static or dynamic time to discover and bind services, and to guarantee that they interact in a protected manner: namely, when a service does not behave as prescribed by its contract, it can be blamed and sanctioned for a contract breach.

In previous work [7] we presented a middleware that uses behavioural contracts to discipline the interactions among distrusting services. Since it supports the COntract-Oriented paradigm, we called it "CO_2 middleware".

Figure 2.1 illustrates the main features of the CO_2 middleware. In (1), the participant A advertises its contract to the middleware, making it available to other participants. In (2), the middleware determines that the contracts of A and B are *compliant*: this means that interactions which respect the contracts are deadlock-free. Upon compliance, the middleware establishes a session through which the two participants can interact. This interaction consists of sending and receiving messages, similarly to a standard message-oriented middleware (MOM): for instance, in (3) participant A

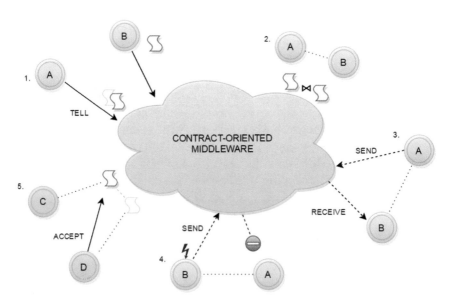

Figure 2.1   Contract-oriented interactions in the CO_2 middleware.

delivers to the middleware a message for B, which can then collect it from the middleware.

Unlike standard MOMs, the interaction happening in each session is monitored by the middleware, which checks whether contracts are respected or not. In particular, the execution monitor verifies that actions occur when prescribed by their contracts, and it detects when some expected action is missing. For instance, in (4) the execution monitor has detected an attempt of participant B to do some illegal action. Upon detection of a contract violation, the middleware punishes the culprit, by suitably decreasing its *reputation*. This is a measure of the trustworthiness of a participant in its past interactions: the lower its reputation is, the lower the probability of being able to establish new sessions with it.

Item (5) shows another mechanism for establishing sessions: here, the participant C advertises a contract, and D just *accepts* it. This means that the middleware associates D with the *canonical compliant* of the contract of C, and it establishes a session between C and D. The interaction happening in this session then proceeds as described previously.

In this chapter we illustrate how to program contract-oriented distributed applications which run on the CO_2 middleware. A public instance of the middleware is accessible from co2.unica.it, together with all examples and experiments we carried out.

2.2 Timed Session Types

The CO_2 middleware currently supports two kinds of contracts:

- first-order binary session types [18];
- timed session types (TSTs) [6].

In this section we illustrate TSTs with the help of a small case study, an online store which receives orders from customers. The use of *un*timed session types in contract-oriented applications is discussed in the literature [3, 4, 8].

2.2.1 Specifying Contracts

Timed session types extend binary session types [18, 26] with clocks and timing constraints, similarly to the way timed automata [1] extend (classic) finite state automata. We informally describe the syntax of TSTs below, and we refer to [5, 6] for the full technical development.

Guards. Guards describe timing constraints, and they are conjunctions of simple guards of the form t ∘ d, where t is a *clock*, d ∈ ℕ, and ∘ is a relation in <, <=, =, >=, >. For instance, the guard t<60,u>10 is true whenever the value of clock t is less than 60, *and* the value of clock u is greater than 10. The value of clocks is in $\mathbb{R}_{\geq 0}$, like for timed automata.

Send and receive. A TST describes the behaviour of a single participant A at the end-point of a session. Participants can perform two kinds of actions:

- a *send action* !a{g;t1,...,tk} stipulates that A will output a message with label a in a time window where the guard g is true. The clocks t1,...,tk will be reset after the output is performed.
- a *receive action* ?a{g;t1,...,tk} stipulates that A will be available to receive a message with label a at *any instant* within the time window where the guard g is true. The clocks t1,...,tk will be reset after the input is received.

When g = true, the guard can be omitted.

For instance, consider the contract store1 between the store and a customer, from the point of view of the store.

```
store1 = "?order{;t} . !price{t<60}"
```

The store declares that it will receive an order at any time. After it has been received, the store will send the corresponding price within 60 seconds.

Internal and external choices. TSTs also feature two forms of choice:

- !a1{g1;R1} + ... + !an{gn;Rn}

  This is an *internal choice*, stipulating that A will decide at run-time which one of the output actions !ai{gi;Ri} (with 1 ≤ i ≤ n) to perform, and at which time instant. After the action is performed, all clocks in the set Ri = {t1,...,tk} are reset.

- ?a1{g1;R1} & ...& ?an{gn;Rn}

  This is an *external choice*, stipulating that A will be able to receive any of the inputs !ai{gi;Ri}, in the declared time windows. The actual choice of the action, and of the instant when it is performed, will be made by the participant at the other endpoint of the session. After the action is performed, all clocks in the set Ri = {t1,...,tk} are reset.

With these ingredients, we can refine the contract of our store as follows:

```
store2 = "?order{;t} . (!price{t<60} + !unavailable{t<10})"
```

This version of the contract deals with the case where the store receives an unknown or invalid product code. In this case, the internal choice allows the store to inform the buyer that the requested item is unavailable.

Recursion. The contracts shown so far can only handle a bounded (statically known) number of interactions. We can overcome this limitation by using recursive TSTs. For instance, the contract store3 below models a store which handles an arbitrary number of orders from a buyer:

```
store3 = "REC 'x' [?addtocart{t<60;t}.'x'
              & ?checkout{t<60;t}.(
                   !price{t<20;t}.(
                        ?accept{t<10} & ?reject{t<10})
              + !unavailable{t<20})]"
```

The contract store3 allows buyers to add some item to the cart, or checkout. When a buyer chooses addtocart, the store must allow him to add more items: this is done recursively. After a checkout, the store must send the overall price, or inform the buyer that the requested items are unavailable. If the store sends a price, it must expect a response from the buyer, who can either accept or reject the price.

Context. Action labels are grouped into *contexts*, which can be created and made public through the middleware APIs. Each context defines the labels related to an application domain, and it associates each label with a *type* and a *verification link*. The type (e.g., int, string) is that of the messages exchanged with that label. The verification link is used by the runtime monitor (described later on in this section) to delegate the verification of messages to a trusted third party. For instance, the middleware supports Paypal as a verification link for online payments [7].

2.2.2 Compliance

Besides being used to specify the interaction protocols between pairs of services, TSTs feature the following primitives:

- a decidable notion of *compliance* between two TSTs;
- an algorithm to detect if a TST admits a compliant one;
- a computable *canonical compliant* construction.

These primitives are exploited by the CO_2 middleware to establish sessions between services: more specifically, the middleware only allows interactions between services with compliant contracts. Intuitively, compliance guarantees that, if *all* services respect *all* their contracts, then the overall distributed application (obtained by composing the services) will not deadlock.

Below we illustrate the primitives of TSTs by examples; a comprehensive formal treatment is in [5].

Informally, two TSTs are *compliant* if, in the interactions where both participants respect their contract, the deadlock state is not reachable (see [5] for details). For instance, recall the simple version of the store contract:

```
store1 = "?order{;t} . !price{t<60}"
```

and consider the following buyer contracts:

```
buyer1 = "!order{;u} . ?price{u<70}"
buyer2 = "!order{;u} . (?price{u<70} & ?unavailable)"
buyer3 = "!order{;u} . (?price{u<30} & ?unavailable)"
buyer4 = "!order{u<20} . ?price{u<70}"
```

We have that:

- `store1` and `buyer1` are compliant: indeed, the time frame where `buyer1` is available to receive `price` is larger than the one where the store can send;
- `store1` and `buyer2` are compliant: although the action `?unavailable` enables a further interaction, this is never chosen by the store `store1`.
- `store1` and `buyer3` are *not* compliant, because the store may choose to send `price` 60 seconds after he got the order, while `buyer2` is only able to receive within 30 seconds.
- `store1` and `buyer4` are *not* compliant. Here the reason is more subtle: assume that the buyer sends the order at time 19: at that point, the store receives the order and resets the clock `t`; after that, the store has 60 seconds more to send `price`. Now, assume that the store chooses to send `price` after 59 seconds (which fits within the declared time window of 60 seconds). The total elapsed time is 19+59=78 seconds, but the buyer is only able to receive before 70 seconds.

We can check if two contracts are compliant through the middleware Java APIs[1]. We show how to do this through the Groovy[2] interactive shell[3].

```
cS1 = new TST(store1)
cS1.isCompliantWith(new TST(buyer1))
>>> true
cS1.isCompliantWith(new TST(buyer3))
>>> false
```

Consider now the second version of the store contract:

```
store2 = "?order{;t} . (!price{t<60} + !unavailable{t<10})"
```

The contract `store2` is compliant with the buyer contract `buyer2` discussed before, while it is *not* compliant with:

```
buyer5 = "!order{;u} . (?price{u<90})"
buyer6 = "!order{;u} . (?price{u<90} + ?unavailable{u>5,u<12})"
```

The problem with `buyer5` is that the buyer is only accepting a message labelled `price`, while `store2` can also choose to send `unavailable`. Although this option is present in `buyer6`, the latter contract is not compliant with `store2` as well. In this case the reason is that the time window for receiving `unavailable` does not include that for sending it (recall that the sender can choose any instant satisfying the guard in its output action). To illustrate some less obvious aspects of compliance, consider the following buyer contract:

```
buyer7 = "!order{u<100} . ?price{u<70}"
```

This contract stipulates that the buyer can wait up to 100 seconds for sending an order, and then she can wait until 60 seconds (from the *start* of the session), to receive the price from the store.

Now, assume that some store contract is compliant with `buyer7`. Then, the store must be able to receive the `order` at least until time 100. If the buyer chooses to send the `order` at time 90 (which is allowed by contract `buyer7`), then the store would never be able to send `price` before time 70. Therefore, no contract can be compliant with `buyer7`.

The issue highlighted by the previous example must be dealt with care: if one publishes a service whose contract does not admit a compliant one, then

---

[1] co2.unica.it/downloads/co2api/

[2] groovy-lang.org/download.html

[3] On Unix-like systems, copy the API's jar in $HOME/.groovy/lib/. Then, add import co2api.* to $HOME/.groovy/groovysh.rc, and run groovysh.

the middleware will never connect that service with others. To check whether a contract admits a compliant one, we can query the middleware APIs:

```
cB7 = new TST(buyer7)
>>> !order{u<100} . ?price{u<70}

cB7.hasCompliant()
>>> false
```

Recall from Section 2.1 that the CO_2 middleware also allows a service to *accept* another service's contract, as per item (5) in Figure 2.1. E.g., assume that the store has advertised the contract store2 above. When the buyer uses the primitive accept, the middleware associates the buyer with the *canonical compliant* of store2, constructed through the method dualOf, i.e.:

```
cS2 = new TST(store2)
>>> ?order{;t} . (!price{t<60} + !unavailable{t<10})

cB2 = cS2.dualOf()
>>> !order{;t} . (?price{t<60} & ?unavailable{t<10})
```

Intuitively, if a TST admits a compliant one, then its canonical compliant is constructed as follows:

1. output labels !a are translated into input labels ?a, and *vice versa*;
2. internal choices are translated into external choices, and *vice versa*;
3. prefixes and recursive calls are preserved;
4. guards are suitably adjusted in order to ensure compliance.

Consider now the following contract of a store which receives an order and a coupon, and then sends a discounted price to the buyer:

```
store4 = "?order{t<60} . ?coupon{t<30;t} . !price{t<60}"
```

In this case store4 admits a compliant one, but this cannot be obtained by simply swapping input/output actions and internal/external choices.

```
cS4 = new TST(store4)
cB4 = new TST("!order{t<60} . !coupon{t<30;t} . ?price{t<60})")
cS4.isCompliantWith(cB4)
>>> false
```

Indeed, the canonical compliant construction gives:

```
cB5 = cS4.dualOf()
>>> !order{t<30} . ?coupon{t<30;t} . ?price{t<60}
```

2.2.3 Run-Time Monitoring of Contracts

In order to detect (and sanction) contract violations, the CO_2 middleware monitors all the interactions that happen through sessions. The monitor guarantees that, in each reachable configuration, only one participant can be "on duty" (i.e., she has to perform some actions); and if no one is on duty nor culpable, then both participants have reached success. Here we illustrate how runtime monitoring works, by making a store and a buyer interact.

To this purpose, we split the paper in two columns: in the left column we show the store behaviour, while in the right column we show the buyer. We assume that both participants call the middleware APIs through the Groovy shell, as shown before. Note that the interaction between the two participants is asynchronous: when needed, we will highlight the points where one of the participants performs a time delay.

Both participants start by creating a connection co2 with the middleware:

```
usr = "testuser1@gmail.com"       usr = "testuser2@gmail.com"
pwd = "testuser1"                 pwd = "testuser2"
co2 = new CO2ServerConnection(    co2 = new CO2ServerConnection(
    usr,pwd)                          usr,pwd)
```

Then, the participants create their contracts, and advertise them to the middleware through the primitive tell. The variables pS and pB are the handles to the published contracts.

```
cS = new TST(store2)              cB = new TST(buyer2)
pS = cS.toPrivate(co2).tell()     pB = cB.toPrivate(co2).tell()
```

Now the middleware has two compliant contracts in its collection, hence it can establish a session between the store and the buyer. To obtain a handle to the session, both participants use the blocking primitive waitForSession:

```
sS = pS.waitForSession()          sB = pB.waitForSession()
```

At this point, participants can query the session to see who is "on duty" (namely, one is on duty if the contract prescribes her to perform the next action), and to check if they have violated the contract:

```
sS.amIOnDuty()                    sB.amIOnDuty()
>>> false                         >>> true
sS.amICulpable()                  sB.amICulpable()
>>> false                         >>> false
```

Note that the first action must be performed by the buyer, who must send the `order`. This is accomplished by the `send` primitive. Dually, the store waits for the receipt of the message, using the `waitForReceive` primitive:

```
msg = sS.waitForReceive()          // send at an arbitrary time
msg.getStringValue()               sB.send("order", "0123")
>>> 0123
sS.amIOnDuty()                     sB.amIOnDuty()
>>> true                           >>> false
```

Since there are no time constraints on sending `order`, this action can be successfully performed at any time; once this is done, the `waitForReceive` unlocks the store. The store is now on duty, and it must send `price` within 60 seconds, or `unavailable` within 10 seconds. Now, assume that the store tries to send `unavailable` after the deadline:

```
// wait more than 10 seconds    msg = sB.waitForReceive()

sS.send("unavailable")          >>> ContractViolationException:
>>> ContractException           "The other participant is culpable"
```

On the store's side, the `send` throws a `ContractException`; on the buyer side, the `waitForReceive` throws an exception which reports the violation of the store. At this point, if the two participants check the state of the session, they find that none of them is still on duty, and that the store is culpable:

```
session.amIOnDuty()             session.amIOnDuty()
>>> false                       >>> false
session.amICulpable()           session.amICulpable()
>>> true                        >>> false
```

At this point, the session is terminated, and the reputation of the store is suitably decreased.

2.3 Contract-Oriented Programming

In this section we develop some simple contract-oriented services, using the middleware APIs via their Java binding[4].

---

[4]Full code listings are available at `co2.unica.it`.

2.3.1 A Simple Store

We start with a basic store service, which advertises the contract `store2`:

```
1 String store2 ="?order{;t}.(!price{t<60} + !unavailable{t<10})";
2 TST c = new TST(store2);

4 CO2ServerConnection co2 =
5   new CO2ServerConnection("testuser@co2.unica.it", "pa55w0rd");
6 Private r = c.toPrivate(co2);
7 Public  p = r.tell();              //advertises the contract store2

9 Session s = p.waitForSession();//blocks until session is created
10 String id = s.waitForReceive().getStringValue();

12 if(isAvailable(id)) { s.send("price", getPrice(id)); }
13 else { s.send("unavailable"); }
```

At lines 1-2, the store constructs a TST `c` for contract `store2`. At lines 4-5, the store connects to the middleware, providing its credentials. At line 6, the `Private` object represents the contract in a state where it has not been advertised to the middleware yet. To advertise the contract, we invoke the `tell` method at line 7. This call returns a `Public` object, modelling a latent contract that can be "fused" with a compliant one to establish a new session. At line 9, the store waits for a session to be established; the returned `Session` object allows the store to interact with a buyer. At line 10, the store waits for the receipt of a message, containing the code of the product requested by the buyer. At lines 12-13, the store sends the message `price` (with the corresponding value) if the item is available, otherwise it sends `unavailable`.

2.3.2 A Simple Buyer

We now show a buyer that can interact with the store. This buyer just accepts the already published contract `store2`. The contract is identified by its hash, which is obtained from `Public.getContractID()`.

```
1 CO2ServerConnection co2 = new CO2ServerConnection(...);

3 String storeCID = "0x...";
4 Integer desiredPrice = 10;

6 Public  p = Public.accept(co2, storeCID, TST.class);
7 Session s = p.waitForSession();

9 s.send("order", "11235811");
```

```
11 try {
12     Message m = s.waitForReceive();
13     switch (m.getLabel()) {
14     case "unavailable": break;
15     case "price":
16         Integer price = Integer.parseInt(m.getStringValue());
17         if (price > desiredPrice) { /*abort the purchase*/ }
18         else { /*proceed with the purchase*/ }
19     }
20 } catch(ContractViolationException e){/*The store is culpable*/}
```

At line 6, the buyer accepts the store's contract, identified by `storeCID`. The call to `Public.accept` returns a `Public` object. At this point a session with the store is already established, and `waitForSession` just returns the corresponding `Session` object (line 7). Now, the buyer sends the item code (line 9), waits for the store response (line 12), and finally in the try-catch statement it handles the messages `price` and `unavailable`.

Note that the `accept` primitive allows a participant to establish sessions with a chosen counterpart; instead, this is not allowed by the `tell` primitive, which can establish a session whenever two contracts are compliant.

2.3.3 A Dishonest Store

Consider now a more complex store, which relies on external distributors to retrieve items. As before, the store takes an order from the buyer; however, now it invokes an external distributor if the requested item is not in stock. If the distributor can provide the item, then the store confirms the order to the buyer; otherwise, it informs the buyer that the item is unavailable.

Our first attempt to implement this refined store is the following.

```
1    TST cB = new TST(store2);
2    TST cD = new TST("!req{;t}.(?ok{t<10} & ?no{t<10})");

4    Public  pB = cB.toPrivate(co2).tell();
5    Session sB = pB.waitForSession();
6    String  id = sB.waitForReceive().getStringValue();

8    if (isAvailable(id)) { // handled internally
9        sB.send("price", getPrice(id));
10   }
11   else { // handled with a distributor
12       Public  pD = cD.toPrivate(co2).tell();
13       Session sD = pD.waitForSession();

15       sD.send("req", id);
```

```
16        Message mD = sD.waitForReceive();

18        switch (mD.getLabel()) {
19        case "no" : sB.send("unavailable"); break;
20        case "ok" : sB.send("price", getPrice(id)); break;
21    }
22  }
```

At lines 1-2 we construct two TSTs: cB for interacting with buyers, and cD for interacting with distributors. In cD, the store first sends a request for some item to the distributor, and then waits for an ok or no answer, according to whether the distributor is able to provide the requested item or not. At lines 4-6, the store advertises cB, and it waits for a buyer to join the session; then, it receives the order, and checks if the requested item is in stock (line 8). If so, the store sends the price of the item to the buyer (line 9).

If the item is not in stock, the store advertises cD to find a distributor (lines 12-13). When a session sD is established, the store forwards the item identifier to the distributor (line 15), and then it waits for a reply. If the reply is no, the store sends unavailable to the buyer, otherwise it sends a price.

Note that this implementation of the store is *dishonest*, namely it may violate contracts [11]. This happens in the following two cases:

1. Assume that the store has received the buyer's order, but the requested item is not in stock. Then, the store advertises the contract cD to find a distributor. Note that there is no guarantee that the session sD will be established within a given deadline, nor that it will be established at all. If more than 60 seconds pass on the waitForSession at line 13, the store becomes culpable with respect to the contract cB. Indeed, such contract requires the store to perform an action before 60 seconds (10 seconds if the action is unavailable).

2. Moreover, if the session sD is established in timely fashion, a slow or unresponsive distributor could make the store violate the contract cB. For instance, assume that the distributor sends message no after nearly 10 seconds. In this case, the store may not have enough time to send unavailable to the buyer within 10 seconds, and so it becomes culpable at session sB.

We have simulated the scenario described in Item 1, by making the store interact with slow or unresponsive distributors (see Figure 2.2).

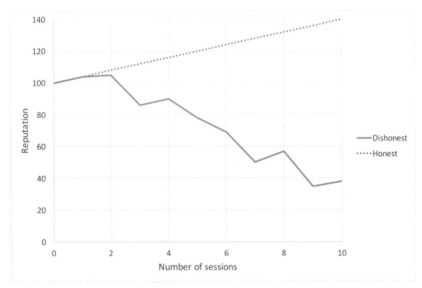

Figure 2.2   Reputation of the dishonest and honest stores as a function of the number of sessions with malicious distributors.

The experimental results show that, although the store is not culpable in all the sessions, its reputation decreases over time. Recovering from such situation is not straightforward, since the reputation system of the CO_2 middleware features defensive techniques against self-promoting attacks [25].

2.3.4 An Honest Store

In order to implement an honest store, we must address the fact that, if the distributor delays its message to the maximum allowed time, the store may not have enough time to respond to the buyer. To cope with this scenario, we adjust the timing constraints in the contract between the store and the distributor, and we implement a revised version of the store as follows.

```
1    TST  cB = new  TST(store2);
2    TST  cD = new  TST("!req{;t} . (?ok{t<5} & ?no{t<5})");

4    Public  pB = cB.toPrivate(co2).tell();
5    Session sB = pB.waitForSession();
6    String  id = sB.waitForReceive().getStringValue();

8    if (isAvailable(id)) { // handled internally
9        sB.send("price", getPrice(id));
10   }
```

```
11   else { // handled with the distributor
12       Public pD = cD.toPrivate(co2).tell(3 * 1000);
13       try {
14           Session sD = pD.waitForSession();
15           sD.send("req", id);

17           try{
18               Message mD = sD.waitForReceive();

20               switch (mD.getLabel()) {
21               case "no": sB.send("unavailable"); break;
22               case "ok": sB.send("price", getPrice(id)); break;
23               }
24           } catch(ContractViolationException e){
25               //the distributor did not respect its contract
26               sB.send("unavailable");
27           }
28       } catch(ContractExpiredException e) {
29           //no distributor found
30           sB.send("unavailable");
31       }
32   }
```

The parameter in the `tell` at line 12 specifies a deadline of 3 seconds: if the session sD is not established within the deadline, the contract cD is retracted from the middleware, and a `ContractExpiredException` is thrown. The store catches the exception at line 28, sending unavailable to the buyer.

Instead, if the session sD is established, the store forwards the item identifier to the distributor (line 15), and then waits for the receipt of a response from it. If the distributor sends neither ok nor no within the deadline specified in cD (5 seconds), the middleware assigns the blame to the distributor for a contract breach, and unblocks the `waitForReceive` in the store with a `ContractViolationException` (line 24). In the exception handler, the store fulfils the contract cB by sending unavailable to the buyer.

2.3.5 A Recursive Honest Store

We now present another version of the store, which uses the recursive contract store3 on page 31. As in the previous version, if the buyer requests an item that is not in stock, the store resorts to an external distributor.

```
1 TST cB = new TST(store3);
2 TST cD = new TST("!req{;t}.(?ok{t<5} & ?no{t<5})");

4 Public  pB = cB.toPrivate(co2).tell();
5 Session sB = pB.waitForSession();
```

```
6 List<String> orders = new ArrayList<>();
7 Message mB;

9 try {
10     do {
11         mB = sB.waitForReceive();
12         if (mB.getLabel().equals("addtocart")){
13             orders.add(mB.getStringValue());
14         }
15     } while(!mB.getLabel().equals("checkout"));

17     if (isAvailable(orders)) { // handled internally
18         sB.send("price", getPrice(orders));
19         String res = sB.waitForReceive().getLabel();
20         switch (res){
21             case "accept": // handle the order
22             case "reject": // terminate
23         }
24     }
25     else { // handled with the distributor
26         Public pD = cD.toPrivate(co2).tell(5 * 1000);
27         try {
28             Session sD = pD.waitForSession();
29             sD.send("req", getOutOfStockItems(orders));
30             try{
31                 switch (sD.waitForReceive().getLabel()) {
32                 case "no": sB.send("unavailable"); break;
33                 case "ok":
34                     sB.send("price", getPrice(orders));
35                     try{
36                         String res =
37                           sB.waitForReceive().getLabel();
38                         switch (res) {
39                         case "accept": // handle the order
40                         case "reject": // terminate
41                         }
42                     }
43                     catch (ContractViolationException e) {
44                         //the buyer is culpable, terminate
45                     }
46                 }
47             } catch (ContractViolationException e){
48                 //the distributor did not respect its contract
49                 sB.send("unavailable");
50             }
51         }
52         catch (ContractExpiredException e) {
53             //no distributor found
54             sB.send("unavailable");
55         }
56     }
57 } catch(ContractViolationException e){/*the buyer is culpable*/}
```

After advertising the contract cB, the store waits for a session sB with the buyer (lines 4-5). After the session is established, the store can receive addtocart multiple times: for each addtocart, it saves the corresponding item identifier in a list. The loop terminates when the buyer selects checkout. If all requested items are available, the store sends the total price to the buyer (line 18). After that, the store expects either accept or reject from the buyer. If the buyer does not respect his deadlines, an exception is thrown, and it is caught at line 57. If the buyer replies on time, the store advertises the contract cD, and waits for a session sD with the distributor (lines 26-28). If the session is not established within 5 seconds, an exception is thrown. The store handles the exception at line 52, by sending unavailable to the buyer. If a session with the distributor is established within the deadline, the store requests the unavailable items, and waits for a response (line 31). If the distributor sends no, the store answers unavailable to the buyer (line 32). If the distributor sends ok, then the interaction between store and buyer proceeds as if the items were in stock. If the distributor does not reply within the deadline, an exception is thrown. The store handles it at line 47, by sending unavailable to the buyer. An untimed specification of this store is proved honest in [4]. We conjecture that also this timed version of the store respects contracts in all possible contexts.

2.4 Conclusions

We have explored the use of behavioural contracts as service-level agreements among the components of a distributed application. In particular, we have considered a middleware where services can advertise contracts (in the form of timed session types, TSTs), and interact through sessions, which are only created between services with compliant contracts. The primitives of the middleware exploit the theory of TSTs: in particular, a decidable notion of compliance between TSTs, a decidable procedure to detect when a TST admits a compliant one, and a decidable runtime monitoring. The middleware has been validated in [7] through a series of experiments, which measure the scalability of the approach when the number of exchanged contracts grows, and the effectiveness of the reputation system.

Although the current version of the middleware only features binary (either timed or untimed) session types as contracts, the underlying idea can be extended to other contract models. Indeed, the middleware only makes mild assumptions about the nature of contracts, e.g., that they feature:

(i) monitorable send and receive actions, (ii) some notion of accepting a contract or a role, or (iii) some notion of compliance with a sound (but not necessarily complete) verification algorithm. Other timed models of contracts would be ideal candidates for extensions of the middleware. For instance, communicating timed automata [13] (which are timed automata with unbounded communication channels) would allow for multi-party sessions.

Security issues should be seriously taken into account when developing contract-oriented applications. As we have shown for the dishonest online store in Section 2.3, adversaries could make a service sanctioned by exploiting discrepancies between its contracts and its actual behaviour. Since these mismatches are not always easy to spot, analysis techniques are needed in order to ensure that a service will not be susceptible to this kind of attacks. A starting point could be the analyses in [8, 9], that can detect whether a contract-oriented specification is honest; the Diogenes toolchain [3] extends this check to Java code. Since these analyses do not take into account time constraints, further work is needed to extend these techniques to timed applications.

2.4.1 Related Work

The theoretical foundations of our middleware are timed session types and CO_2 [12, 10], a specification language for contract-oriented services. The middleware implements the main primitives of CO_2 (tell, send, receive), and it introduces new concepts, such as the accept primitive, time constraints, and reputation.

From the theoretical viewpoint, the idea of constraint-based interactions has been investigated in other process calculi, such as Concurrent Constraint Programming (CCP) [24], and cc-pi [16]. The kind of interactions they induce is quite different from ours. In CCP, processes can interact by telling and asking for the validity of constraints on a global constraint store. In cc-pi, interaction is a mix of name communication *à la* π-calculus [21] and tell *à la* CCP (which is used to put constraints on names). In cc-pi consistency plays a crucial role: tells *restrict* the future interactions with other processes, since adding constraints can lead to more inconsistencies; by contrast, in our middleware advertising a contract *enables* interaction with other services, so consistency is immaterial, but compliance is a key notion.

Several formalisms for expressing timed communication protocols have been proposed over the years. The work [14] addresses a timed extension of multi-party asynchronous session types [19]. Unlike ours, the approach pursued in [14] is top-down: a *global type*, specifying the overall communication protocol of a set of services, is projected onto a set of *local types*. Then, a composition of services preserves the properties of the global type (e.g., deadlock-freedom) if each service type-checks against the associated local type. The CO_2 middleware, instead, fosters a bottom-up approach to service composition. Both our approach and [14, 23] use runtime monitoring to detect contract violations and assign the blame to the party that is responsible for a contract violation. The CO_2 middleware also exploits these data in its reputation system.

The work [13] studies *communicating timed automata*, a timed version of communicating finite-state machines [15]. In this model, participants in a network communicate asynchronously through bi-directional FIFO channels; similarly to [14], clocks, guards and resets are used to impose time constraints on when communications can happen. An approximate (sound, but not complete) decidable technique allows one to check when a system of automata enjoys progress. This technique is based on *multiparty compatibility*, a condition that guarantees deadlock-freedom of untimed systems [20].

From the application viewpoint, several works have investigated the problem of *service selection* in open dynamic environments [2, 22, 27, 28]. This problem consists in matching client requests with service offers, in a way that, among the services respecting the given functional constraints, the one that maximises some *non-functional* constraints is selected. These non-functional constraints are often based on quality of service (QoS) metrics, e.g. cost, reputation, guaranteed throughput or availability, etc. The selection mechanism featured in our middleware does not search for the "best" contract that is compliant with a given one (actually, typical compliance relations in behavioural contracts are qualitative, rather than quantitative); the only QoS parameter we take into account is the reputation of services. In some approaches [2, 28] clients can require a sequence of tasks together with a set of non-functional constraints, and the goal is to find an assignment of tasks to services that optimises all the given constraints. There are two main differences between these approaches and ours. First, unlike behavioural contracts, tasks are considered as atomic activities, not requiring any interaction between clients and services. Second, unlike ours, these approaches do not consider the possibility that a service may not fulfil the required task.

Some works have explored service selection mechanisms where functional constraints can be required in addition to QoS constraints [22]: the first are described by a web service ontology, while the others are defined as requested and offered ranges of basic QoS attributes. Runtime monitor and reputation systems are also implemented, which, similarly to ours, help to marginalise those services that do not respect the advertised QoS constraints. Some kinds of QoS constraints cannot be verified by the service broker, so their verification is delegated to clients. This can be easily exploited by malicious participants to carry on *slandering attacks* to the reputation system [17]: an attacker could destroy another participant's reputation by involving it in many sessions, and each time declare that the required QoS constraints have been violated. In our middleware there is no need to assume that participants are trusted, as the verification of contracts is delegated to the middleware itself and to trusted third parties.

Acknowledgments This work is partially supported by Aut. Reg. of Sardinia grants L.R.7/2007 CRP-17285 (TRICS), P.I.A. 2013 ("NOMAD"), and by EU COST Action IC1201 "Behavioural Types for Reliable Large-Scale Software Systems" (BETTY). Alessandro Sebastian Podda gratefully acknowledges Sardinia Regional Government for the financial support of her PhD scholarship (P.O.R. Sardegna F.S.E. Operational Programme of the Autonomous Region of Sardinia, European Social Fund 2007–2013 – Axis IV Human Resources, Objective 1.3, Line of Activity 1.3.1).

References

[1] Rajeev Alur and David L. Dill. A theory of timed automata. *Theor. Comput. Sci.*, 126(2):183–235, 1994.

[2] Danilo Ardagna and Barbara Pernici. Adaptive service composition in flexible processes. *IEEE Trans. Software Eng.*, 33(6):369–384, 2007.

[3] Nicola Atzei and Massimo Bartoletti. Developing honest Java programs with Diogenes. In *Formal Techniques for Distributed Objects, Components, and Systems (FORTE)*, volume 9688 of *LNCS*, pages 52–61. Springer, 2016.

[4] Nicola Atzei, Massimo Bartoletti, Maurizio Murgia, Emilio Tuosto, and Roberto Zunino. Contract-oriented design of distributed applications: a tutorial. tcs.unica.it/papers/diogenes-tutorial.pdf, 2016.

[5] Massimo Bartoletti, Tiziana Cimoli, and Maurizio Murgia. Timed session types, 2015. Pre-print available at tcs.unica.it/papers/tst.pdf

[6] Massimo Bartoletti, Tiziana Cimoli, Maurizio Murgia, Alessandro Sebastian Podda, and Livio Pompianu. Compliance and subtyping in timed session types. In *Formal Techniques for Distributed Objects, Components, and Systems (FORTE)*, volume 9039 of *LNCS*, pages 161–177. Springer, 2015.

[7] Massimo Bartoletti, Tiziana Cimoli, Maurizio Murgia, Alessandro Sebastian Podda, and Livio Pompianu. A contract-oriented middleware. In *Formal Aspects of Component Software (FACS)*, volume 9539 of *LNCS*, pages 86–104. Springer, 2015. co2.unica.it

[8] Massimo Bartoletti, Maurizio Murgia, Alceste Scalas, and Roberto Zunino. Verifiable abstractions for contract-oriented systems. *Journal of Logical and Algebraic Methods in Programming (JLAMP)*, 86:159–207, 2017.

[9] Massimo Bartoletti, Alceste Scalas, Emilio Tuosto, and Roberto Zunino. Honesty by typing. *Logical Methods in Computer Science*, 12(4), 2016. Pre-print available at: arxiv.org/abs/1211.2609

[10] Massimo Bartoletti, Emilio Tuosto, and Roberto Zunino. Contract-oriented computing in CO_2. *Sci. Ann. Comp. Sci.*, 22(1):5–60, 2012.

[11] Massimo Bartoletti, Emilio Tuosto, and Roberto Zunino. On the realizability of contracts in dishonest systems. In *COORDINATION*, volume 7274 of *LNCS*, pages 245–260. Springer, 2012.

[12] Massimo Bartoletti and Roberto Zunino. A calculus of contracting processes. In *IEEE Symposium on Logic in Computer Science (LICS)*, pages 332–341. IEEE Computer Society, 2010.

[13] Laura Bocchi, Julien Lange, and Nobuko Yoshida. Meeting deadlines together. In *CONCUR*, volume 42 of *LIPIcs*, pages 283–296. Schloss Dagstuhl – Leibniz-Zentrum fuer Informatik, 2015.

[14] Laura Bocchi, Weizhen Yang, and Nobuko Yoshida. Timed multiparty session types. In *CONCUR*, volume 8704 of *LNCS*, pages 419–434. Springer, 2014.

[15] Daniel Brand and Pitro Zafiropulo. On communicating finite-state machines. *J. ACM*, 30(2):323–342, 1983.

[16] Maria Grazia Buscemi and Ugo Montanari. CC-Pi: A constraint-based language for specifying service level agreements. In *European Symposium on Programming (ESOP)*, volume 4421 of *LNCS*, pages 18–32. Springer, 2007.

[17] Kevin J. Hoffman, David Zage, and Cristina Nita-Rotaru. A survey of attack and defense techniques for reputation systems. *ACM Comput. Surv.*, 42(1), 2009.

[18] Kohei Honda, Vasco T. Vasconcelos, and Makoto Kubo. Language primitives and type disciplines for structured communication-based programming. In *European Symposium on Programming (ESOP)*, volume 1381 of *LNCS*, pages 22–138. Springer, 1998.

[19] Kohei Honda, Nobuko Yoshida, and Marco Carbone. Multiparty asynchronous session types. In *ACM SIGPLAN-SIGACT Symposium on Principles of Programming Languages (POPL)*, pages 273–284. ACM, 2008.

[20] Julien Lange, Emilio Tuosto, and Nobuko Yoshida. From communicating machines to graphical choreographies. In *ACM SIGPLAN-SIGACT Symposium on Principles of Programming Languages (POPL)*, pages 221–232. ACM, 2015.

[21] Robin Milner, Joachim Parrow, and David Walker. A Calculus of Mobile Processes, I and II. *Information and Computation*, 100(1):1–40,41–77, September 1992.

[22] A. Mukhija, Andrew Dingwall-Smith, and D.S. Rosenblum. QoS-aware service composition in Dino. In *ECOWS*, volume 5900 of *LNCS*, pages 3–12. Springer, 2007.

[23] Rumyana Neykova, Laura Bocchi, and Nobuko Yoshida. Timed runtime monitoring for multiparty conversations. In *BEAT*, volume 162 of *EPTCS*, pages 19–26, 2014.

[24] Vijay A. Saraswat and Martin C. Rinard. Concurrent constraint programming. In *ACM SIGPLAN-SIGACT Symposium on Principles of Programming Languages (POPL)*, pages 232–245. ACM, 1990.

[25] Mudhakar Srivatsa, Li Xiong, and Ling Liu. TrustGuard: countering vulnerabilities in reputation management for decentralized overlay networks. In *International Conference on World Wide Web (WWW)*, pages 422–431. ACM, 2005.

[26] Kaku Takeuchi, Kohei Honda, and Makoto Kubo. An interaction-based language and its typing system. In *PARLE*, pages 398–413, 1994.

[27] Tao Yu, Yue Zhang, and Kwei-Jay Lin. Efficient algorithms for Web services selection with end-to-end QoS constraints. *ACM Transactions on the Web*, 1(1):6, 2007.

[28] Liangzhao Zeng, Boualem Benatallah, Anne HH Ngu, Marlon Dumas, Jayant Kalagnanam, and Henry Chang. QoS-aware middleware for Web services composition. *IEEE Transactions on Software Engineering*, 30(5):311–327, 2004.

3

A Runtime Monitoring Tool for Actor-Based Systems

Duncan Paul Attard[1], Ian Cassar[1], Adrian Francalanza[1], Luca Aceto[2] and Anna Ingólfsdóttir[2]

[1]Department of Computer Science, Faculty of ICT, University of Malta, Malta
[2]School of Computer Science, Reykjavík University, Iceland

Abstract

This chapter discusses detectEr, an experimental runtime monitoring tool that can be used to formally verify concurrent systems developed in Erlang. Formal correctness properties in detectEr are expressed using a monitorable subset of Hennessy-Milner Logic with recursion, and synthesised into actor-based runtime monitors. Our exposition focusses on how the specification logic is enriched and extended with pattern-matching and conditional constructs which allow monitors to be adept at processing the data obtained dynamically from the system's execution trace. The tool leverages the native tracing functionality provided by the Erlang language platform so as to produce asynchronous monitors that can be instrumented to run alongside the system with minimal effort. To demonstrate how detectEr can be used in practice, this material also provides a hands-on guide that is especially aimed at users wishing to use our tool to monitor Erlang applications.

3.1 Introduction

Concurrency [30] refers to software systems whose functionality is expressed in terms of multiple *components* or processes that are specifically designed to work simultaneously with each other. In recent years, a *concurrency-oriented* [3] approach to software development has become increasingly commonplace, and is greatly favoured over monolithic-style approaches. This is, in part, owed to the rigidity that the latter types of architectures are synonymous with, where attempts at addressing scalability concerns usually lead to notoriously complex and often, inadequate solutions. Instead, concurrency recasts the notion of system design in a way that makes it possible to avail oneself of the multi-processor and multi-core platforms that are prevalent nowadays.

Formally ensuring the correctness of concurrent systems is an arduous, albeit necessary, task, especially since the interactions between fine-grained computational components can easily harbour subtle software bugs. Despite several success stories in their application to real-life applications, static verification techniques such as Model Checking (MC) scale poorly in concurrent scenarios, particularly because the system state space that needs to be *exhaustively* verified grows exponentially with respect to the size of the system [13, 14] – this is on account of the considerable number of possible execution paths that result from process interleaving. Moreover, situations often arise whereby verification cannot be performed statically (*i.e., pre-deployment*), as certain application components might not always be available for inspection before the system starts executing (*e.g.* in systems where functional components such as add-ons are downloaded and installed dynamically at runtime). There are also cases where the internal workings of a component (*e.g.* source code or execution graph) are not accessible and need to be treated as a black box. In these cases, Runtime Verification (RV) presents an appealing compromise towards ensuring the correctness of component-based applications. It is a *lightweight* verification technique that analyses the current runtime execution path of the system under scrutiny by considering partial executions incrementally, up to the current execution point [17, 26]. Its nature inherently circumvents the scalability issues attributed to MC and provides a means for post-deployment verification. Despite these advantages, RV has limited expressiveness and cannot be used to verify arbitrary specifications such as (general) liveness properties [27].

This chapter discusses the implementation of a prototype RV tool called detectEr, that targets concurrent, component-based applications written in Erlang. The presented material aspires to introduce this tool from a pragmatic standpoint, and thus omits technical details that may be abstruse to users of the tool. Interested readers should consult previous work [4, 19, 21] for details regarding the monitor synthesis and runtime behaviour of the monitoring tool.

The content that follows is organised into three sections. Section 3.2 gives a concise overview of the ideas behind RV and monitoring; this is followed by a review of mHML, the logic used for specifying correctness properties in our tool. Although this section helps to make the presentation self-contained, it may be safely skipped by readers familiar with the subject or merely interested in using the tool. Section 3.3 revisits the logic mHML from Section 3.2, and examines how it was adapted to address the practical requirements of users wishing to define correctness properties for Erlang concurrent programs. It also very briefly touches on the compilation process that transforms mHML specification scripts into executable runtime monitors. The final section takes the form of a hands-on tutorial that guides readers through the basic steps that need to be performed in order to instrument an Erlang application with runtime monitors using the tool.

3.2 Background

An executing system results in the generation of a (possibly infinite) sequence of events known as a *trace*. These events are the upshot of internal or external system behaviours, such as message exchanges between processes or function invocations. An *execution, i.e., a finite prefix* of an infinite trace, is consumed and processed by a software entity known as a *monitor*, tasked with the job of checking whether the execution provides enough evidence so as to determine whether a property is satisfied or violated. *Correctness specifications (properties)* serve to unambiguously describe the behaviour to which the executing system should adhere to. *Verdicts* denote monitoring outcomes and are assumed to be *definite* and non-retractable (*i.e.,* once given, cannot change). These typically consist of judgements relating to property violations and satisfactions, but may also include *inconclusive* verdicts for when the exhibited execution trace does not permit any definite judgement in relation to the property being monitored for [4, 6, 17, 19, 26]. A RV monitor for some correctness property is typically synthesised automatically from a high-level specification that finitely describes the property. Property specifications are given in terms of formal logics [4, 6, 7, 19] or other formalisms such as regular expressions [20] or automata [5, 15, 29]. Figure 3.1 depicts a correctness specification (denoted by φ) that is translated into an executable monitor, Monitor_{φ}, and instrumented with the running system. Trace events are sequentially analysed by the monitor whenever these are generated by the system through the instrumentation mechanism. Once the monitor reaches a verdict, it typically stops executing.

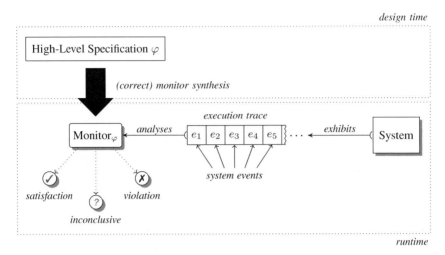

Figure 3.1   Runtime monitor synthesis and operational set-up.

3.2.1 Runtime Monitoring Criteria

Monitor synthesis, *i.e.,* the translation procedure from specifications to monitors and the associated system instrumentation, should ideally provide some *guarantees* of correctness. This covers both aspects that relate to how monitor verdicts correspond to the semantics of the property being monitored for (*e.g.* a monitor trace *rejection* should correspond to the system *violating* the property being monitored for), as well as requirements that the monitors instrumented with the executing system under scrutiny do *not* introduce fresh bugs *themselves* (consult our previous work [9, 18, 19, 21] for a detailed rendition on the subject). Equally important is the *efficiency* with which monitors execute, as this can adversely affect the monitored system or even alter its functional behaviour (*e.g.* slowdown due to inefficient monitors might cause the system to violate time-dependent properties that would not have been violated in the unmonitored system). A monitoring set-up that induces considerable levels of performance overhead may be deemed too costly to be feasibly used in practice.

3.2.2 A Branching-Time Logic for Specifying Correctness Properties

Specification logics can be categorised into two classes. *Linear-time* logics [6, 13, 26] treat time as having one possible future, and regard the behaviour of a system under observation in terms of execution traces or paths. On the other hand, *branching-time* logics [1, 13] make it possible to perceive time instances as potentially having more than one future, thereby giving rise to a *tree* of possible execution paths that may be (non-deterministically) taken by the executing system at runtime.

μHML [1, 25] is a branching-time logic that can be used to specify correctness properties over *Labelled Transition Systems* (LTSs) — graphs modelling the possible behaviours that can be exhibited by executing processes (see Figure 3.3 for a depiction of two LTSs). A LTS consists of a set of system states $p, q \in$ SYS, a set of actions $\alpha \in$ ACT, and finally, a ternary transition relation between states labelled by actions, $p \xrightarrow{\alpha} q$. When $p \xrightarrow{\alpha} q$ for no process q, the notation $p \xslashed{\alpha}$ is used. Additionally, $p \Longrightarrow q$ denotes $p(\xrightarrow{\tau})^* q$, whereas $p \xRightarrow{\alpha} q$, is written in place of $p \Longrightarrow \cdot \xrightarrow{\alpha} \cdot \Longrightarrow q$. Actions labelled by τ are used to denote unobservable (silent) actions that are performed by the system internally.

The μHML syntax, given in Figure 3.2, assumes a countable set of logical variables $X, Y \in$ LVAR, thereby allowing formulae to recursively express largest and least fixpoints using **max** $X.\varphi$ and **min** $X.\varphi$ respectively; these constructs bind free instances of the variable X in φ. In addition to the standard constructs for truth, falsity, conjunction and disjunction, the syntax also includes the necessity and possibility modalities.

The semantics of the logic is defined in terms of the function mapping μHML formulae φ to the set of LTS states $S \subseteq$ SYS satisfying them. Figure 3.2 describes the semantics for both open and closed formulae, and uses a map $\rho \in$ LVAR $\rightharpoonup 2^{\text{SYS}}$ from variables to sets of system states to enable an inductive definition on the structure of the formula φ. The formula **tt** is satisfied by all processes, while **ff** is satisfied by none; conjunctions and disjunctions bear the standard set-theoretic meaning of intersection and union. Necessity formulae $[\alpha]\varphi$ state that *for all* system executions producing event α (possibly none), the subsequent system state must then satisfy φ (*i.e.*, $\forall p', p \xRightarrow{\alpha} p'$ implies $p' \in [\![\varphi, \rho]\!]$ must hold). Possibility formulae $\langle \alpha \rangle \varphi$ require the existence of *at least* one system execution with event α whereby the subsequent state then satisfies φ (*i.e.*, $\exists p', p \xRightarrow{\alpha} p'$ and $p' \in [\![\varphi, \rho]\!]$ must hold). The recursive formulae **max** $X.\varphi$ and **min** $X.\varphi$ are respectively satisfied by the largest and least set of system states satisfying φ. The semantics of recursive variables X with respect to an environment instance ρ is given by the mapping of X in ρ, *i.e.*, the set of processes associated with X. *Closed* formulae (*i.e.*, formulae containing no free variables) are interpreted

Syntax

$$\varphi, \phi \in \mu\text{HML} ::= \; \textbf{ff} \;\; (\text{falsity}) \quad | \quad \textbf{tt} \;\; (\text{truth})$$
$$| \quad \varphi \wedge \phi \;\; (\text{conjunction}) \quad | \quad \varphi \vee \phi \;\; (\text{disjunction})$$
$$| \quad [\alpha]\varphi \;\; (\text{necessity}) \quad | \quad \langle \alpha \rangle \varphi \;\; (\text{possibility})$$
$$| \quad \textbf{max } X.\varphi \;\; (\text{max. fixpoint}) \quad | \quad \textbf{min } X.\varphi \;\; (\text{min. fixpoint})$$
$$| \quad X \;\; (\text{recursive variable})$$

Semantics

$$\llbracket \textbf{ff}, \rho \rrbracket \stackrel{\text{def}}{=} \emptyset \qquad\qquad \llbracket \textbf{tt}, \rho \rrbracket \stackrel{\text{def}}{=} \text{Sys}$$

$$\llbracket \varphi \wedge \phi, \rho \rrbracket \stackrel{\text{def}}{=} \llbracket \varphi, \rho \rrbracket \cap \llbracket \phi, \rho \rrbracket \qquad \llbracket \varphi \vee \phi, \rho \rrbracket \stackrel{\text{def}}{=} \llbracket \varphi, \rho \rrbracket \cup \llbracket \phi, \rho \rrbracket$$

$$\llbracket [\alpha]\varphi, \rho \rrbracket \stackrel{\text{def}}{=} \left\{ p \mid \forall p'. \; p \stackrel{\alpha}{\Longrightarrow} p' \text{ implies } p' \in \llbracket \varphi, \rho \rrbracket \right\} \qquad \llbracket \langle \alpha \rangle \varphi, \rho \rrbracket \stackrel{\text{def}}{=} \left\{ p \mid \exists p'. \; p \stackrel{\alpha}{\Longrightarrow} p' \text{ and } p' \in \llbracket \varphi, \rho \rrbracket \right\}$$

$$\llbracket \textbf{max } X.\varphi, \rho \rrbracket \stackrel{\text{def}}{=} \bigcup \left\{ S \mid S \subseteq \llbracket \varphi, \rho[X \mapsto S] \rrbracket \right\} \qquad \llbracket \textbf{min } X.\varphi, \rho \rrbracket \stackrel{\text{def}}{=} \bigcap \left\{ S \mid \llbracket \varphi, \rho[X \mapsto S] \rrbracket \subseteq S \right\}$$

$$\llbracket X, \rho \rrbracket \stackrel{\text{def}}{=} \rho(X)$$

Figure 3.2   The syntax and semantics of μHML.

independently of the environment ρ, and the shorthand $[\![\varphi]\!]$ is used to denote $[\![\varphi, \rho]\!]$, *i.e.,* the set of system states in SYS that satisfy φ. In view of this, we say that a system (state) p satisfies some closed formula φ whenever $p \in [\![\varphi]\!]$, and conversely, that it violates φ whenever $p \notin [\![\varphi]\!]$.

Example 3.2.1. The μHML formula $\langle\alpha\rangle$tt describes systems that *can* produce action α, while $[\alpha]$ff describes systems that *cannot* produce action α.

$$\varphi_1 = \mathbf{max}\, X.\big([\texttt{req}]([\texttt{resp}]X \wedge [\texttt{resp}][\texttt{resp}]\mathbf{ff})\big)$$
$$\varphi_2 = \mathbf{min}\, X.(\langle\texttt{req}\rangle\langle\texttt{resp}\rangle X \vee \langle\texttt{lim}\rangle\mathbf{tt})$$

Formula φ_1 describes a property that prohibits a system from producing duplicate responses in answer to client requests. System p whose LTS is depicted in Figure 3.3a *violates* φ_1 through any trace in the regular language $(\texttt{req.resp})^+.\texttt{resp}$. Formula φ_2 describes systems that *can* reach a service limit after a number (possibly zero) of request and response interactions; system q depicted in Figure 3.3b *satisfies* φ_2 through any trace in the regular language $(\texttt{req.resp})^*.\texttt{lim}$. ∎

3.2.3 Monitoring μHML

Despite its limitations (*i.e.,* monitors can only analyse single execution traces), RV can be still effectively applied in cases where correctness properties can be shown to be satisfied (or violated) by analysing a *single* finite execution. As explained previously, the formula $[\alpha]$ff states that *all* α-actions performed by a satisfying system state should satisfy property **ff** afterwards. Since no system state can satisfy **ff**, the only way how to satisfy $[\alpha]$ff is for a system *not* to perform α. From a RV perspective, for a monitor to detect a violation of this requirement, observing *one negative witness* execution trace that

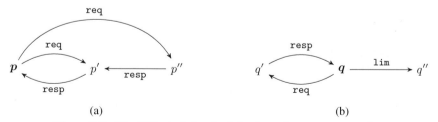

(a)   (b)

Figure 3.3   The LTSs depicting the behaviour of two servers p and q.

starts with action α suffices to show that property φ is infringed. Dually, when monitoring for the formula $\langle\alpha\rangle$**tt**, observing *one positive witness* that starts with action α suffices to show that property φ is satisfied.

Example 3.2.2. The μHML formula $\varphi_3 = \langle \text{lim} \rangle$**tt** requires that "*a process can perform action* lim". System q in Figure 3.3b can exhibit the trace lim.ϵ which suffices to show that system q satisfies φ_3. Yet, q may also exhibit other traces, such as those matching $(\text{req.resp})^*$, that all start with the event req. These traces do not provide enough evidence that system q satisfies φ_3. Stated otherwise, the monitor for formula φ_3 can reach an *acceptance* verdict *only* when a trace starting with event lim is observed. Otherwise, no verdict relating to the satisfaction or violation of the formula can be reached; in our specific case, the monitors we consider will reach an inconclusive verdict. ∎

The availability of a single finite runtime trace *does* however restrict the applicability of RV in cases such as those involving correctness properties describing *infinite* or *branching* executions. In view of this, certain properties expressed using the full expressive power of a branching-time logic such as μHML cannot be monitored for at runtime. The work by Francalanza *et al.* [19] explores the limits of monitorability for μHML, identifies a syntactic logical subset called mHML, and shows it to be monitorable and maximally expressive with respect to the constraints of runtime monitoring. The syntax of mHML, given in Figure 3.4, consists of two syntactic classes, *Safety* HML (sHML), describing *invariant* properties stipulating that bad things do *not* happen, and *Co-Safety* HML (cHML), describing properties that *eventually* hold after a *finite* number of events [2, 6, 23]. Formulae φ_1 and φ_2 from Example 3.2.1 are instances of sHML and cHML specifications respectively.

Monitorable Logic Syntax

$$\psi \in \text{mHML} \overset{\text{def}}{=} \text{sHML} \cup \text{cHML} \text{ where:}$$

$$\theta, \vartheta \in \text{sHML} ::= \textbf{tt} \quad | \quad \textbf{ff} \quad | \quad \theta \wedge \vartheta \quad | \quad [\alpha]\theta \quad | \quad \textbf{max}\, X.\theta \quad | \quad X$$
$$\pi, \varpi \in \text{cHML} ::= \textbf{tt} \quad | \quad \textbf{ff} \quad | \quad \pi \vee \varpi \quad | \quad \langle\alpha\rangle\pi \quad | \quad \textbf{min}\, X.\pi \quad | \quad X$$

Figure 3.4   The syntax of mHML.

3.3 A Tool for Monitoring Erlang Applications

We briefly review the implementation of our RV tool detectEr that analyses the correctness of concurrent programs developed in Erlang. It builds on the work by Francalanza *et al.* [19] which specifies a synthesis procedure that generates *correct* monitor descriptions from formulae written in mHML. We adapt this synthesis procedure so as to produce *concurrent* monitors in the form of Erlang actors that are instrumented with the running system via the tracing mechanism exposed by the VM of the host language. The synthesis procedure exploits the compositional semantics of mHML formulae to generate a choreography of monitor (actor) components that independently analyse the individual subformulae constituting a global formula, while still guaranteeing the correctness of the overall monitoring process.

In the sequel we refrain from delving into the specifics of how these concurrent monitors are synthesised; readers are encouraged to consult our previous work [4, 21], where the synthesis procedure is discussed at length. Instead, we limit ourselves to a high-level description of the main concepts and technologies required by readers to be able to adequately use the monitoring tool. In particular, we discuss the mechanisms of the host language used by the tool, the adaptations to the specification logic that facilitate the handling of data, and finally, give an overview of the tool's compilation process.

3.3.1 Concurrency-Oriented Development Using Erlang

Erlang is a general-purpose, concurrent programming language suitable for the development of fault-tolerant and distributed systems [3, 12, 22]. It adopts the actor model for concurrency as the primary means for structuring its applications. An *actor* is a concurrency unit of decomposition that represents a processing entity sharing no mutable memory with other actors. It interacts with other actors by sending (asynchronous) messages, and changes its internal state based on the messages received from other actors. In Erlang, actors are implemented as *lightweight* processes that are uniquely identified via their process PID (a number triple). Each process owns a message queue, known as a *mailbox*, to which messages from other processes can be sent in a non-blocking fashion; these can be consumed selectively at a later stage by the recipient process. Messages are comprised of elements of Erlang data types, including integers, floats, atoms, functions, binaries, *etc.*. Since process PIDs are allocated dynamically to newly spawned processes, Erlang provides a mechanism for registering a PID with a fixed alias name. This allows external entities to refer to a specific process statically via the registered name alias [3, 12].

The Erlang Virtual Machine (EVM) offers a powerful and flexible *tracing* mechanism that makes it possible to observe process behaviour *without* modifying the system source code through commonly used instrumentation techniques such as Aspect Oriented Programming (AOP) [3, 12]. Its flexibility stems from the fact that it can be *selectively* applied on specific processes as required, thereby fine tuning the tracing effort to the desired level of granularity. When traced, processes generate *action* messages that are directed by the Erlang runtime to a specially designated *tracer* process. Trace messages assume the form of Erlang *tuples* that describe the nature of trace events (*e.g.* function calls, message sends and receives, garbage collection triggers, *etc.*) and are deposited (like any other message) asynchronously inside the tracer's mailbox. Tracing serves as the basis for a number of utilities, including Erlang's text-based tracing facility dbg, and trace tool builder ttb [3]. Our tool, detectEr, employs this tracing mechanism to achieve *lightweight* trace event extraction for monitoring purposes; refer to the work by Attard *et al.* [4] for further details.

3.3.2 Reasoning about Data

Adapting mHML to be used for specifying the behaviour of Erlang programs adequately requires auxiliary functionality that describes system events carrying *data*; this involves mechanisms for generalising over specific data values and for expressing data dependencies. detectEr assumes a richer set of system events that carry data. Our account focusses on two types of events, namely outputs $i\,!\,d$ and inputs $i\,?\,d$, where i ranges over process PIDs, and d denotes the data payload associated with the action in the form of Erlang data values (*e.g.* PID, lists, tuples, atoms, *etc.*). In addition, our tool enriches the syntax of Figure 3.4 by introducing *pattern-matching* extensions for event actions (see Figure 3.5). Necessity and possibility formulae may contain *event patterns* instead of specific events: these possess the *same structure* of the aforementioned data-carrying events, but may also employ *variables* (Erlang-style alphanumeric identifiers starting with an upper-case letter) in place of values. Variables denote quantifications over data and are dynamically bound to values when they are *pattern-matched* to specific system events at runtime. Event patterns also allow us to express data dependencies across multiple events. Intuitively, whenever a variable is used in a pattern inside a necessity or possibility formula and again in the ensuing guarded subformula, the *first*

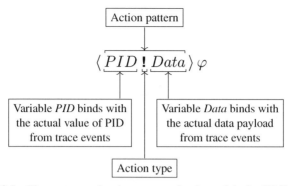

Figure 3.5 The anatomy of action patterns for the enriched mHML syntax.

variable instance acts as a binder for subsequent variable uses. The next example illustrates this concept.

Example 3.3.1. The client-server set-up shown in Figure 3.6 consists of a successor server process (with PID <0.33.0>) that increments the numeric payloads it receives from requesting clients by 1. Client requests should adhere to the following protocol. A client sends a tuple of the form $\{tag, return\_addr, value\_to\_increment\}$ where the first element is a qualifier tag stating that it is a client request (tag = req). The client then awaits for an answer back from the server in the form of a message with format $\{resp, incremented\_value\}$. The server obtains the identity of the client from the client request data $return\_addr$, which should carry the PID of the client sending the request (*e.g.* <0.38.0> in the case of Figure 3.6). One attempt at verifying the correctness of the executing system is by specifying a safety property stating that

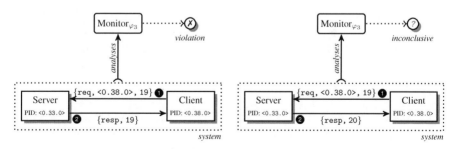

(a) The incorrect server implementation.   (b) The correct server implementation.

Figure 3.6 Runtime verifying the correctness of a client-server system.

> *"the numeric payload contained in the server's response cannot equal the one sent in the original client request."*

This requirement can be expressed as follows:

$$\varphi_3 = [Srv \text{ ? } \{\texttt{req}, Clt, Num\}] [Clt \text{ ! } \{\texttt{resp}, Num\}] \textbf{ff}$$

The two necessity constructs in the sHML formula φ_3 describe a request-response interaction between the client and server processes. The first necessity $[Srv \text{ ? } \{\texttt{req}, Clt, Num\}]$ specifies an *input* event data pattern that conforms to the structure of the data sent by the client when initiating its interaction with the server (*i.e.*, the action labelled by ❶ in Figure 3.6); meanwhile, the second necessity $[Clt \text{ ! } \{\texttt{resp}, Num\}] \textbf{ff}$ specifies an *output* action data pattern that conforms to the structure of the data sent by the server in reply to the client's request (*i.e.*, action ❷ in Figure 3.6). Formula φ_3 matches events in the execution trace whenever the server Srv receives a request with numeric payload Num from client Clt, and replies back to the same client Clt with an unchanged value Num. Note the dependency between the patterns in the two necessities: the values matched to the variable Clt and Num in first pattern are then instantiated in the subsequent necessity pattern.

To illustrate concretely how binding actually works, we can consider how the two different executions of client-server system depicted in Figures 3.6a and 3.6b are monitored at runtime. When the event pattern Srv ? $\{\texttt{req}, Clt, Num\}$ from the first necessity is matched to the first trace event <0.33.0> ? $\{\texttt{req}, \texttt{<0.38.0>}, 19\}$ (resulting from the execution of action ❶), the free pattern variables Srv, Clt and Num become *bound* to the runtime values <0.33.0>, <0.38.0> and 19 respectively. The runtime binding of variables Srv, Clt and Num in turn, also instantiates subsequent (guarded) patterns in the second necessity — this leaves us with the (continuation) residual formula $[\texttt{<0.38.0>} \text{ ! } \{\texttt{resp}, 19\}] \textbf{ff}$ to check for. This closed formula can now match the *second* trace event (due to action ❷), *only if* an *incorrectly* implemented server responds to the initial client request with the *same* numeric payload sent to it, as is the case in Figure 3.6a. This leads to a *violation* detection. Contrastingly, Figure 3.6b shows the case where the server's reply sent back to the client contains the value '20' that does not match the runtime binding for the subformula $[Clt \text{ ! } \{\texttt{resp}, Num\}] \textbf{ff}$ of φ_3. After the first pattern-match, Num is bound to '19', and this does not match with event $\{\texttt{resp}, 20\}$ of action ❷ in Figure 3.6b (Clt is bound to <0.38.0> as before), thus leading to an *inconclusive* verdict.   ■

3.3.2.1 Properties with specific PIDs

Since process PIDs are allocated at runtime, there is no direct way for a correctness property to refer to a *specific* process. Nevertheless, the tool still provides an indirect method how to specify this via the process PID registering mechanism offered by the host language. For instance, in the case of formula φ_3 from Example 3.3.1, one could refer to a particular process (instead of *any* arbitrary process that is dynamically bound to variable *Srv* in the pattern $[Srv\ ?\ \{\texttt{req},\ Clt,\ Num\}]$) using the notation @ *srv* in place of *Srv*. This would then map to the process that is registered with the fixed (atom) name *srv* in the system and, subsequently, the respective event analysis would only match events sent specifically to the process whose PID is registered as *srv*.

3.3.2.2 Further reasoning about data

Readers might have been wary of the fact that formula φ_3 in Example 3.3.1 only guards against cases where the server merely *echoes* back the same numeric payload sent to it by clients. This only partially addresses the ideal correctness requirements, because it does not capture the full behaviour expected of the successor server in Figure 3.6. Reformulating the property from Example 3.3.1 to read as

> *"the numeric payload contained in the server's response must be equal to the successor of the one sent in the original client request."*

while more specific, requires the monitor to check whether *all* responses issued by the server in reply to client requests do in fact contain the successor of the number enclosed in said requests.

Our logic handles this expressiveness requirement by extending the enriched mHML syntax from this section with conditional constructs and predicates, thus enabling it to perform complex reasoning on data values acquired dynamically through pattern matching. Data predicates[1], together with boolean expressions, are evaluated to values $b \in \{\text{false}, \text{true}\}$. Conditionals, written as **if** b **then** θ **else** ϑ for sHML formulae and **if** b **then** π **else** ϖ for cHML formulae, evaluate to θ and π respectively when b evaluates to true, and to ϑ and ϖ otherwise. The **else** clause may be omitted if not required. Correctness formulae of the latter form are given

---

[1]Data predicates are assumed to be decidable (*i.e.,* guaranteed to terminate). Our implementation makes use of a restricted subset of Erlang side effect-free functions employed in standard guard expressions (*e.g.* is_list/1, is_number/1, is_pid/1, *etc.*) [12].

an *inconclusive* interpretation whenever the boolean condition inside the **if** clause evaluates to false. Conditional constructs increase the expressiveness of mHML, because they make it possible to formalise properties that are otherwise hard to express using the basic form of the logic. When compiled, conditional formulae are translated into monitors whose runtime analysis branches depending on dynamic decisions made on data obtained at runtime.

Example 3.3.2. The reformulated safety property *"the numeric payload contained in the server's response must be equal to the successor of the one sent in the original client request"* can be specified as follows using the extended sHML syntax:

$$\varphi_4 = [Srv \; \textbf{?} \; \{\texttt{req}, Clt, Num\}] \, [Clt \; \textbf{!} \; \{\texttt{resp}, Succ\}]$$
$$\textbf{if}(Succ \neq Num + 1) \, \textbf{then ff}$$

Formula φ_4 differs slightly from the one specified in Example 3.3.1. It introduces a new variable *Succ* that binds to the server's return value. This, in turn, enables the conditional construct to determine whether the successor operation is correctly implemented by the server, thus ensuring that φ_4 is violated *only* when the value bound to *Succ* is not the successor of *Num*. An inconclusive verdict is assumed by the formula whenever $(Succ \neq Num + 1)$ does not hold, *i.e., Succ is* indeed the successor of *Num*, as in the case of Figure 3.6b.   ∎

3.3.3 Monitor Compilation

Following closely the synthesis function of [4], our tool is able to parse mHML formulae and generate Erlang code that monitors for the input formulae. The inherent concurrency features offered by Erlang, together with the modular structure of the synthesis function are used to translate formulae into choreographed collections of (sub)monitors. These are expressed as concurrent *processes* that execute independently of one another and analyse different parts of the exhibited system trace (*e.g.* one submonitor may be analysing the second event in an execution trace of length five, whereas another may forge ahead and analyse the fourth event in the trace). In order to ensure that submonitors have access to the same trace events, they are organised as supervision trees [3, 12]: the (parent) monitor to which the submonitors are attached *forks* (*i.e.,* replicates and forwards) individual trace events to its children. The moment a verdict is reached by any submonitor process, all monitoring processes are terminated, and said verdict is used to declare the final monitoring outcome. Interested readers are referred to our previous work [4, 21] for details on how these monitor choreographies are organised.

Figure 3.7 outlines the compilation steps required to transform a formula script file (*e.g.* `script.hml`) into a corresponding Erlang source code implementing the monitor functionality (*e.g.* `monitor.erl`). The tool instruments the synthesised monitors to run asynchronously with the system to be analysed using the native tracing functionality provided by the EVM. Crucially, this type of instrumentation requires *no changes to the monitor source code* (or the target system binaries). In Figure 3.7, the file packaging component of the compiler leaves the system source files unchanged; this increases confidence in the correctness of the resulting monitoring set-up. In addition to the monitor source file, Figure 3.7 shows also a second module, `launcher.erl`, that is generated automatically based on the specified system start up configuration. The launcher is tasked with the responsibility of starting the system and corresponding monitors in tandem. Said modules, together with other supporting tool-related source code files are afterwards compiled into executable modules (`.beam` files), which are then packaged and placed alongside other system binary files.

3.4 detectEr in Practice

We revisit the runtime monitoring tool depicted in Figure 3.7 from a user's perspective, and present a brief guide showcasing its main functionality. This guide, presented in the form of a tutorial, goes through the steps required

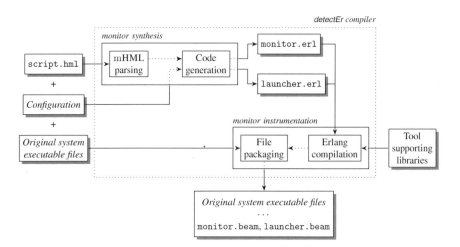

Figure 3.7 The monitor synthesis process and instrumentation pipeline.

to apply our tool to monitor an Erlang implementation of the client-server system seen earlier in Example 3.3.1. It shows how a simple (but useful) safety property can be scripted as a sHML formula, and compiled into a runtime monitor that is used to verify the incorrect and correct behaviour of the successor server illustrated in Figures 3.6a and 3.6b. cHML properties from Figure 3.4 can also be monitored for using the same sequence of steps.

The current prototype tool implementation is capable of instrumenting only one monitor inside the target system. Nevertheless, the tool's compilation and instrumentation processes were developed with extensibility in mind, and the steps that are outlined in the following tutorial will remain valid once the tool is extended to support multiple monitors. Although the example presented in this guide is fairly basic, it conveys the essence of how the tool should be applied in practice; more complex properties [8, 10] would be approached following the same instructions and procedures outlined in the coming sections.

3.4.1 Creating the Target System

The initial distribution of the tool is available from `https://bitbucket.org/duncanatt/detecter-lite`, and requires a working installation of Erlang. This guide assumes that GNU make is installed on the host system. OSX users can acquire make by installing the XCode Command Line Tools; Windows users can install the MinGW suite of tools. Although Linux was used to create this tutorial, the steps below can be replicated on any other operating system.

3.4.1.1 Setting up the Erlang project

To facilitate the development of Erlang applications, detectEr includes a generic makefile which we use in this guide. The following make targets are provided:

- `init`: Creates the standard Erlang project structure;
- `clean`: Removes Erlang `.beam` and other temporary files;
- `all`: Cleans and compiles the Erlang project;
- `instrument`: Synthesises and instruments monitors into the target system, given the HML script, target system binary directory, and application entry point configuration.

We begin by creating a target directory called `example`. This contains the client-server system Erlang project and all its associated source code files. At the root of the example directory, we also place the aforementioned makefile, since this is used to manage the build process of our simple Erlang application. The latest version of the makefile can be downloaded directly from the project site using wget:

```
duncan@term:/$ mkdir example
duncan@term:/$ cd example
duncan@term:/example$ wget https://bitbucket.org/duncanatt/detecter-lite\
/raw/detecter-lite-1.0/Makefile
```

Once the makefile is downloaded, the standard Erlang directory structure is created using the init target:

```
duncan@term:/example$ make init
duncan@term:/example$ ls -l
drwxrwxr-x 2 duncan duncan 4096 May 15 16:53 include
-rw-rw-r-- 1 duncan duncan 5463 May 15 16:53 Makefile
drwxrwxr-x 2 duncan duncan 4096 May 15 16:53 src
drwxrwxr-x 2 duncan duncan 4096 May 15 16:53 test
```

To avoid writing the Erlang server manually from scratch, the guide borrows a number of sample source code files that are included in the tool's distribution. For simplicity, we assume that the tool is set up in the same directory as our example project directory. The plus_one module that forms part of the tool distribution, implements a version of the successor server as described in Figure 3.6. This file, together with its dependencies should be copied into the src and include directories as shown below; these commands result in the creation of a directory structure that corresponds to the one shown in Figure 3.8a.

```
duncan@term:/example$ cd src
duncan@term:/example/src$ cp ../../detecter-lite/test/plus_one.erl .
duncan@term:/example/src$ cp ../../detecter-lite/src/mon/log.erl .
duncan@term:/example/src$ cd ../include/
duncan@term:/example/include$ cp ../../detecter-lite/include/* .
```

After the files have been copied successfully into their respective directories, the Erlang project can be built by invoking make:

```
duncan@term:/example/include$ cd ..
duncan@term:/example$ make

Compiling Erlang source file: src/log.erl to ebin/log.beam
```

Compiling Erlang source file: src/plus_one.erl to ebin/plus_one.beam

```
>-----------------------------------<
  Build completed successfully!
>-----------------------------------<
```

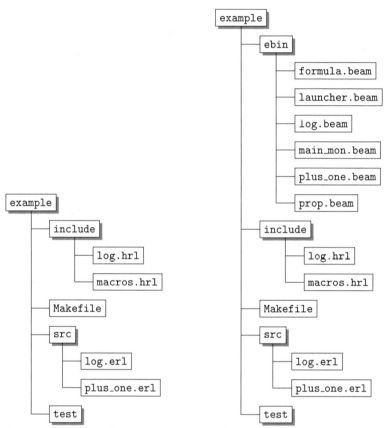

(a) The example project directory sturuc- (b) The example project directory sturuc-
ture before compilation. ture after compilation and instrumentation.

Figure 3.8   Creating the Erlang project directory structure.

3.4.1.2 Running and testing the server

With the build now completed, the plus_one successor server can be
launched and tested. Since we have not developed a complete application,

but only the server part, testing is conducted using the Erlang shell in place of a full client implementation. For illustrative purposes, the plus_one server may exhibit different behaviours at runtime depending on the flag it is started up with. Concretely, the plus_one server and shell can be launched from the terminal as follows:

```
1  duncan@term:/example$ erl -pa ebin -eval "plus_one:start(bad)"
2
3  Erlang/OTP 18 [erts-7.2] [source] [smp:4:4] [async-threads:10] [kernel-poll:false]
4  Eshell V7.2  (abort with ^G)
5
6  [<0.33.0> - plus_one:22] - Started PLUS ONE server with initial value'0'\,and\,mode 'bad'.
7  1> _
```

The plus_one server is intentionally started using the startup flag bad, in order to simulate the *incorrect* server behaviour depicted in Figure 3.6a. This serves its purpose later when scripting the formula used to verify the server's behaviour. We can confirm that the server started up successfully by ensuring that the plus_one start up log (line 6) shows up in the terminal. Once loaded, the server can be tested by submitting requests to it using the Erlang ! (send) operator (line 8 below). Following the protocol outlined in Example 3.3.1, the test request is sent to the process identified by the Erlang registered process name plus_one. This test request observes the tuple format {req, *return_addr*, *value_to_increment*}, where *return_addr* corresponds to the PID of the sender actor (in this case, the Erlang shell), and *value_to_increment* contains the actual numeric data payload, *i.e.,* the number the client wishes to increment. In Erlang, a process may obtain its own PID through the function call self(). Note that commands typed in the Erlang shell must terminate with a period symbol, otherwise these will not be processed.

```
8   1> plus_one ! {req, self(), 19}.
9
10  [<0.33.0> - plus_one:41] - Received request with value '19'.
11  [<0.33.0> - plus_one:46] - Sending response with value '{resp,19}', Current cnt '1'.
12  {req,<0.38.0>,19}
13  2> _
```

As can be gleaned from the logs above, the plus_one server receives the number '19' as payload, and echoes back that same value to the shell (lines 10–11). A correct implementation of the server should have replied with a value of '20', that corresponds to the client's request being incremented by '1'. The server's response can be extracted from the Erlang shell by invoking the `flush()` function to empty the shell's mailbox (line 14). After confirming that the server is working (incorrectly) as intended, the Erlang shell can be closed by typing "`q().`" at the terminal.

```
14  2> flush().
15  Shell got {resp,19}
16  ok
17  3> _
```

3.4.2 Instrumenting the Test System

We are now in a position to generate a monitor that verifies the safety property below, a generalisation of the property discussed earlier in Example 3.3.1:

> *"After any sequence of request-response interactions with arbitrary clients, the numeric payload contained in the server's response following a client request must never equal the one sent in the original client request."*

The monitor synthesised for this property should detect the violating behaviour exhibited by the `plus_one` server.

3.4.2.1 Property specification

Properties using our tool are specified in plain text files that are processed to produce monitors in the form of Erlang code. These, together with other supporting source files, are compiled to executable Erlang `.beam` files and copied into the target system's binary directory, `ebin`. As explained in Section 3.3.3, the tool also creates a `launcher` module that is used to bootstrap the system together with the synthesised monitor. Once loaded, the system executes as it normally would, while concurrently, the monitor passively observes the system's behaviour expressed in terms of the messages exchanged between it and its environment. A violation will be promptly flagged when discovered by the monitor analysing the trace generated by our successor server. The aforestated safety property can be scripted by pasting the sHML formula given below into a plain text editor, and saving it as `prop.hml` *in* the `example` directory.

```
1  max('X',
2    [Srv ? {req, Clt, Num}][Clt ! {resp, Num}] ff
3    &&
4    [Srv ? {req, Clt, Num}][Clt ! {resp, Other}] 'X')
```

This *recursive* sHML formula makes use of a conjunction (&&) construct to express the two possible behaviours expected of the system. The violating behaviour, specified using [*Srv* ? {req, *Clt*, *Num*}][*Clt* ! {resp, *Num*}] ff, demands that a violation be flagged when the server *Srv* receives a request containing *Num* from client *Clt*, and returns to *Clt* the same value *Num*. The recursive (non-violating) behaviour, expressed by [*Srv* ? {req, *Clt*, *Num*}][*Clt* ! {resp, *Other*}] 'X', requires the monitor to recurse whenever a request received from *Clt* is answered with some value *Other*, *i.e.,* not just the successor of *Num*. This is in line with the property above, as it requires the monitor to detect violations *only* when the same value of *Num* is returned by a server in reply to a client's request. Recursion, made possible by the maximal fixpoint construct max('X', ...) and the recursive variable 'X', allows the monitor to unfold repeatedly, thereby *continuously* analysing the system trace until the violating behaviour is detected. Note that the formula in prop.hml is an extension of the simpler property φ_3 from Example 3.3.1. In φ_3, the absence of recursion restricts the corresponding monitor to analyse, at most, two trace events before terminating. Note also that a more comprehensive interpretation of the aforementioned correctness property would of course require the formula to check that each number in the server's response is actually the successor of the one sent in the client's request, as discussed earlier in Example 3.3.2. This can be expressed by modifying line 4 in the above script to

```
max('X',  ...  &&
   [Srv ? {req, Clt, Num}][Clt ! {resp, Other}] if Other =:= Num + 1 then 'X')
```

In what follows, we stick to the weaker variant of the property to simplify our presentation.

3.4.2.2 Monitor synthesis and instrumentation

The monitor corresponding to the sHML script created above is synthesised using the instrument target from the application makefile:

```
duncan@term:/example$ cd ../detecter-lite
duncan@term:/detecter-lite$ make instrument hml="../example/prop.hml"\
app-bin-dir="../example/ebin"\
MFA="{plus_one,start,[bad]}"
```

The `instrument` target requires the following command line arguments:

- `hml`: The relative or absolute path that leads to the formula script file;
- `app-bin-dir`: The target application's binary base directory;
- `MFA`: The target application's entry point function, encoded as a {Mod, Fun, [Args]} tuple, where we specify the plus_one module's start function passing bad as argument, like previously.

Monitor synthesis and instrumentation (refer to Figure 3.7) results in the Erlang project directory structure shown in Figure 3.8b. All the original target system binaries remain untouched, and the plus_one server application can be still run without monitors, as before (see Section 3.4.1.2).

3.4.2.3 Running the monitored system

The instrumented system can be started up by using the automatically generated `launcher` module as shown:

```
1 duncan@term:/example$ erl -pa ebin -eval "launcher:start()"
2
3 Erlang/OTP 18 [erts-7.2] [smp:4:4] [async-threads:10] [kernel-poll:false]
4 Eshell V7.2 (abort with ^G)
5
6 [<0.31.0> - main_mon:38] - Started main monitor for processes/PIDs [].
7 [<0.33.0> - plus_one:22] - Started PLUS ONE server with initial cnt value '0' and mode 'bad'.
8
9 [<0.32.0> - main_mon:24] - System to be monitored started.
10 [<0.34.0> - main_mon:62] - Resolved procs [].
11 [<0.40.0> - formula:152] - mon_max adding var 'X' to formula env.
12 [<0.40.0> - formula:91] - mon_and spawned processes '<0.41.0>' and '<0.42.0>'.
13 [<0.34.0> - main_mon:84] - Starting main monitor loop.
14 1> _
```

As indicated by the above logs, the plus_one server and corresponding monitor are now executing in parallel with PIDs <0.33.0> and <0.34.0> that are dynamically assigned at runtime once the respective processes are spawned (lines 6–7). The synthesised monitor corresponding to the recursion in the formula of Section 3.4.2.1 eagerly unfolds one iteration of the formula (lines 10–11) exposing a conjunction construct at top level (see Francalanza *et al.* [21] for a detailed discussion of how recursion is handled in the synthesised monitors). The "conjunction monitor" mon_and spawns its two submonitor actors once it starts executing (line 12); these correspond to the violation submonitor created from subformula [*Srv* ? {req, *Clt*, *Num*}] [*Clt* ! {resp, *Num*}] ff and the recursive submonitor created from [*Srv* ? {req, *Clt*, *Num*}] [*Clt* ! {resp, *Other*}] 'X'. As before, the server is tested using the same request, sent from the Erlang shell (line 15):

```
15  1> plus_one ! {req, self(), 19}.
16
17  [<0.33.0> - plus_one:41] - Received request with value '19'.
18  [<0.41.0> - formula:120] - mon_nec evaluating action: {recv,<0.33.0>,{req,<0.38.0>,19}}.
19  [<0.42.0> - formula:120] - mon_nec evaluating action: {recv,<0.33.0>,{req,<0.38.0>,19}}.
20  [<0.33.0> - plus_one:46] - Sending response with value '{resp,19}', Current cnt '1'.
21
22  {req,<0.38.0>,19}
23  [<0.41.0> - formula:120] - mon_nec evaluating action: {send,<0.38.0>,{resp,19}}.
24  [<0.42.0> - formula:120] - mon_nec evaluating action: {send,<0.38.0>,{resp,19}}.
25  [<0.41.0> - formula:67] - mon_ff matched 'ff' action.
26  [<0.42.0> - formula:180] - mon_var retrieving var 'X' from formula env and recursing.
27  [<0.34.0> - main_mon:113] -
28
29  Main monitor/tracer received 'ff' - *** Violation detected! ***
30
31  2> _
```

The violation (PID <0.41.0>) and recursive (PID <0.42.0>) submonitor processes acquire trace events from their parent "conjunction monitor" process mon_and as soon as new trace events are reported by the EVM. For instance, the trace event generated by the message {req, self(), 19} sent from the shell is forwarded by mon_and to its child submonitors (lines 18–19). Next, the plus_one server computes the result and sends it back to the Erlang shell (line 20). This causes the second trace event to be generated by the system and reported by the EVM's tracing mechanism; once again this trace event is forwarded to, and processed by both submonitors (lines 23–24). At this point, the recursive submonitor tries to unfold in preparation for the next computation (line 26), while the violation submonitor flags a violation verdict **ff** (line 25), which is in turn sent to the main monitor. As a *single* detection suffices to ensure a global verdict, the main monitor terminates accordingly with **ff** (line 29); consult the work by Attard *et al.* [4] for reasons on why this is the case.

3.4.2.4 Running the correct server

So far, the plus_one successor server has been intentionally launched in bad mode in order to demonstrate how violations are handled by our monitor. We now re-instrument the system in order to emulate the correct successor server behaviour depicted in Figure 3.6b; invoking the instrument target differs only in the MFA tuple used to start the server, where instead of bad, the flag good is used:

```
duncan@term:/detecter-lite$ make instrument hml="../example/prop.hml"\
app-bin-dir="../example/ebin"\
MFA="{plus_one,start,[good]}"
```

The server should now behave correctly, and return the successor value of any numeric payload that we choose to send to it from the Erlang shell.

```
 1  duncan@term:/example$ erl -pa ebin -eval "launcher:start()"
 2
 3  Erlang/OTP 18 [erts-7.2] [source] [smp:4:4] [async-threads:10] [kernel-poll:false]
 4  Eshell V7.2  (abort with ^G)
 5
 6  [<0.34.0> - main_mon:38] - Started main monitor for processes/PIDs [].
 7  [<0.33.0> - plus_one:22] - Started PLUS ONE server with initial cnt value '0' and mode 'good'.
 8
 9  [<0.32.0> - main_mon:24] - System to be monitored started.
10  [<0.34.0> - main_mon:62] - Resolved procs [].
11  [<0.40.0> - formula:152] - mon_max adding var 'X' to formula environment.
12  [<0.40.0> - formula:91] - mon_and spawned processes '<0.41.0>' and '<0.42.0>'.
13  [<0.34.0> - main_mon:84] - Starting main monitor loop.
14  1> _
15  1> plus_one ! {req, self(), 19}.
16
17  [<0.33.0> - plus_one:41] - Received request with value '19'.
18  [<0.41.0> - formula:120] - mon_nec evaluating action: {recv,<0.33.0>,{req,<0.38.0>,19}}.
19  [<0.42.0> - formula:120] - mon_nec evaluating action: {recv,<0.33.0>,{req,<0.38.0>,19}}.
20  [<0.33.0> - plus_one:46] - Sending response with value '{resp,20}', Current cnt '1'.
21
22  {req,<0.38.0>,19}
23  [<0.41.0> - formula:120] - mon_nec evaluating action: {send,<0.38.0>,{resp,20}}.
24  [<0.42.0> - formula:120] - mon_nec evaluating action: {send,<0.38.0>,{resp,20}}.
25  [<0.41.0> - formula:59] - mon_id no match.
26  [<0.42.0> - formula:180] - mon_var retrieving var 'X' from formula env and recursing.
27  [<0.42.0> - formula:91] - mon_and spawned processes '<0.44.0>' and '<0.45.0>'.
28  2> _
```

When the client request {req, self(), 19} is submitted to the server from the Erlang shell (line 15), this again generates a response from the server answering back with the tuple {resp,20}. Although the sequence of trace events is similar to the ones in Section 3.4.2.3, the data in these events is *different*: the server response now carries value '20' as opposed to '19'. This causes the violation submonitor to terminate with an inconclusive verdict (line 25) and the recursive submonitor to unfold (line 26) in preparation for the next trace events. Stated otherwise, no violation is detected by the monitor up to the current point of execution.

3.5 Conclusion

We have presented an overview of detectEr from the perspective of a user wishing to employ this tool to verify Erlang systems at runtime. The tool automatically synthesises monitoring code from specifications written in the monitorable subset of the Hennessy-Milner Logic with maximal and minimal fixpoints [25, 19]. The monitoring code which is then instrumented to run alongside the system under scrutiny infers specification satisfactions or violations by analysing the runtime execution trace exhibited by the system. One salient aspect of the tool is that the instrumentation employs the tracing facility of the host language virtual machine. It therefore requires no access to system source code and relies only on the application's binary files. The execution of the monitor and the system being analysed is decoupled — this may lead to late (satisfaction or violation) detections from the monitor. In spite of this, the lightweight instrumentation approach adopted by detectEr leaves the target system binaries untouched, thus making it possible to employ our tool in cases where (commercial) software with licenses and/or support agreements explicitly forbid the modification of binary code.

3.5.1 Related and Future Work

Apart from being a manifestation of the work due to Francalanza *et al.* [19], the tool detectEr was also used as a starting point for a number of other investigations. Cassar *et al.* [11] explored choreographed reconfigurations for submonitors as means to lower the monitoring computational overhead, whereas in subsequent work [8], the authors also explored modifications to the tool to be able to synchronise more closely the executions of the system and the monitor, thereby avoiding problems associated with late detections. In other work by Cassar *et al.* [10], the investigators consider extensions to the tool that enable the runtime analysis to administer adaptation actions to the system once a violation is detected. Following this work, the authors also developed a type-based approach [9] to ensure that runtime adaptations are administered correctly by the tool. We are presently considering tool extensions that enable monitoring analysis to be distributed across sites and also alternative monitor synthesis procedures that guarantee a degree of property enforcement.

There has also been an extensive body of work [16, 28] on the runtime checking of session types. Lange *et al.* [24] demonstrate the correspondence between session types and a fragment of the modal μ-calculus, which has been previously shown by Larsen [25] to be a reformulation of the logic

μHML. Crucially, the monitors we study consider the system from a global level. By contrast, the aforementioned works project global multiparty session types to local endpoint types, which are then synthesised into local monitors that analyse traffic at individual channel endpoints.

Acknowledgments This work was partly supported by the project "Theo-FoMon: Theoretical Foundations for Monitorability" (nr.163406-051) of the Icelandic Research Fund.

References

[1] Luca Aceto, Anna Ingólfsdóttir, Kim Guldstrand Larsen, and Jiri Srba. *Reactive Systems: Modelling, Specification and Verification.* Cambridge Univ. Press, New York, NY, USA, first edition, 2007.

[2] Bowen Alpern and Fred B. Schneider. Recognizing Safety and Liveness. *Distributed Computing*, 2(3):117–126, 1987.

[3] Joe Armstrong. *Programming Erlang: Software for a Concurrent World.* Pragmatic Bookshelf, first edition, 2007.

[4] Duncan Paul Attard and Adrian Francalanza. A Monitoring Tool for a Branching-Time Logic. In *RV*, volume 10012 of *LNCS*, pages 473–481. Springer, 2016.

[5] Howard Barringer, Yliès Falcone, Klaus Havelund, Giles Reger, and David E. Rydeheard. Quantified Event Automata: Towards Expressive and Efficient Runtime Monitors. In *FM*, volume 7436 of *LNCS*, pages 68–84. Springer, 2012.

[6] Andreas Bauer, Martin Leucker, and Christian Schallhart. Runtime Verification for LTL and TLTL. *ACM Trans. Softw. Eng. Methodol.*, 20(4):14, 2011.

[7] Andreas Klaus Bauer and Yliès Falcone. Decentralised LTL Monitoring. In *FM*, volume 7436 of *LNCS*, pages 85–100. Springer, 2012.

[8] Ian Cassar and Adrian Francalanza. On Synchronous and Asynchronous Monitor Instrumentation for Actor-Based Systems. In *FOCLASA*, volume 175 of *EPTCS*, pages 54–68, 2014.

[9] Ian Cassar and Adrian Francalanza. Runtime Adaptation for Actor Systems. In *RV*, volume 9333 of *LNCS*, pages 38–54. Springer, 2015.

[10] Ian Cassar and Adrian Francalanza. On Implementing a Monitor-Oriented Programming Framework for Actor Systems. In *IFM*, volume 9681 of *LNCS*, pages 176–192. Springer, 2016.

[11] Ian Cassar, Adrian Francalanza, and Simon Said. Improving Runtime Overheads for detectEr. In *FESCA*, volume 178 of *EPTCS*, pages 1–8, 2015.

[12] Francesco Cesarini and Simon Thompson. *Erlang Programming*. O'Reilly Media, first edition, 2009.

[13] Edmund M. Clarke, Orna Grumberg, and Doron A. Peled. *Model Checking*. The MIT Press, first edition, 1999.

[14] Edmund M. Clarke, William Klieber, Milos Nováček, and Paolo Zuliani. Model Checking and the State Explosion Problem. In *LASER*, volume 7682 of *LNCS*, pages 1–30. Springer, 2011.

[15] Christian Colombo, Adrian Francalanza, and Rudolph Gatt. Elarva: A Monitoring Tool for Erlang. In *RV*, volume 7186 of *LNCS*, pages 370–374. Springer, 2011.

[16] Romain Demangeon, Kohei Honda, Raymond Hu, Rumyana Neykova, and Nobuko Yoshida. Practical interruptible conversations: Distributed dynamic verification with multiparty session types and Python. *Formal Methods in System Design*, 46(3):197–225, 2015.

[17] Yliès Falcone, Jean-Claude Fernandez, and Laurent Mounier. What can you verify and enforce at runtime? *STTT*, 14(3):349–382, 2012.

[18] Adrian Francalanza. A Theory of Monitors. In *FoSSaCS*, volume 9634 of *LNCS*, pages 145–161. Springer, 2016.

[19] Adrian Francalanza, Luca Aceto, and Anna Ingólfsdóttir. On Verifying Hennessy-Milner Logic with Recursion at Runtime. In *RV*, volume 9333 of *LNCS*, pages 71–86. Springer, 2015.

[20] Adrian Francalanza, Andrew Gauci, and Gordon J. Pace. Distributed System Contract Monitoring. *J. Log. Algebr. Program.*, 82(5–7):186–215, 2013.

[21] Adrian Francalanza and Aldrin Seychell. Synthesising Correct Concurrent Runtime Monitors. *Formal Methods in System Design*, 46(3):226–261, 2015.

[22] Fred Hebert. *Learn You Some Erlang for Great Good!: A Beginner's Guide*. No Starch Press, first edition, 2013.

[23] Orna Kupferman. Variations on Safety. In *TACAS*, volume 8413 of *LNCS*, pages 1–14. Springer, 2014.

[24] Julien Lange and Nobuko Yoshida. Characteristic Formulae for Session Types. In *TACAS*, volume 9636 of *LNCS*, pages 833–850. Springer, 2016.

[25] Kim Guldstrand Larsen. Proof Systems for Satisfiability in Hennessy-Milner Logic with Recursion. *Theor. Comput. Sci.*, 72(2&3):265–288, 1990.

[26] Martin Leucker and Christian Schallhart. A Brief Account of Runtime Verification. *J. Log. Algebr. Program.*, 78(5):293–303, 2009.

[27] Zohar Manna and Amir Pnueli. Completing the Temporal Picture. *Theor. Comput. Sci.*, 83(1):91–130, 1991.

[28] Rumyana Neykova, Laura Bocchi, and Nobuko Yoshida. Timed Runtime Monitoring for Multiparty Conversations. In *BEAT*, volume 162 of *EPTCS*, pages 19–26, 2014.

[29] Giles Reger, Helena Cuenca Cruz, and David E. Rydeheard. Marq: Monitoring at Runtime with QEA. In *TACAS*, volume 9035 of *LNCS*, pages 596–610. Springer, 2015.

[30] A. Roscoe. *Theory and Practice of Concurrency*. Prentice Hall, first edition, 1997.

4

How to Verify Your Python Conversations

Rumyana Neykova and Nobuko Yoshida

Imperial College London, UK

Abstract

In large-scale distributed systems, each application is realised through inter-actions among distributed components. To guarantee safe communication (no deadlocks and communication mismatches) we need programming languages and tools that structure, manage, and policy-check these interactions. Multiparty session types (MSPT), a typing discipline for structured interactions between communicating processes, offers a promising approach. To date, however, session types applications have been limited to static verification, which is not always feasible and is often restrictive in terms of programming API and specifying policies. This chapter investigates the design and implementation of a runtime verification framework, ensuring conformance between programs and specifications. Specifications are written in Scribble, a protocol description language formally founded on MPST. The central idea of the approach is a runtime monitor, which takes a form of a communicating finite state machine, automatically generated from Scribble specifications, and a communication runtime stipulating a message format. We demonstrate Scribble-based runtime verification in manifold ways. First, we present a Python library, facilitated with session primitives and verification runtime. Second, we show examples from a large cyber-infrastructure project for oceanography, which uses the library as a communication medium. Third, we examine communication patterns, featuring advanced Scribble primitives for verification of exception handling behaviours.

4.1 Framework Overview

Figure 4.1 illustrates the methodology of our framework. The development of a communication-oriented application starts with the specification of the intended interactions (the choreography) as a *global protocol* using the Scribble protocol description language [9]. From a global protocol, the toolchain mechanically generates (*projects*) a Scribble *local protocol*, represented as a finite state machine (FSM), for each participant (abstracted as a *role*). As a session is conducted at run-time, the monitor at each endpoint validates the communication actions performed by the local endpoint, and the messages that arrive from the other endpoints, against the transitions permitted by the monitor's FSM. Each monitor thus works to protect (1) the endpoint from invalid actions by the network environment, and (2) the network from incorrectly implemented endpoints. Our runtime multiparty session types (MPST) [6] framework is designed in this way to ensure, using the decentralised monitoring of each local endpoint, that the session as a whole conforms to the original global protocol [1], and that unsafe actions by a bad endpoint cannot corrupt the protocol state of other compliant endpoints.

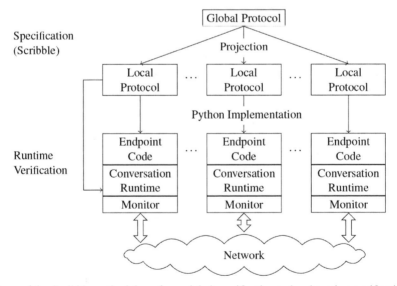

Figure 4.1   Scribble methodology from global specification to local runtime verification.

Outline. We outline the structure of this chapter. Section 4.2 demonstrates our runtime MPST framework through an example. We present a global-to-local projection of Scribble protocols, endpoint implementations, and local FSM generation. Section 4.3 demonstrates an API for conversation programming in Python that supports standard socket-like operations, as well as event-driven interface. The API decorates conversation messages with meta information required by the monitors to perform runtime verification. Section 4.4 discusses the monitor implementation, focusing on key architectural requirements of our framework. Section 4.5 presents an extension of Scribble with asynchronous session interrupts. This is a feature of MPST, giving a general mechanism for handling nested, multiparty exception handling. We present the extension of the Python API with a new construct for scopes, endpoint implementations, and local FSM generation for monitoring. Section 4.6 explains the correspondence between a theoretical model and our implementation.

4.2 Scribble-Based Runtime Verification

This section illustrates the stages of our framework and its implementation through a use case, emphasising the properties verified at each verification stage. The presented use case is obtained from our industrial partners Ocean Observatories Institute (OOI) [7] (use case UC.R2.13 "Acquire Data From Instrument"). The OOI is a project to establish a cyberinfrastructure for the delivery, management and analysis of scientific data from a large network of ocean sensor systems. Its architecture relies on the combination of high-level protocol specifications of network services (expressed as Scribble protocols [8]) and distributed runtime monitoring to regulate the behaviour of third-party applications within the system. We have integrated our framework into the Python-based runtime platform developed by OOI [7].

4.2.1 Verification Steps

Global Protocol Correctness. The first level of verification is when designing a global protocol. A Scribble global protocol for the use case is listed in Figure 4.2 (left). Scribble describes interactions between session participants through message passing sequences, branches and recursion. Each message has a label (an operator) and a payload. The Scribble protocol in Figure 4.2 starts by protocol declaration, which specifies the name of the protocol, Data

```
1  global protocol DataAcquisition(        local protocol DataAcquisition
2  role U, role A, role I) {                 at A (role U, role A, role I)}
3  Request(string:info) from U to A;       Request(string:info) from U;
4  Request(string:info) from A to I;       Request(string:info) to I;
5    choice at I {                           choice at I {
6      Supported() from I to A;                Supported() from I;
7      rec Poll {                              rec Poll {
8        Poll() from A to I;                     Poll() to I;
9        choice at I {                           choice at I {
10       Raw(data) from I to A                   Raw(data) from I
11       @{size(data) <= 512};                   @{size(data) <= 512};
12       Formatted(data) from A to U;            Formatted(data) to U;
13       continue Poll;                          continue Poll;
14       } or {                                  } or {
15       Stop() from I to A;                     Stop() from I;
16       Stop() from A to U;}}                   Stop() to U;}}
17   } or {                                  } or {
18     NotSupported from I to A;               NotSupported from I;
19     Stop() from A to I;                     Stop() to I;
20     Stop from A to U;}}                     Stop to U;}}
```

Figure 4.2   Global Protocol (left) and Local Protocol (right).

Acquisition, and its participating roles – a User (U), an Agent service (A) and an Instrument (I). The overall scenario is as follows: U requests through A to start streaming a list of resources from I (line 3–4). At line 5 I makes a choice whether to continue the interaction or not. If I supports the requested resource, I sends a message Supported and the communication continues by A sending a Poll request to I. The raw resource data is sent from I to A, at A the data is formatted and forwarded to U (line 10–12). Line 11 demonstrates an *assertion* construct specifying that I is allowed to send data packages that are less than 512MB.

The Scribble toolchain validates that a protocol is well-formed and thus projectable for each role. For example, in each case of a choice construct, the deciding party (e.g. at I) must correctly communicate the decision outcome unambiguously to all other roles involved; a choice is badly-formed if the actions of the deciding party would cause a race condition on the selected case between the other roles, or if it is ambiguous to another role whether the decision has already been made or is still pending.

Local protocol conformance.   The second level of verification is performed at runtime and ensures that each endpoint program conforms to the local protocol structure. Local protocols specify the communication behaviour for each conversation participant. Local protocols are mechanically projected from a global protocol. A local protocol is essentially a view of the global protocol from the perspective of one participant role. Projection works by identifying the message exchanges where the participant is involved, and disregarding the rest, while preserving the overall interaction structure of the global protocol.

From the local protocols, an FSM is generated. At runtime, the endpoint program is validated against the FSM states. There are two main checks that are performed. First, we verify that the type (a label and payloads) of each message matches its specification (labels can be mapped directly to message headers, or to method calls, class names or other relevant artefacts in the program). Second, we verify that the flow of interactions is correct, i.e. interaction sequences, branches and recursions proceed as expected, respecting the explicit dependencies (e.g. m1() from A to B; m2() from B to C; imposes a causality at B, which is obliged to receive the messages from A before sending a message to C).

Policy validation. The final level of verification enables the elaboration of Scribble protocols using annotations (@{} in Figure 4.2). Annotations function as API hooks to the verification framework: they are not verified by the MPST monitor itself, but are, instead, delegated to a third-party library. Our current implementation uses a Python library for evaluating basic predicates (e.g. the size check in Figure 4.2). At runtime, the monitor passes the annotated information, along with the FSM state information, to the appropriate library to perform the additional checks or calculations. To plug in an external validation engine, our toolchain API requires modules for parsing and evaluating the annotation expressions specified in the protocol.

4.2.2 Monitoring Requirements

Positioning. In order to guarantee global safety, our monitoring framework imposes complete mediation of communications: communication actions should not have an effect unless the message is mediated by the monitor. The tool implements this principal for outline monitor configurations, i.e. the monitor is running as a separate application. Outline monitoring is realised by dynamically modifying the application-level network configuration to (asynchronously) route every message through a monitor. Our prototype is built over an Advanced Message Queuing Protocol (AMQP) [1] transport. An AMQP is a publish-subscribe middleware. An AMQP network consists of a federation of distributed virtual routers (called brokers) and queues. A monitor dispatcher is assigned to each network endpoint as a conversation gateway. The dispatcher can create new routes and spawn new monitor processes if needed, to ensure the scalability of this approach.

Message format. To monitor Scribble conversations, our toolchain relies on a small amount of message meta data, referred to as Scribble header, and embedded into the message payload. Messages are processed depending on their kind, as recorded in the first field of the Scribble header. There are two kinds of conversation messages: initialisation (exchanged when a session is started, carrying information such as the protocol name and the role of the monitored process) and in-session (carrying the message operation and the sender/receiver roles). Initialisation messages are used for routing reconfiguration, while in-session messages are the ones checked for protocol conformance.

Principals and Conversation runtime. A principal (an application) implements a protocol behaviour using the Conversation API. The API is built on top of a Conversation Runtime. The runtime provides a library for instantiating, managing and programming Scribble protocols and serialising and deserializing conversation messages. The library is implemented as a thin wrapper over an existing transport library. The API provides primitives for creating and joining a conversation, as well as primitives for sending and receiving messages.

4.3 Conversation Programming in Python

The Python Conversation API is a message passing API, which offers a high-level interface for safe conversation programming, mapping the interaction primitives of session types to lower-level communication actions on concrete transports. The API primitives are displayed in Figure 4.3. In summary, the API provides functionality for (1) session initiation and joining, (2) basic send/receive.

Conversation Initiation. A session is initiated using the `create` method. It creates a fresh conversation id and the required AMQP objects (principal exchange and queue), and sends an invitation for each role specified in the protocol. Invitations are sent to principals.

Conversation API operation	Purpose
`create(protocol_name, config.yml)`	Initiate conversation, send invitations
`join(self, role, principal_name)`	Accept invitation
`send(role, op, payload)`	Send a message
`recv(role)`	Receive message from role
`recv_async(self, role, callback)`	Asynchronous receive

Figure 4.3   The core Python Conversation API operations.

We use a configuration file to provide the mapping between roles and principals. We give on the right an example of the configuration file (invitation section) for the `DataAcquisition` protocol. Principal names direct the routing of invitation message to the right endpoint. Each invitation carries a role, a principal name and a name for a Scribble local specification file. An invitation is accepted using the `Conversation.join` method. It establishes an AMQP connection and, if one does not exist, creates an invitation queue for receiving invitations.

```
invitations:
  -role: U
   principal name: bob
   local capability: DataAcquisition.spr
  -role: A
   principal name: allice
   local capability: DataAcquisition.spr
  -role: I
   principal name: carol
   local capability: DataAcquisition.spr
```

We demonstrate the usage of the API in a Python implementation of the local protocol projected for the Agent role. The local protocol is given in Figure 4.2 (right). Figure 4.4 (left) gives the Agent role implementation. First, the `create` method of the Conversation API initiates a new conversation instance of the `DataAcquisition` protocol, and returns a token that is used to join the conversation locally. The `config.yml` file specifies which network principals will play which roles in this session and the runtime sends invitation messages to each principal. The `join` method confirms that the endpoint is joining the conversation as the principal `alice` playing role `A`. Once the invitations are sent and accepted (via `Conversation.join`), the conversation is established and the intended message exchange can proceed. As a result of the initiation procedure, the runtime at every participant has a mapping (conversation table) between each role and their AMQP addresses.

Conversation Message Passing. The API provides standard send/receive primitives. Send is asynchronous, meaning that a basic send does not block on the corresponding receive; however, the basic receive does block until the message has been received. In addition, an asynchronous receive method, called `recv_async`, is provided to support event-driven usage of the conversation API. These asynchronous features map closely to those supported by

```
class ClientApp(BaseApp):               class ClientApp(BaseApp):
  def start(self):                        def start(self):
    c = Conversation.create(                c = Conversation.create(
    'DataAcquisition', 'config.yml')        'DataAcquisition', 'config.yml')
    c.join('A', 'alice')                    c.join('A', 'alice')
                                            c.recv_async('U', on_request)
    resource_request = c.recv('U')
    c.send('I', resource_request)         def on_request(self, conv, op, msg):
    req_result = c.recv('I')                if (op == SUPPORTED):
                                              conv.send('I', 'Poll')
    if (req_result == 'Supported'):           conv.recv_async('I', 'on_data')
      c.send('I', 'Poll')                   else: conv.send([I, U], 'Stop')
      op, data = c.recv('I')

      while (op != 'Stop'):               def on_data(self, conv, op, payload):
        formatted_data = format(data)       if (op != 'Stop'):
        c.send('U', formatted_data)           formatted_data = format(payload)
      c.send('U', stop)                       c.send('U', formatted_data)
    else:                                   else:
      c.send([U, I], stop)                    conv.send('U', 'Stop')
      c.stop()                                conv.stop()
```

Figure 4.4 Python program for A: synchronous implementation (left) and event-driven implementation (right).

Pika[1], a Python transport library used as an underlying transport library in our implementations.

Each message signature in a Scribble specification contains an operation and payloads (message arguments). The API does not mandate how the operation field should be treated, allowing the flexibility to interpret the operation name in various ways, e.g. as a plain message label, an RMI method name, etc. We treat the operation name as a plain label.

Following its local protocol, the program for A receives a request from U and forwards the message to I. The recv returns a tuple, (label, payload) of the message. When the message does not have a payload, only the label is returned (req_result = c.recv('I')). The recv method can also take the source role as a single argument (c.recv('I')), or additionally the label of the desired message (c.recv('I', 'Request')). The send method called on the conversation channel c takes, in this order, the destination role, message operator and payload values as arguments. In our example, the received payload resource_request is forwarded without modifications to I. After A receives the reply from I, the program checks the label value req_result using conditional statements, if (req_result== 'Supported'). If I replies with 'Supported', A enters a loop, where it continuously sends a 'Poll' requests to I and after receiving the result from I, formats the received data (format(data)) and resends the formatted result to U.

---

[1]http://pika.readthedocs.org/

Event-driven conversations. For asynchronous, non-blocking receives, the Conversation API provides `recv_async` to be used in an event-driven style. Figure 4.4 (right) shows an alternative implementation of the `user` role using callbacks. The method `recv_async` accepts as arguments a callback to be invoked when a message is received.

We first create a conversation variable similar to the synchronous implementation. After joining the conversation, A registers a callback to be invoked when a message from U is received (`on_request`). The callback executions are linked to the flow of the protocol by taking the conversation id as an argument (e.g. `conv`). It also accepts as arguments the label for the message (`op`) and the payload (`msg`). In the message handler for `Request`, the role A forwards the received payload to I and registers a new message handler for the next message. Although the event-driven API promotes a notably different programming style, our framework monitors both implementations in Figure 4.4 transparently without any modifications.

4.4 Monitor Implementation

Figure 4.5 depicts our outline monitor configuration. The interception mechanism is based on message forwarding. A principal has at least one queue for consuming messages, although the number of queues can be tuned to

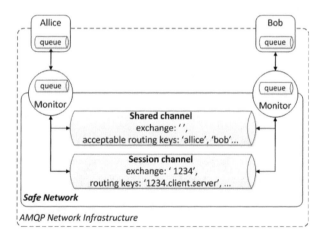

Figure 4.5 Configuration of distributed session monitors for an AMQP-based network.

use separate queues for invitations and roles. We outline a concrete scenario. Principal Alice is authenticated and connected to her local broker.

1. Authentication creates a network access point for Alice (the Monitor circle in Figure 4.5). The access point consists of a new conversation monitor instance, monitor queues (monitor as a consumer), and an exchange. Alice is only permitted to send messages to this exchange.
2. Alice initiates a new session (creates an exchange with id 1234 in Figure 4.5) and dispatches an invitation to principal Bob. The invitation is received and checked by Alice's monitor and then dispatched on the shared channel, from where it is rerouted to Bob's Monitor.
3. Bob's monitor checks the invitation, generates the local FSM and session context for Bob and Bob's role (for example client), and allocates a session channel (with exchange: 1234 and routing keys matching Bob's role (1234.*client.** and 1234. * .*client*). The invitation is delivered to Bob's queue.
4. Any message sent by Alice (e.g. to Bob) in this session is similarly passed by the monitor and validated. If valid, the message is forwarded to the session channel to be routed. The receiver's monitor will similarly but independently validate the message.

Figure 4.6 depicts the main components and internal workflow of our prototype monitor. The lower part relates to conversation initiation. The *invitation* message carries (a reference to) the local protocol for the invitee and the conversation id. We use a parser generator (ANTLR[2]) to produce, from a Scribble local protocol, an abstract syntax tree with MPST constructs as nodes. The tree is traversed to generate a finite state machine, represented in Python as a hash table, where each entry has the shape:

$$(current\_state,\ transition) \mapsto (next\_state,\ assertion,\ var)$$

where *transition* is a quadruple (*interaction type, label, sender, receiver*), *interaction type* is either *send* or *receive* and *var* is a variable binder for a message payload. We number the states using a generator of natural numbers. The FSM generation is based on the translation of local Scribble protocols to FSMs, presented in [5].

The upper part of Figure 4.6 relates to *in-conversation* messages, which carry the conversation id (matching an entry in the FSM hash table), sender and receiver fields, and the message label and payload. This information

---

[2]http://www.antlr.org/

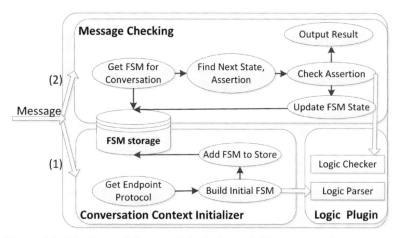

Figure 4.6 Monitor workflow for (1) invitation and (2) in-conversation messages.

allows the monitor to retrieve the corresponding FSM (by matching the message signature to the FSM's transition function). Assertions associated to communication actions are evaluated by invoking a library for Python predicate evaluation.

4.5 Monitoring Interruptible Systems

This section presents the implementation of a new construct for verifying *asynchronous multiparty session interrupts*. Asynchronous session interrupts express communication patterns in which the behaviour of the roles following the default flow through a protocol segment may be overruled by one or more other roles concurrently raising asynchronous interrupt messages. Extending MPST with asynchronous interrupts is challenging because the inherent communication race conditions that may arise conflict with the MPST safety properties. Taking a continuous stream of messages from a producer to a consumer as a simple example: if the consumer sends an interrupt message to the producer to pause or end the stream, stream messages (those already in transit or subsequently dispatched before the interrupt arrives at the producer) may well continue arriving at the consumer for some time after the interrupt is dispatched. This scenario is in contrast to the patterns permitted by standard session types, where the safety properties guarantee that no message is ever lost or redundant by virtue of disallowing all protocols with potential races.

This section introduces a novel approach based on reifying the concept of *scopes* within a protocol at the runtime level when an instance of the protocol is executed. A scope designates a sub-region of the protocol, derived from its syntactic structure, on which certain communication actions, such as interrupts, may act on the region as a whole. At run-time, every message identifies the scope to which it belongs as part of its meta data. From this information and by tracking the local progress in the protocol, the runtime at each endpoint in the session is able to resolve discrepancies in a protocol state by discarding incoming messages that have become irrelevant due to an asynchronous interrupt. This mechanism is transparent to the user process, and although performed independently by each distributed endpoint, preserves global safety for the session.

We integrate the new interrupt construct in our framework for runtime monitoring. The FSM generation is extended to support interruptible protocol scopes. We treat interruptible scopes by generating nested FSM structures. In the case of scopes that may be entered multiple times by recursive protocols, we use dynamic FSM nesting (conceptually, a new sub-FSM is created each time the scope is entered, and the sub-FSM is terminated once it reaches its end state or when an interrupt message is received) corresponding to the generation of fresh scope names in the syntactic model.

4.5.1 Use Case: Resource Access Control (RAC)

This section expands on how we extend Scribble to support the specification and verification of asynchronous session interrupts, henceforth referred to as just interrupts. Our running example is based on an OOI project use case, which we have distilled to focus on session interrupts.

Figure 4.7 (left) gives an abridged version of a sequence diagram given in the OOI documentation for the Resource Access Control (RAC) use case [8], regarding access control of users to sensor devices in the OOI cyberinfrastucture for data acquisition. In the OOI setting, a User interacts with a sensor device via its Agent proxy (which interacts with the device using a separate protocol outside of this example). OOI Controller agents manage concerns such as authentication of users and metering of service usage.

For brevity, we omit from the diagram some of the data types to be carried in the messages and focus on the *structure* of the protocol. The depicted interaction can be summarised as follows. The protocol starts at the top of the left-hand diagram. User sends Controller a request message to use a sensor

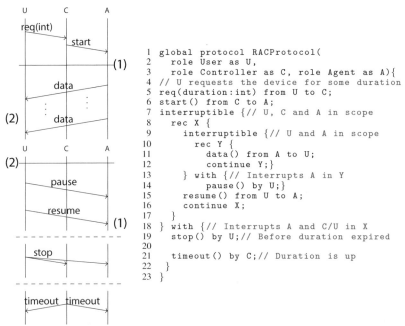

```
1  global protocol RACProtocol(
2      role User as U,
3      role Controller as C, role Agent as A){
4  // U requests the device for some duration
5  req(duration:int) from U to C;
6  start() from C to A;
7  interruptible {// U, C and A in scope
8      rec X {
9          interruptible {// U and A in scope
10             rec Y {
11                 data() from A to U;
12                 continue Y;}
13             } with {// Interrupts A in Y
14                 pause() by U;}
15         resume() from U to A;
16         continue X;
17     }
18 } with {// Interrupts A and C/U in X
19     stop() by U;// Before duration expired
20
21     timeout() by C;// Duration is up
22 }
23 }
```

Figure 4.7 Sequence diagram (left) and Scribble protocol (right) for the RAC use case.

for a certain amount of time (the int in parentheses), and Controller sends a start to Agent. The protocol then enters a phase (denoted by the horizontal line) that we label (1), in which Agent streams data messages (acquired from the sensor) to User. The vertical dots signify that Agent produces the stream of data freely under its own control, i.e. without application-level control from User. User and Controller, however, have the option at any point in phase (1) to move the protocol to the phase labelled (2), below.

Phase (2) comprises three alternatives, separated by dashed lines. In the upper case, User *interrupts* the stream from Agent by sending Agent a pause message. At some subsequent point, User sends a resume and the protocol returns to phase (1). In the middle case, User interrupts the stream, sending both Agent and Controller a stop message. This is the case where User does not want any more sensor data, and ends the protocol for all three participants. Finally, in the lower case, Controller interrupts the stream by sending a timeout message to User and Agent. This is the case where, from Controller's view, the session has exceeded the requested duration, so Controller interrupts the other two participants to end the protocol. Note this diagram actually intends that stop (and timeout) can arise anytime after (1), e.g. between pause and resume (a notational ambiguity that is compensated by additional prose comments in the specification).

4.5.2 Interruptible Multiparty Session Types

Figure 4.7 (right) shows a Scribble protocol that formally captures the structure of interaction in the Resource Access Control (RAC) use case and demonstrates the uses of our extension for asynchronous interrupts. Besides the formal foundations, we find the Scribble specification more explicit and precise, particularly regarding the combination of compound constructs such as choice and recursion, than the sequence diagram format, and provides firmer implementation guidelines for the programmer.

The protocol starts with a header declaring the protocol name (given as RACProtocol in Figure 4.7) and role names for the participants (three roles, aliased in the scope of this protocol definition as U, C and A). Lines 5 and 6 straightforwardly correspond to the first two communications in the sequence diagram, a User sends a request message, carrying an int *payload*, to the Controller and then the Controller replies with a start() message and an empty payload.

Then the intended communication in "phase" (1) and (2) in the diagram, is clarified in Scribble as two nested interruptible statements. The outer statement, on lines 7–22, corresponds to the options for User and Controller to end the protocol by sending the stop and timeout interrupts. An interruptibleconsists of a main body of protocol actions, here lines 8–17, and a set of interrupt message signatures, lines 18–22. The statement stipulates that each participant behaves by either (a) following the protocol specified in the body until finished for their role, or (b) raising or detecting a specified interrupt at any point during (a) and exiting the statement. Thus, the outer interruptible states that U can interrupt the body (and end the protocol) by a stop() message, and C by a timeout().

The body of the outer interruptible is a labelled recursion statement with label X. The continue X; inside the recursion (line 16) causes the flow of the protocol to return to the top of the recursion (line 8). This recursion corresponds to the loop implied by the sequence diagram that allows User to pause and resume repeatedly. Since the recursion body always leads to the continue, Scribble protocols of this form state that the loop should be driven indefinitely by one role, until one of the interrupts is raised by *another* role. This communication pattern cannot be expressed in multiparty session types without interruptible.

The body of the X-recursion is the inner `interruptible`, which corresponds to the option for User to pause the stream. The stream itself is specified by the Y-recursion, in which A continuously sends `data()` messages to U. The inner `interruptible` specifies that U may interrupt the Y-recursion by a `pause()` message, which is followed by the `resume()` message from U before the protocol returns to the top of the X-recursion.

4.5.3 Programming and Verification of Interruptible Systems

We extend the Python API, presented in Section 4.3, to provide functionality for *scope* management for handling interrupt messages. We demonstrate the usage of the construct through an implementation of the local protocol projected for the `User` role. Figure 4.8 gives the local protocol and its implementation.

Similarly to the previous example from Section 4.3, the implementation starts by creating a conversation instance c of the Resource Access Control protocol (Figure 4.7) using method `create` (line 6, left) and `join`. The latter returns a conversation channel object for performing the subsequent communication operations.

Interrupt handling. The implementation of the User program demonstrates a way of handling conversation interrupts by combining conversation scopes

```
 1 class UserApp(BaseApp):
 2   user, controller, agent =
 3     ['User', 'Controller', 'Agent']
 4   def start(self):
 5     self.buffer = buffer(MAX_SIZE)
 6     conv = Conversation.create(
 7       'RACProtocol', 'config.yml')
 8     c = conv.join(user, 'alice')
 9     c.send(controller, 'req', 3600)
10     with c.scope('timeout', 'stop') as c_x:
11       while not self.should_stop():
12         with c_x.scope('pause') as c_y:
13           while not self.buffer.is_full():
14             data = c_y.recv(agent)
15             self.buffer.append(data)
16           c_y.send_interrupt('pause')
17           use_data(self.buffer)
18           self.buffer.clear()
19           c_x.send(agent, 'resume')
20       c_x.send_interrupt('stop')
21     c.close()
```

```
local protocol RACProtocol
  at U (role C, role A){
  req(duration:int) to C;
  interruptible {
    rec X {
      interruptible {
        rec Y {
          data() from A;
          continue Y;
        }
      } with {
        pause() by U;
      }
      resume() to A;
      continue X;
    }
  } with {
    stop() by U;
    timeout() by C;
  }
}
```

Figure 4.8 Python implementation (left) and Scribble local protocol (right) for the User role for the global protocol from Figure 4.7.

with the Python `with` statement (an enhanced try-finally construct). We use `with` to conveniently capture interruptible conversation flows and the nesting of interruptible scopes, as well as automatic `close` of interrupted channels in the standard manner, as follows. The API provides the `c.scope()` method, as in line 10, to create and enter the scope of an `interruptible` Scribble block (here, the outer interruptible of the RAC protocol). The `timeout` and `stop` arguments associate these message signatures as interrupts with this scope. The conversation channel `c_x` returned by `scope` is a wrapper of the parent channel `c` that (1) records the current scope of every message sent in its meta data, (2) ensures every send and receive operation is guarded by a check on the local interrupt queue, and (3) tracks the nesting of scope contexts through nested `with` statements. The interruptible scope of `c_x` is given by the enclosing `with` (lines 10–20); if, e.g., a `timeout` is received within this scope, the control flow will exit the `with` block to line 21. The inner `with` (lines 12–16), corresponding to the inner interruptible block, is associated with the `pause` interrupt. When an interrupt, e.g. `pause` in line 16, is thrown (`send_interrupt`) to the other conversation participants, the local and receiver runtimes each raise an internal exception that is either handled or propagated up, depending on the interrupts declared at the current scope level, to direct the interrupted control flow accordingly. The delineation of interruptible scopes by the global protocol, and its projection to each local protocol, thus allows interrupted control flows to be coordinated between distributed participants in a structured manner.

The scope wrapper channels are closed (using the Python construct `with`) after throwing or handling an interrupt message. Since we assume asynchronous communication, there is a delay from the time when an interrupt mesasage is sent untill the time when the interrupt message is received by all participants. Hence, the monitor reacts differently when checking message sending (a check driven by the monitored participant) and message receive (an action driven by a message arriving in the queue of the monitor); the monitor discards the message in the latter case and marks the message as wrong in the former case. More precisely, when a monitor receives a message from a closed scope, it discards it as to accommodate for the delay in receiving of an interrupt message. However, if a participant that is monitoried attepts to send a message on a scope that is already closed (after an interrupt message has been recieved or after the participant has thrown interrupt himself) then the monitor flagges the interaction as an error. For example, using `c_x` after a `timeout` is received (i.e. outside its parent scope) will be flagged as an error. However, receiving messages on that scope will be

```
1   class UserApp(BaseApp):
2     def start(self):
3       self.buffer = buffer(MAX_SIZE)
4       conv = Conversation.create(
5           'RACProtocol', config.yml)
6       c = conv.join(user, 'alice')
7       # request 1 hour access
8       c.send(controller, 'req', 3600)
9       c_x = c.scope('timeout', 'stop')
10      c_y = c_x.scope('pause')
11      c_y.recv_async(agent, recv_handler)
12
13    def recv_handler(self, c, op, payload):
14      with c:
15        if self.should_stop():
16          c.send_interrupt('stop')
17        elif self.buffer.is_full():
18          self.process_buffer(c, payload)
19        else:
20          self.buffer.append(payload)
21        c.recv_async(agent, recv_handler)
22
23    def process_buffer(self, c, payload):
24      with c:
25        c_x = c.send_interrupt('pause')
26        use_data(self.buffer, payload)
27        self.buffer.clear()
28        c_x.send(agent, 'resume')
29        c_y = c_x.scope('pause')
30        c_y.recv_async(agent, recv_handler)
```

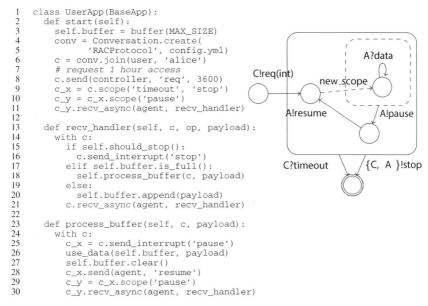

Figure 4.9 Event-driven conversation implementation for the User role (left) and Nested FSM generated from the User local protocol (right).

discarded and will not be dispatched to the application. In our example, the User runtime discards `data` messages that arrive after `pause` is thrown. The API can also make the discarded data available to the programmer through secondary (non-monitored) operations.

Message handlers with scopes. As demonstrated in Section 4.3, our Python API supports asynchronous receive through the primitive `recv_async`. The construct is used to register a method that should be invoked on message receive. To support event-driven programming with interrupts, we extend the implementation presented in Section 4.3. The difference is in the semantics of event handlers. More precisely, each event handler is associated with a scope. Therefore, if an interrupt is received, but the protocol state is not in the same scope as the scope written in the conversation header of the interrupt message, the interrupt will be discarded.

Figure 4.9 (left) shows an alternative implementation of the `User` role using callbacks. We first enter the nested conversation scopes according to the potential interrupt messages (lines 9 and 10). The callback method is then registered using the `recv_async` operation (line 11). The callback executions are linked to the flow of the protocol by taking the scoped channel as an argument (e.g. c on line 13). Note that if the stop and pause interrupts were not declared for these scopes, line 16 and line 25 would be considered invalid by the monitor. When the buffer is full (line 17), the user sends the pause interrupt. After raising an interrupt, the current scope becomes obsolete and the channel object for the parent scope is returned. After the data is processed and the buffer is cleared, the resume message is sent (line 28) and a fresh scope is created and again registered for receiving data events (line 29). Our framework monitors both this implementation and that in Figure 4.8 transparently without any modifications.

4.5.4 Monitoring Interrupts

FSM generation for interruptible local protocols makes use of nested FSMs. Each interruptible induces a nested FSM given by the main interruptible block, as illustrated in Figure 4.9 (right) for the User local protocol. The monitor internally augments the nested FSM with a scope id, derived from the signature of the interruptible block, and an interrupt table, which records the interrupt message signatures that may be thrown or received in this scope. Interrupt messages are marked via the same meta data field used to designate invitation and in-conversation messages, and are validated in a similar way except that they are checked against the interrupt table. However, if an interrupt arrives that does not have a match in the interrupt table of the immediate FSM(s), the check searches upwards through the parent FSMs; the interrupt is invalid if it cannot be matched after reaching the outermost FSM.

4.6 Formal Foundations of MPST-Based Runtime Verification

In this section, we explain the correspondence between a theoretical model for MPST-based monitoring and the implementation, presented in this chapter. Our implementation is formalised in a theory for MPST-based verification of networks, first proposed in [3], and later extended in [1], and in [2]. The interrupt extension is formalised in [4]. [3] only gives an overview of the desired properties, and requires all local processes to be dynamically verified through the protections of system monitors, while [1] presents a framework for semantically precise decentralised run-time verification, supporting statically and dynamically verified components. In addition, the routing mechanism of AMPQ networks is explicitly presented in [1], while in [3] it is implicit.

A delicate technical difference between the theory and the implementation lies in handling of out-of-order delivery of messages when messages are sent from different senders to the same receiver. Asynchrony poses a challenge in the treatment of out-of-order asynchronous message monitoring, and thus, to prevent false positive results, in the theoretical model, a type-level permutations of actions is required, e.g a monitor checks messages up to permutations. The use of global queues and local permutations is inefficient in practice, and thus we have implemented the theoretical model of a global queue as different physical queues. Specifically, we introduce a queue per pair of roles, which ensures messages from the same receivers are delivered in order and are not mixed with messages from other roles. This model is semantically equivalent to a model of a global indexed queue, permitting permutation of messages.

Next we explain the correspondence between the asynchronous π-calculus with fine-grained primitives for session initiation and our Python API. Also in [1] specifications are given as local types. Instead of using local types, for efficient checking, we use communicating finite state machines (CFSMs) generated from local Scribble protocols, which are equivalent to local types, as has been shown in [5].

Processes. Our Python API embodies the primitives of the asynchronous π-calculus with fine grained primitives for session initiation, presented in [1]. The correspondence is given in Figure 4.10. Note that the API does not stipulate the use of a recursion and a conditional, which appear in the syntax of session π-calculus, since these constructs are handled by native Python constructs. The `create` method, which, we remind, creates a fresh conversation id and the required AMQP objects (principal exchanges and queues), and sends an invitation for each role specified in the protocol, corresponds to the action $\overline{a}\langle s[r] : T\rangle$, which sends on the shared channel a, an

Conversation API operation	Purpose
`create(protocol_name, config.yml)`	$\overline{a}\langle s[r] : T\rangle$
`join(self, role, principal_name)`	$a(y[r] : T).P$
`send(role, op, payload)`	$k[r_1, r_2]!l\langle e\rangle$
`recv(role)`	$k[r_1, r_2]?\{l_i(x_i).P_i\}_{i \in I}$
`recv_async(self, role, callback)`	$-$

Figure 4.10 The core Python Conversation API operations and their session π-calulus counterparts.

invitation to join the fresh conversation s as the role of r with a specification T. In the implementation, this information is codified in the message header, which as we have explained contains the new session id (abstracted as s), the name of the local Scribble protocol (i.e. T) and the role (i.e. r). The invitation action $a(y[r] : T).P$ models session join. As a result of join new queues and a routing bindings are created. For example, when Bob joins a conversation with id of 1234 as the role of client, as shown in Figure 4.5, an AMQP binding 1234.client.* is created, which ensures that all messages to the role of a client are delivered to Bob. The reduction rule for $a(y[r] : T).P$, in the semantics in [1], reflects this behaviour by adding a record in the routing table. The primitive for sending a message $k[r_1, r_2]!l\langle e \rangle$ corresponds to the API call send, and results in sending a message of type $s[r_1, t_2]!l\langle e \rangle$, which in the implementation is codified in the message header, consisting of session id s, sender r_1, receiver r_2, label l and a payload e.

Properties of monitored networks. Finally, we give an overview of the properties of monitored networks as presented in [1]. Due to the correspondence explained above, these properties are preserved in the context of the monitor implementation, presented in this chapter.

Local safety states that a monitored process respects its local protocol, i.e. that dynamic verification by monitoring is sound.

Local transparency states that a monitored process has equivalent behaviour to an unmonitored but well-behaved process, e.g. statically verified against the same local protocol.

Global safety states that a system satisfies the global protocol, provided that each participant behaves as if monitored.

Global transparency states that a fully monitored network has equivalent behaviour to an unmonitored but well-behaved network, i.e. in which all local processes are well-behaved against the same local protocols.

Session fidelity states that, as all message flows of a network satisfy global specifications, whenever the network changes because some local processes take actions, all message flows continue to satisfy global specifications.

4.7 Concluding Remarks

We have presented a runtime verification framework for Python programs based on Scribble protocols. We discuss the core design elements of the implementation of a conversation-based API in a dynamically typed language,

Python. Through a runtime layer of protocol management Scribble protocols are loaded and translated to CFSMs such that during a program execution, messages emitted by the program are checked against a corresponding CFSM. We also introduce a construct for expressing exception-like patterns in Scribble, which syntactically splits the protocol into sub-regions, allowing certain messaging to act on the regions as a whole and thus permitting controllable races, traditionally disallowed by the theory of session types.

Acknowledgements We thank the anonymous reviewers for their insightful comments, which helped us to improve the article. This work is partially supported by EPSRC projects EP/K034413/1, EP/K011715/1, EP/L00058X/1, EP/N027833/1 and EP/N028201/1; by EU FP7 612985 (UP-SCALE).

References

[1] Laura Bocchi, Tzu-Chun Chen, Romain Demangeon, Kohei Honda, and Nobuko Yoshida. Monitoring networks through multiparty session types. In *FMOODS*, volume 7892 of *LNCS*, pages 50–65, 2013.

[2] Tzu-Chun Chen. *Theories for Session-based Governance for Large-scale Distributed Systems*. PhD thesis, Queen Mary, University of London, 2013.

[3] Tzu-Chun Chen, Laura Bocchi, Pierre-Malo Deniélou, Kohei Honda, and Nobuko Yoshida. Asynchronous distributed monitoring for multiparty session enforcement. In *TGC'11*, volume 7173 of *LNCS*, pages 25–45, 2012.

[4] Romain Demangeon, Kohei Honda, Raymond Hu, Rumyana Neykova, and Nobuko Yoshida. Practical interruptible conversations: Distributed dynamic verication with multiparty session types and python. *FMSD*, pages 1–29, 2015.

[5] Pierre-Malo Deniélou and Nobuko Yoshida. Multiparty session types meet communicating automata. In *ESOP*, volume 7211 of *LNCS*, pages 194–213. Springer, 2012.

[6] Kohei Honda, Nobuko Yoshida, and Marco Carbone. Multiparty Asynchronous Session Types. *Journal of the ACM*, 63, 2016.

[7] Ocean Observatories Initiative. `http://www.oceanobservatories.org/`

[8] OOIExamples. `http://confluence.oceanobservatories.org/display/CIDev/Identify+required+Scribble+extensions+for+advanced+scenarios+of+R3+COI`

[9] Scribble project home page. `http://www.scribble.org`

5

The DCR Workbench: Declarative Choreographies for Collaborative Processes

Søren Debois and Thomas T. Hildebrandt

Department of Computer Science, IT University of Copenhagen, Rued Langgaards Vej 7, 2300 Copenhagen S, Denmark

Abstract

The *DCR Workbench* is an online tool for simulation and analysis of collaborative distributed processes specified as *DCR graphs*. The Workbench is a robust and comprehensive implementation of DCR graphs, providing concrete syntax, specification by refinement, visualisation, simulation, static analysis, time analysis, enforcement, declarative subprocesses, data dependencies, translation to other declarative models, and more. This chapter introduces the Workbench and, through the features of the Workbench, surveys the DCR formalism. The Workbench is available on-line at `http://dcr.tools`.

5.1 Introduction

Citizens, businesses and public organisations increasingly rely on distributed business processes. Many such processes involve at the same time information systems, humans and mechanical artefacts, and are thus highly unpredictable. Moreover, such processes are constantly evolving due to advances in technology, improvement in business practices, and changes in legislation.

In this climate of distribution and continuous change, the traditional vision of verifying a system once and for all against a final formal description has little hope of realisation. Instead, we need tools and techniques for describing, building, and analysing systems of continuously changing distributed collaborative processes. Dynamic Condition Response graphs, DCR graphs, is a formal model developed in response to this need.

Developed through a series of research projects, DCR graphs today stands on three pillars: a substantial body of academic publications on both case studies [8, 10, 15, 17, 20, 38] and formal aspects [2, 6, 7, 12–14, 16, 18, 21–25, 29–31, 34, 37]; the DCR Workbench, implementing most major advances of the formalism (the subject of this chapter); and a commercial adaptive case-management system developed by independent vendor Exformatics A/S [9, 11, 15, 19, 23, 28].

Declarative process notations such as DCR graphs, DECLARE [33, 40] and GSM [26] generally support *specification and analysis* of requirements, whereas imperative notations such as Workflow Nets [1] and BPMN [32] generally support *implementation* of requirements. DCR graphs have the advantage of serving as *both* the specification of requirements *and* the runtime representation of a process instance, which can be adapted dynamically if the requirements change.

The DCR Workbench is a comprehensive tool for modelling with DCR graphs and analysing DCR models. The Workbench serves the dual purposes of being a communication and teaching tool, used both in classroom settings and in discussions with industry, as well as a test-bed for experimentation with new analysis and variants.

This chapter gives an introduction to DCR graphs in general and the Workbench in particular. As we shall see, the Workbench implements a majority of published DCR graph variants and analysis methods, as well as some work-in-progress experimental additions and algorithms that have yet to be published. Through the features of the Workbench, the chapter also provides a survey of the state-of-the-art of DCR graphs variants, their technical properties, and their published analysis methods and algorithms.

5.1.1 History of the DCR Workbench

DCR graphs were introduced in 2010 [18,29] by Thomas Hildebrandt and his group at the ITU. Soon after, Danish vendor of adaptive case-management solutions, Exformatics A/S, entered into a long-term collaboration with the ITU group; a collaboration which continues to this day. The continued financial support and interest of Exformatics A/S has been instrumental in the development of the formalism.

DCR graphs were implemented repeatedly as the formalism evolved. Notably, an early implementation created by industrial PhD Tijs Slaats in collaboration with Exformatics A/S [37] eventually grew into that company's current commercial DCR tool [9, 15, 28], available at dcrgraphs.net. In

2013 this tool was solidifying into a commercial offering. While the backing of commercial vendor was *extremely* helpful to DCR graph, the Exformatics tool was becoming too heavyweight for quick academic experiments. Accordingly, the ITU group in 2013 commenced development of a nimbler implementation. This effort was spearheaded by Søren Debois and became the DCR Workbench of the present chapter.

The two tools have different goals: Exformatics' offering is aimed at non-expert commercial users and emphasises stability and usability. Conversely, the DCR Workbench is aimed at academics and prioritises ease-of-experimentation overall. This division has so far been productive: sufficiently good ideas implemented in the Workbench has later been re-implemented by Exformatics in their commercial offering [8–10, 28].

The Workbench made its first appearance in a research paper in 2014 [12], and its first appearance in industry collaborations in 2015 [9]. Subsequently, the Workbench has provided implementation and examples for most major developments of the formalism [2, 6, 12–14, 16].

The DCR Workbench is implemented in F# [39], using the WebSharper library [5] to derive server- and client-side components from the same F# code base. The choice of implementation language and platform is no accident: On the one hand, F# is very well-suited to manipulating formal models; on the other, the web-based platform makes the Workbench *immediately* available to interested researchers: all it takes is a browser.

5.1.2 The DCR Workbench

The DCR Workbench is available at

```
http://dcr.tools/2017chapter
```

This URL leads to a special page supporting this chapter with the collection of examples used on the following pages. We encourage the reader to visit this page and actively try out the examples presented in the remainder of this chapter as he progresses through the it.

Overview In Section 5.2, we introduce a running example, and in Section 5.3, we recall DCR graphs. In Section 5.4 we introduce basic modelling, simulation and analysis of DCR graphs in the Workbench. In Section 5.5 we construct models by *refinement*; in Section 5.6 we discuss timed models; in Section 5.7 we talk about subprocesses; and in Section 5.8 data. In Section 5.9, we mention briefly other tools in the Workbench, before concluding in Section 5.10.

5.2 Running Example

As a running example we consider a stylised mortgage loan application process distilled from real-life cases [9, 13]. Mortgage application processes are in practice *extremely* varied, depending on the type of mortgage, the neighbourhood, the applicant, and the credit institution in question. The purpose of the process is to arrive at a point where the activity Assess loan application can be carried out. This requires in turn:

1. collecting appropriate documentation,
2. collecting a budget from the applicant, and
3. appraising the property.

In practice, applicants' budgets tend to be underspecified, so an intern will screen the budget and request a new one if the submitted one happens to be so. The case worker should not spend time assessing the application if the documentation has not been collected or the budget is underspecified. The caseworker decides if the appraisal can be entirely statistical, i.e., carried out without physical inspection, or if it requires an on-site appraisal. For reasons of cost efficiency, only one appraisal should be carried out.

5.3 Dynamic Condition-Response Graphs

In this section, we recall DCR graphs [6, 9, 14, 15, 18, 29, 37]. We begin by an informal walkthrough, followed by a formal development in Section 5.3.4.

DCR graphs constitute a declarative modelling notation describing at the same time a process and its run-time state. The core notation comprises *labelled events*, *event states*, and five possible *relations* between events. The relations govern: (a) how executability of one event depend on the state of another, and (b) how execution of one event updates the states of another.

5.3.1 Event States

The *event state* consists of three booleans: The *executed, included,* and *pending* states of the event.

- The *executed* state simply registers whether the event has been previously executed (an event may execute more than once). It is updated to true whenever the event executes. It is never updated to false.
- The *included* state indicates whether the event is included, i.e. relevant for the process. Being included is a prerequisite for an event to execute.

- The *pending* state indicates whether the event is required to eventually execute (or become not included).

We give events and initial states for the running example in Figure 5.1. Except Request new budget, which becomes relevant only when a budget has been submitted, all events are included. The Assess loan application and Submit budget events are pending: they are required to complete the process.

5.3.2 Relations

Each pair of events may be related by one of five different relations. Relations regulate (a) which events may execute in a given graph (condition, milestone) and (b) the effect of executing an event (inclusion, exclusion, response).

We give a full DCR model of the running example[1] in Figure 5.2.

Conditions. A condition e →• f causes the target activity f to be not executable whenever the source activity e is included (its "included" state is true) and has not been previously executed (its "executed" state is false). E.g., in Figure 5.2, we must execute Collect documents before Assess loan application can be executed.

Milestones. A milestone e →◇ f causes the target activity f to be not executable whenever the source activity e is included and pending (its "included" and "pending" states are true). In Figure 5.2, whenever Submit budget is pending, Assess loan application is prevented from executing.

Event	Role	Initial state
Collect documents	Caseworker	
Budget screening approve	Intern	
Request new budget	Intern	Excluded
Submit budget	Customer	Pending
On-site appraisal	Mobile consultant	
Statistical appraisal	Caseworker	
Assess loan application	Caseworker	Pending

Figure 5.1   Events and initial states (marking) for the mortgage application process. Where nothing else is indicated, the initial state of an event is not executed, included, and not pending.

---

[1]This graph is in fact the output of the DCR Workbench visualiser; we describe in Section 5.4.2 exactly how the visualiser represents event state.

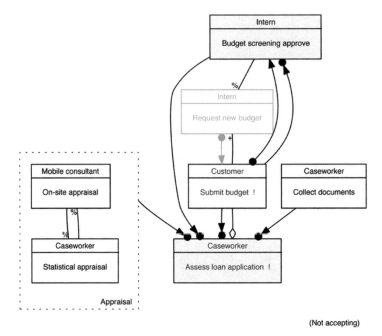

Figure 5.2   DCR graph modelling the mortgage application process.

Responses. A response e •→ f causes the target activity f to be pending (its "pending" state is true) whenever the source activity e is executed. In Figure 5.2, when an applicant executes Submit budget, we require a subsequent screening: there is a response from Submit budget to Budget screening approve.

Inclusions and exclusions. An inclusion e →+ f resp. an exclusion e →% f causes the "included" state of the target activity f to be true resp. false when the source activity e is executed. In Figure 5.2, the activity Submit budget includes the activity Request new budget.

5.3.3 Executing Events

Enabled events are the ones which have their "included" state true, and which do not have their execution prohibited by a condition or milestone as indicated above. Conditions or milestones from excluded events *do not* disable their target events. E.g., in Figure 5.2, both On-site appraisal and Statistical appraisal are conditions for Assess loan application, but once one executes, the other is excluded and thus no longer required for Assess loan application to execute.

Executing an enabled event in a DCR graph updates event states as indicated by inclusion, exclusion, and response relations. Conceptually, we can think of the execution as producing a new DCR graph with the same relations but with updated event states.

The *denotation* of a DCR model is the set of (finite and infinite) sequences of event labels, corresponding to a sequence of such event executions, where at each step, every pending event is eventually executed or excluded at some later step. We refer to such sequences as *accepting traces*. It follows that for finite accepting traces, in the final state, no event is pending and included.

We consider potential traces for Figure 5.2.

- $s_0 = \langle$ Collect documents, Assess loan application$\rangle$. This sequence intuitively corresponds to assessing the loan application *without* getting a budget and appraising the property. After Collect documents, the event Assess loan application is not enabled, so s_0 is not a trace.
- $s_1 = \langle$ Collect documents, Submit budget$\rangle$. This sequence is a trace, but not an accepting one, since Assess loan application is pending and included in the final state. It follows that s_1 sequence is not part of the denotation of Figure 5.2.
- $s_2 = \langle$ Collect documents, Submit budget, Budget screening approve, Statistical appraisal, Assess loan application$\rangle$ is an accepting trace.

Notice that between the notions of enabledness and accepting trace, DCR graphs express both permissions and obligations. We return to expressiveness of DCR graphs in Section 5.7 below.

5.3.4 Formal Development

Definition 1 (DCR Graph [18]). A *DCR graph*, ranged over by G, is a tuple $(\mathsf{E}, \mathsf{R}, \mathsf{M}, \ell)$ where

- E is a finite set of (labelled) *events*, the nodes of the graph.
- R is the edges of the graph. Edges are partitioned into five kinds, named and drawn as follows: The *conditions* ($\rightarrow\bullet$), *responses* ($\bullet\rightarrow$), *milestones* ($\rightarrow\diamond$), *inclusions* ($\rightarrow+$), and *exclusions* ($\rightarrow\%$).
- M is the *marking* of the graph. This is a triple $(\mathsf{Ex}, \mathsf{Re}, \mathsf{In})$ of sets of events, respectively the previously executed (Ex), the currently pending (Re), and the currently included (In) events.
- ℓ is a labelling function assigning to each $e \in \mathsf{E}$ a label comprising an activity name and a set of roles.

When G is a DCR graph, we write, e.g., $E(G)$ for the set of events of G, $Ex(G)$ for the executed events in the marking of G, etc.

Notation. Let $R \subseteq X \times Y$ be a relation. For $y \in Y$ we take $Ry = \{x \in X \mid (x,y) \in R\}$; dually for $x \in X$ we take $xR = \{y \in Y \mid (x,y) \in R\}$. We use this notation for relations, e.g.,, $(\rightarrow\bullet\, e)$ is the set of events that are conditions for e.

Definition 2 (Enabled events). Let $G = (E, R, M, \ell)$ be a DCR graph, with marking $M = (Ex, Re, In)$. An event $e \in E$ is *enabled*, written $e \in \text{enabled}(G)$, iff (a) $e \in In$, (b) $In \cap (\rightarrow\bullet e) \subseteq Ex$, and (c) $In \cap (\rightarrow\diamond e) \subseteq E \backslash Re$.

That is, enabled events (a) are included, (b) have their included conditions executed, and (c) have no included milestone with an unfulfilled responses.

Definition 3 (Execution). Let $G = (E, R, M, \ell)$ be a DCR graph with marking $M = (Ex, Re, In)$. Suppose $e \in \text{enabled}(G)$. We may *execute* e obtaining the DCR graph $G' = (E, R, M', \ell)$ with $M' = (Ex', Re', In')$ defined as follows.

1. $Ex' = Ex \cup e$
2. $Re' = (Re \backslash e) \cup (e \bullet \rightarrow)$
3. $In' = (In \backslash (e \rightarrow \%)) \cup (e \rightarrow +)$

That is, to execute an event e one must: (1) add e to the set Ex of executed events; (2) update the currently required responses Re by first removing e, then adding any responses required by e; and (3) remove from In those events excluded by e, then adding those included by e.

Technically, the operational semantics of a DCR graph is the labelled transition system where states are graphs and transitions are executions.

Definition 4 (Transitions). Let G be a DCR graph. If $e \in \text{enabled}(G)$ and executing e in G yields H, we say that G has *transition on e to H* and write $G \longrightarrow_e H$. A *run* of G is a (finite or infinite) sequence of DCR graphs G_i and events e_i such that: $G = G_0 \longrightarrow_{e_0} G_1 \longrightarrow_{e_1} \dots$. A *trace* of G is a sequence of labels of events e_i associated with a run of G. We write $\text{runs}(G)$ and $\text{traces}(G)$ for the set of runs and traces of G, respectively.

The denotation of a DCR graph is the set of *accepting* finite and infinite traces allowed by its operational semantics.

Definition 5 (Acceptance). A run $G_0 \longrightarrow_{e_0} G_1 \longrightarrow_{e_1} \dots$ is *accepting* iff for all n with $e \in In(G_n) \cap Re(G_n)$ there exists $m > n$ s.t. either $e_m = e$, or $e \notin In(G_m)$. A *trace is accepting* iff it has an underlying run which is.

Acceptance tells us which workflows a DCR graph accepts, its *language*.

Definition 6 (Language). The *language* of a DCR graph G is the set of its accepting traces. We write $\mathsf{lang}(G)$ for the language of G.

We conclude this Section by noting that by Definitions 2 and 3, because the set of events is finite, both the set of enabled events and the result of executing an event are computable in polynomial time.

5.4 Modelling with the Workbench

A typical configuration of the Workbench can be seen in Figure 5.3. The Workbench is divided into panels. In the configuration in Figure 5.3, we see the Visualiser, Parser and Activity Log panels.

The Workbench maintains at all times a current DCR graph and a current trace. Each panel allow the user to interact with this current graph and current trace. A few panels also maintain a DCR graph of their own.

Panels are dynamic: The user is free to remove panels by clicking "close" in the lower-left corner of a panel; or to add panels by selecting a new panel in the "Add a new panel" section of the Workbench panel. At the time of writing, the Workbench implements 22 different panels.

When working with the Workbench, it is customary to have several panels open; e.g., a visualiser and one or more analysis panels. The Workbench panel contains a selection of seven pre-made such panel configurations called "presets". These presets are accessible through the left-hand "Load a preset" section of the Workbench panel.

Finally, the Workbench can function as a process engine, making some of its functionality available programmatically as a REST interface; see the right-hand "REST API" section.

5.4.1 Inputting a Model: The Parser Panel

The parser panel allows input of DCR graphs as plain text. The parser accepts programs written according to the grammar of Figure 5.4.

As an example program, consider the abridged variant of our running example given in Figure 5.6; the corresponding input program is listed in Figure 5.5. The Workbench accepts such source programs as input, producing visualisations automatically. (Visualisations in this chapter was so produced.)

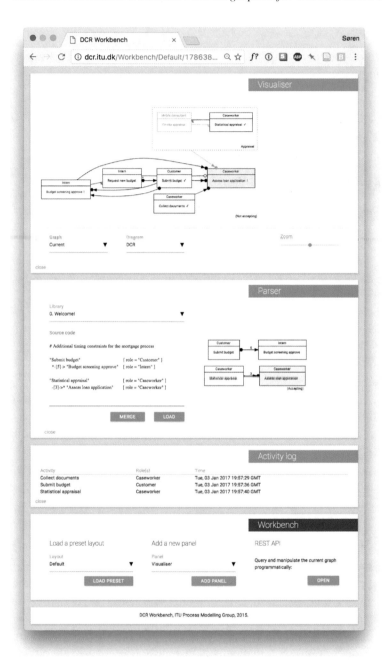

Figure 5.3   The DCR Workbench (http://dcr.tools).

⟨*expressions*⟩ ::= ⟨*expressions*⟩ ⟨*relation*⟩
         | ⟨*expressions*⟩ ⟨*relation*⟩ WHEN ⟨*condition*⟩
         | ⟨*expressions*⟩ ⟨*event*⟩
         | ⟨*expressions*⟩ GROUP { ⟨*expressions*⟩ }
         | ⟨*empty*⟩

⟨*relation*⟩ ::= ⟨*event*⟩ ⟨*arrow*⟩ ⟨*event*⟩
         | ⟨*event*⟩ ⟨*arrow*⟩ ⟨*relation*⟩

⟨*arrow*⟩ ::= -->* | --<> | -->+ | -->% | *-->
         | -[⟨*num*⟩]->*   (Timed condition)
         | *-[⟨*num*⟩]->   (Timed response)

⟨*event*⟩ ::= % ⟨*event*⟩
         | / ⟨*event*⟩
         | ! [⟨*num*⟩] ⟨*event*⟩
         | : [⟨*num*⟩] ⟨*event*⟩
         | (⟨*event*⟩+)
         | ⟨*identifier*⟩ [⟨*meta*⟩] [⟨*sub*⟩]

⟨*meta*⟩ ::= [[⟨*identifier*⟩] [⟨*string*⟩ = ⟨*string*⟩]*]

⟨*sub*⟩ ::= [?] { ⟨*expressions*⟩ }

⟨*condition*⟩ ::= (* ... *)

Figure 5.4   EBNF definition of the language recognised by the Parser panel.

```
( "Collect documents"       [ role = Caseworker ]          3
  "Submit budget"           [ role = Customer   ] )        4
  -->*                                                     5
  !"Assess loan application" [ role = Caseworker ]         6
```

Figure 5.5   Source code for the core process.

The concrete syntax specifies events and relations such as "condition from A to B" with expressions such as "A -->* B". An event state is specified by prefixing an event with modifiers such as ! or %. We see this on line 6 in Figure 5.5. If the event occurs more than once in the program, it is sufficient to prefix the modifier only once. We specify roles by adding a role tag to the event. We see this on line 3 in Figure 5.5. More than one role may be added; in general, the same tag may be added multiple times.

It is occasionally convenient to relate more than one event at the same time. In the present case, Assess loan application needs conditions on both

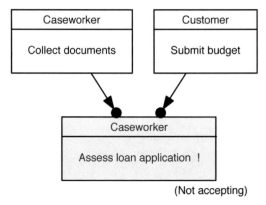

Figure 5.6   Visualisation of core process of Figure 5.5.

Collect documents and Submit budget. We specify these conditions concisely by enclosing the latter two in parenthesis, as seen on line 3–4.

As the user types in the parser, a preview of the graph being input is presented on the right. The input graph substitutes the current graph and resets the current trace when the user clicks "Load".

The parser also understands the XML format output by Exformatics A/S commercial `http://dcrgraphs.net` tool [15, 38].

5.4.2 Visualisation and Simulation: The Visualiser and Activity Log Panels

The Visualiser, top-most in Figure 5.3, displays a visualisation of the current DCR graph. In Figure 5.3, the chosen visualisation is simply the graph layout; alternatively, the underlying transition system may be shown (see below).

The visualisation of the core application process (Figure 5.5) is reproduced in Figure 5.6. The visualiser represent events as boxes, labelled by the activity of the event (centre) and the role or participant executing that activity (top). E.g., the top-left box represents an activity Collect documents which is carried out by a Caseworker.

Activities are coloured according to their state: grey background is not currently executable (Assess loan application in Figure 5.6), red label! with an exclamation mark is pending (ditto); "greyed out" boxes are excluded events (Request new budget in the original Figure 5.2); and finally, executed events have a tick mark after their action label, (Submit budget in Figure 5.3).

Simulation The visualiser allows executing events by clicking. E.g., to execute Submit budget, simply click it. This will extend the current trace with that execution, and replace the current graph with the one obtained by applying the updates to event state resulting from the execution of Submit budget (in this case, setting "executed" of that event true). Use the browser's back buttons to revert to a previous state.

The Activity Log panel, third from the top in Figure 5.3, displays the current trace, analogous to the way the visualiser displays the current graph.

State-space enumeration As mentioned in Definition 4, a DCR graph gives rise to a labelled transition system (LTS), where states are markings and transitions are labelled event executions. The visualiser can be configured to render a visualisation of the state space of the DCR graph rather than the DCR graph itself, through the drop-down button on the left of the panel. The visualiser highlights the current run in that LTS. The visualisation of the full LTS of the full mortgage application process of Figure 5.2.

The visualiser was originally reported in [12], with the transition system generator following in [9].

5.5 Refinement

We proceed to construct step-wise the full mortgage process application by *refinement* [6, 14]. We begin with the core process of Figures 5.5 and 5.6. We first add the process fragments for budget submission and screening given in Figures 5.8 and 5.9.

The Workbench supports step-wise refinement: by using in the parser the "Merge" button rather than the "Load" button. Whereas "Load" replaces the current graph and sets the current trace to empty, the "Merge" button preserves both, replacing the global current graph with (graph) union $G \oplus H$ of the current graph G and the parser's current local graph H.

To refine the core process by the budget fragment, we make sure that the core process is the current graph, then enter the fragment (Figure 5.9) in the parser, and click "Merge". The result is the graph in Figure 5.10. As can be seen, the resulting process is close to the full running example in Figure 5.2, except the process fragment for appraising the property is missing.

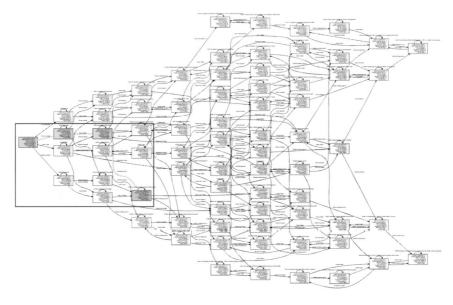

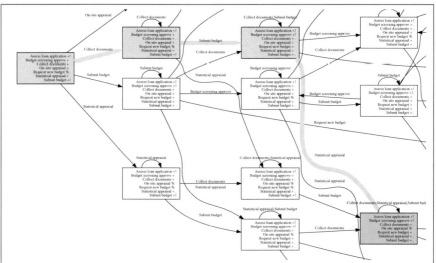

Figure 5.7   Transition system of the full mortgage application process (top), with the red box expanded for readability (bottom).

Repeating the Merge procedure with the process fragment in Figure 5.11 adds the missing bits and leaves us with exactly Figure 5.2—this is how the examples for the present chapter has been constructed.

```
! "Submit budget"                                                   3
    -->* "Budget screening approve"   [ role = Intern ]            4
    -->* "Assess loan application"     [ role = Caseworker ]       5
                                                                    6
"Submit budget"                                                     7
    --<> "Assess loan application"                                 8
                                                                    9
"Submit budget"                                                    10
    -->+ %"Request new budget"        [ role = "Intern" ]         11
    *--> "Submit budget"              [ role = "Customer" ]       12
    *--> "Budget screening approve"                               13
    -->% "Request new budget"                                     14
```

Figure 5.8   Budget process fragment.

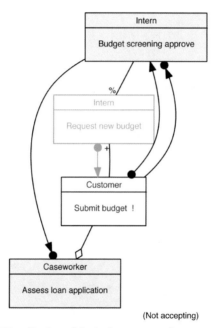

(Not accepting)

Figure 5.9   Visualisation of the budget process fragment of Figure 5.8.

Not every such merge preserves the language of the original graph. Exclusions may void conditions, giving the merged graph behaviour not present in the original graph, even when restricting attention to only the events of that graph. As a very simple example, consider the two graphs $G = a \rightarrow\bullet b$ and $H = c \rightarrow\% a$. The union $G \oplus H = a \rightarrow\bullet b, c \rightarrow\% a$ has the trace $\langle c, b \rangle$; even if we dismiss the new event c, G could not by itself exhibit the trace $\langle b \rangle$.

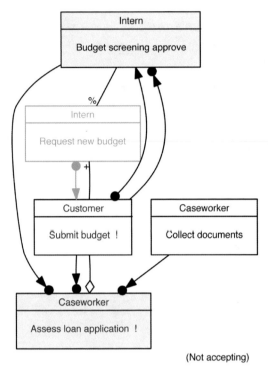

(Not accepting)

Figure 5.10   Visualisation of the core process (Figures 5.5 and 5.6) refined by the budget fragment (Figures 5.8 and 5.9).

```
Group "Appraisal" {                                                        3
    "On-site appraisal"          [ role = "Mobile consultant" ]           4
    "Statistical appraisal"      [ role = "Caseworker" ]                  5
}                                                                          6
                                                                           7
"Statistical appraisal" -->% "On-site appraisal"                          8
"On-site appraisal" -->% "Statistical appraisal"                          9
                                                                          10
"Appraisal"                                                               11
 -->* "Assess loan application" [ role = "Caseworker" ]                   12
```

Figure 5.11   Appraisal process fragment.

This situation was investigated in detail in previous work [6, 14], where a sufficient condition for a refinement to be language preserving in the above sense was established in a much richer setting. For the present notion of DCR graphs, it is sufficient to require that the refining graph does not exclude or include any events of the original graph.

Notation. Given a sequence s and an alphabet Σ, write $s \downharpoonright_\Sigma$ for the largest sub-sequence s' of s s.t. $s'_i \in \Sigma$; e.g, if $s = AABC$ then $s \downharpoonright_{A,C} = AAC$.

Definition 7. Given DCR graphs G and H, we say that *H is a refinement of G* iff $\mathsf{lang}(H) \downharpoonright_{l(\mathsf{E}(G))} \subseteq \mathsf{lang}(G)$.

That is, the language of H restricted to the labels used by events in G must be a subset of the language of G. We can now state the following Proposition [6, Theorem 43]:

Proposition 1. *Let G and G' be DCR processes such that for every $e \in \mathsf{E}(G)$, there is no relation $x \to\% e$ or $x \to+ e$ in G'. Then the graph union $G \oplus G'$ is a refinement of G.*

If the Workbench has current graph G and the parser has graph G' not satisfying (a published [6, 14] generalisation of) the conditions of Proposition 1, the Parser issues a warning and requires confirmation before merging.

DCR refinement was originally suggested in [17, 24] and worked out comprehensively in [6, 14]. The Workbench implements this latter mechanism.

5.6 Time

The Workbench supports the extension of DCR graphs with time [2, 23]. Time is modelled discretely by a special action tick modelling the passage of time; conditions are augmented with an optional delay, $e \xrightarrow{k}\bullet f$, and responses with an optional deadline $e \bullet\xrightarrow{k} f$. Intuitively, the delay in the timed condition requires that at least k ticks have passed after the last execution of e before f may execute; dually, the deadline in the timed response requires that at most k ticks pass after the last execution of e before f must execute.

The former requirement makes it possible to have *timelocks*, i.e., situations where, say, f must execute, but is not allowed to. As a very simple example, consider the DCR graph $G = e \xrightarrow{3}\bullet f, e \bullet\xrightarrow{2} f$. In this graph, once e executes, *at least* 3 ticks must pass before f *can* execute because of the condition delay, but *at most* 2 ticks may pass before f *must* execute because of the response deadline. After the sequence $\langle e, \mathsf{tick}, \mathsf{tick} \rangle$, the graph is said to be *time-locked*: Time cannot advance without a constraint being violated.

For our running example, suppose that (a) the initial screening of the customer's budget must be completed within 5 days, and (b) that the final assessment of the loan application must wait a 3-day "grace period" after a statistical appraisal (in order to prevent caseworkers from doing overly optimistic statistical appraisals). We model these constraints as a timed DCR graph directly using a timed response and condition in Figures 5.12 and 5.13.

Timed DCR graphs are still in finite state [23], but deciding time-lock freedom naively by exploring in the state space is infeasible. Recent research [2] established a sufficient condition for a graph to be time-lock–free and gave a generic "enforcement mechanism" for time-lock–free graphs, that is, a device which monitors the progression of time and a DCR graph and proactively causes events to execute to avoid missing deadlines.

The Workbench implements time as defined in [23], and time-lock–analysis and enforcement as defined in [2].

```
"Submit budget"                        [ role = "Customer" ]      3
  *-[5]-> "Budget screening approve"   [ role = "Intern" ]        4
                                                                  5
"Statistical appraisal"                [ role = "Caseworker" ]    6
  -[3]->* "Assess loan application"    [ role = "Caseworker" ]    7
```

Figure 5.12   Additional timing constraints for the mortgage application process in Figure 5.2.

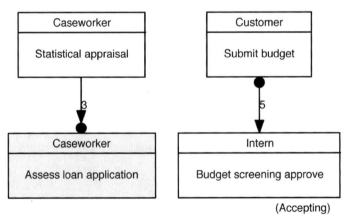

(Accepting)

Figure 5.13   Visualisation of additional timing constraints for the mortgage application process in Figure 5.2.

5.7 Subprocesses

It may happen that a customer *during the application process* applies for pre-approval of an expected increase in property value due to, e.g., on-going kitchen remodellings. In this case, the caseworker must assess the limit extension before deciding on the mortgage application itself. At the caseworker's discretion, an intern may or may not collect bank statements from the customer for the limit extension assessment; however, collecting that statement requires the customer's explicit consent.

Such limit extensions in practice may happen several times during a mortgage application due to, e.g., expanded scope of a kitchen remodelling project. Thus, the limit extension fragment is a *subprocess*: A process that may be added to the main process when necessary, and possibly repeatedly.

Note that since subprocesses may be added repeatedly, each such addition must duplicate the events of the subprocess. This situation is akin to bound names under replication being duplicated in in the π-calculus [35]. The subprocess may contain both events local to the subprocess, *bound* events, and references to events of the containing graph. The former are indicated syntactically with a / prefix as seen in lines 6–8.

The Workbench supports subprocesses; we add the above limit extension process in Figures 5.14 and 5.15. Note that in visualisation, the subprocess is not visible until it has been expanded once.

The visualisation shows the triggering event Apply for limit extension—singled out as spawning a subprocess by the ⊞ following contemporary business process notations, e.g., BPMN [32]. The bound events in a subprocess are shown with round corners and inside a dashed box[2].

```
"Assess loan application"                                              3
                                                                       4
"Apply for limit extension"            [ role = Customer ]            5
    { /"Assess limit extension"        [ role = Caseworker ]         6
     /"Collect consent"                [ role = Intern ]             7
      -->* /"Collect bank statement"   [ role = Intern ]             8
      "Submit budget"                  [ role = Customer ]           9
      --<> !"Assess limit extension"   [ role = Caseworker ]        10
      -->* "Assess loan application"   [ role = Caseworker ]        11
    }                                                               12
    *--> "Submit budget"                                           13
```

Figure 5.14 Additional subprocess constraints (credit limit extension) for the full mortgage application process of Figure 5.2.

---

[2]If a subprocess adds new global events—as opposed to bound ones—these would appear with square corners inside the box.

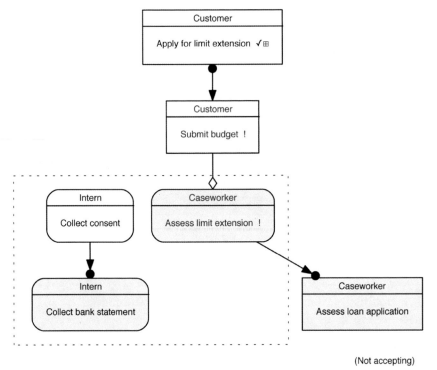

(Not accepting)

Figure 5.15 Visualisation of additional subprocess constraints (credit limit extension) for the full mortgage application process of Figure 5.2; after one execution of Apply for limit extension.

Subprocess semantics is based on graph union; if $G = \ldots, e\{H\}$ contains a subprocess-spawning event e, then executing e will form the graph $G \oplus H$, then apply the effects of e. Note the importance of bound names here: If events in H were not bound, then repeated instantiation of the subprocess would not change the graph, i.e., $G \oplus H \oplus H = G \oplus H$. This equation emphatically does not hold under the current semantics, where events in H may be bound, and thus replicated. In the running example, executing Apply for limit extension *twice* would result in all rounded-box event to be replicated *twice*—we invite the reader to try this out in the Workbench.

Adding subprocesses and bound events significantly increase the expressive power of DCR graphs [6, Theorem 9]:

Theorem 1. *DCR graphs express the union of regular and ω-regular languages. Graphs with subprocesses and bound events are Turing complete.*

In particular, while event-reachability and refinement is decidable for plain and timed DCR graphs, they are undecidable for DCR graphs with subprocesses and bound events.

DCR graphs was extended with a notion of subprocesses and bound events in [12], followed by an investigation of expressive power in [6, 14]. The Workbench implements subprocesses in the sense of [12].

5.8 Data

The Workbench augments DCR graphs with a notion of "input events" and relations conditional on data. Suppose for our running example that if the amount applied for in a credit limit extension exceeds EUR 10.000, then having a bank statement becomes a condition for evaluating the loan application.

Technically, this is accomplished by: (a) adding the option of inputting a value when a subprocess is spawned; and (b) adding data-guards on select relations. When the subprocess is instantiated, condition on the input value dictates whether, e.g., the relation e $\rightarrow\bullet$ f takes effect or not.

We extend the running example with such a conditional condition in Figure 5.16. Note that since the "variable" associated with an event is simply the name of the event, it becomes convenient to specify separately the name and label of the event. This is done in line 3, where it is specified that the event limit has label Apply for a limit extension and role Customer. (Without an explicit specification, the Workbench identifies event and label.)

The visualiser does not show data-guarded relations, and the formal semantics of DCR graphs with data have yet to be published.

```
limit["Apply for limit extension" role = Customer ] ?              3
     { /"Assess limit extension"                                   4
       /"Collect consent" -->* /"Collect bank statement"           5
         "Submit budget"                                           6
              --<> !"Assess limit extension"                       7
              -->* "Assess loan application"                       8
       "Collect bank statement"                                    9
         -->* "Assess loan application"                           10
         when "$limit > 10000"                                    11
     }                                                            12
     *--> "Submit budget"                                        13
```

Figure 5.16 Alternative subprocess-with-data constraints (credit limit extension) for the full mortgage application process of Figure 5.2.

An interesting application of data is that of specifying user-input "forms" (think Web forms) via DCR graph, associating with each event in a graph an input field in such a form. This idea was implemented in the Actions panel and later realised [27] in collaboration with Exformatics A/S.

5.9 Other Panels

We mention here briefly a few panels of the Workbench not discussed so far.

1. An encoding from DCR to the GSM model [26] was defined in recent research [16]; the Workbench implements this encoding an outputs CMMN [4] XML.
2. Notions of concurrency and independence of DCR events, following the standard notion of labelled asynchronous transition systems [3, 36], was recently investigated [13]. The Workbench' Concurrency panel implements these notions, automatically identifying concurrent events.
3. Work on applications of DCR in practical settings suggested a need for simplifying process models when presented to end users [9]. The Workbench contains a number of such simplifying views, most notably a "swimlane" view of the current trace in the panel of the same name, and a mechanism for projecting a graph in various ways to sub-graphs of interest.

5.10 Conclusion

We have given an introduction to DCR graphs, and an overview of the DCR Workbench. Since the Workbench implements most major variations of DCR graphs, this Chapter has also served as a survey of the state-of-the-art of DCR graphs as modelling and analysis tool for continuously changing distributed collaborative processes.

The Workbench has been instrumental for scientific research, providing a test-bed for quick experiments with new ideas; for teaching, providing students the opportunities for hands-on learning of abstract concepts; and for collaborations with industry and knowledge transfer to industry. In all these instances, providing a practical platform on which to demonstrate sometimes difficult-to-communicate abstract concepts helps to cement the reality and applicability of DCR as a modelling methodology. In particular, the Workbench has paved the way for academic results [9, 12, 18, 21, 27] to find their way to implementation in commercial tools [8, 28].

We invite the reader to use the Workbench for research and teaching. It is available at `http://dcr.tools`.

References

[1] Wil M. P. van der Aalst. Verification of Workflow Nets. In *Proc. of the 18th Int. Conf. on Application and Theory of Petri Nets, ICATPN*, pages 407–426, 1997.

[2] David A. Basin, Søren Debois, and Thomas T. Hildebrandt. In the nick of time: Proactive prevention of obligation violations. In *IEEE 29th Computer Security Foundations Symposium, CSF 2016*, pages 120–134. IEEE Computer Society, 2016.

[3] Marek Bednarczyk. *Categories of asynchronous systems*. PhD thesis, U. Sussex, 1988.

[4] BizAgi and others. Case Management Model and Notation (CMMN), v1, May 2014. OMG Document Number formal/2014-05-05, Object Management Group.

[5] Joel Bjornson, Anton Tayanovskyy, and Adam Granicz. *Composing Reactive GUIs in F# Using WebSharper*, pages 203–216. Springer, 2011.

[6] Søren Debois, Thomas Hildebrandt, and Tijs Slaats. Replication, refinement & reachability: Complexity in Dynamic Condition-Response graphs. *Acta Informatica*, 2017. Accepted for publication.

[7] Søren Debois, Thomas T. Hildebrandt, Paw Høsgaard Larsen, and Kenneth Ry Ulrik. Declarative process mining for DCR graphs. In *SAC '17*, 2017. Accepted for publication.

[8] Søren Debois, Thomas T. Hildebrandt, Morten Marquard, and Tijs Slaats. Bridging the valley of death: A success story on danish funding schemes paving a path from technology readiness level 1 to 9. In *SER&IP 2015*, pages 54–57. IEEE, 2015.

[9] Søren Debois, Thomas T. Hildebrandt, Morten Marquard, and Tijs Slaats. Hybrid process technologies in the financial sector. In *BPM 2015, Industry track*, volume 1439 of *CEUR Workshop Proceedings*, pages 107–119. CEUR-WS.org, 2015.

[10] Søren Debois, Thomas T. Hildebrandt, Morten Marquard, and Tijs Slaats. The DCR graphs process portal. In *BPM 2016*, volume 1789 of *CEUR Workshop Proceedings*, pages 7–11. CEUR-WS.org, 2016.

[11] Søren Debois, Thomas T. Hildebrandt, and Lene Sandberg. Experience report: Constraint-based modelling and simulation of railway emergency response plans. In *ANT 2016 / SEIT-2016*, volume 83 of *Procedia Computer Science*, pages 1295–1300. Elsevier, 2016.

[12] Søren Debois, Thomas T. Hildebrandt, and Tijs Slaats. Hierarchical declarative modelling with refinement and sub-processes. In *BPM 2014*, volume 8659 of *LNCS*, pages 18–33. Springer, 2014.

[13] Søren Debois, Thomas T. Hildebrandt, and Tijs Slaats. Concurrency and asynchrony in declarative workflows. In *BPM 2015*, volume 9253 of LNCS, pages 72–89. Springer, 2015.

[14] Søren Debois, Thomas T. Hildebrandt, and Tijs Slaats. Safety, liveness and run-time refinement for modular process-aware information systems with dynamic sub processes. In *FM 2015*, pages 143–160, 2015.

[15] Søren Debois, Thomas T. Hildebrandt, Tijs Slaats, and Morten Marquard. A case for declarative process modelling: Agile development of a grant application system. In *EDOC Workshops '14*, pages 126–133. IEEE Computer Society, 2014.

[16] Rik Eshuis, Søren Debois, Tijs Slaats, and Thomas T. Hildebrandt. Deriving consistent GSM schemas from DCR graphs. In *ICSOC 2016*, volume 9936 of *Lecture Notes in Computer Science*, pages 467–482. Springer, 2016.

[17] Thomas T. Hildebrandt, Morten Marquard, Raghava Rao Mukkamala, and Tijs Slaats. Dynamic condition response graphs for trustworthy adaptive case management. In *OTM 2013 Workshops*, volume 8186 of *LNCS*, pages 166–171. Springer, 2013.

[18] Thomas T. Hildebrandt and Raghava Rao Mukkamala. Declarative Event-Based Workflow as Distributed Dynamic Condition Response Graphs. In *PLACES 2010*, volume 69 of *EPTCS*, pages 59–73, 2010.

[19] Thomas T. Hildebrandt, Raghava Rao Mukkamala, and Tijs Slaats. Declarative modelling and safe distribution of healthcare workflows. In *FHIES 2011*, volume 7151 of *LNCS*, pages 39–56. Springer, 2011.

[20] Thomas T. Hildebrandt, Raghava Rao Mukkamala, and Tijs Slaats. Designing a cross-organizational case management system using dynamic condition response graphs. In *EDOC 2011*, pages 161–170. IEEE Computer Society, 2011.

[21] Thomas T. Hildebrandt, Raghava Rao Mukkamala, and Tijs Slaats. Nested dynamic condition response graphs. In *FSEN 2011, Revised Selected Papers*, volume 7141 of *Lecture Notes in Computer Science*, pages 343–350. Springer, 2011.

[22] Thomas T. Hildebrandt, Raghava Rao Mukkamala, and Tijs Slaats. Safe distribution of declarative processes. In *SEFM 2011*, volume 7041 of *LNCS*, pages 237–252. Springer, 2011.

[23] Thomas T. Hildebrandt, Raghava Rao Mukkamala, Tijs Slaats, and Francesco Zanitti. Contracts for cross-organizational workflows as timed dynamic condition response graphs. *J. Log. Algebr. Program.*, 82(5–7):164–185, 2013.

[24] Thomas T. Hildebrandt, Raghava Rao Mukkamala, Tijs Slaats, and Francesco Zanitti. Modular context-sensitive and aspect-oriented processes with dynamic condition response graphs. In *FOAL 2013*, pages 19–24. ACM, 2013.

[25] Thomas T. Hildebrandt and Francesco Zanitti. A process-oriented event-based programming language. In *DEBS 2012*, pages 377–378. ACM, 2012.

[26] R. Hull, E. Damaggio, F. Fournier, M. Gupta, F. Heath III, S. Hobson, M. Linehan, S. Maradugu, A. Nigam, P. Sukaviriya, and R. Vaculín. Introducing the guard-stage-milestone approach for specifying business entity lifecycles. In *WS-FM 2010*, LNCS. Springer, 2010.

[27] Morten Marquard, Søren Debois, Tijs Slaats, and Thomas T. Hildebrandt. Forms are declarative processes! In *BPM 2016 Industry Track (to appear)*, 2016.

[28] Morten Marquard, Muhammad Shahzad, and Tijs Slaats. Web-based modelling and collaborative simulation of declarative processes. In *BPM 2015*, volume 9253 of LNCS, pages 209–225. Springer, 2015.

[29] Raghava Rao Mukkamala. *A Formal Model For Declarative Workflows: Dynamic Condition Response Graphs*. PhD thesis, IT University of Copenhagen, 2012.

[30] Raghava Rao Mukkamala and Thomas T. Hildebrandt. From dynamic condition response structures to büchi automata. In *TASE 2010*, pages 187–190. IEEE, 2010.

[31] Raghava Rao Mukkamala, Thomas T. Hildebrandt, and Tijs Slaats. Towards trustworthy adaptive case management with dynamic condition response graphs. In *EDOC 2013*, pages 127–136. IEEE Computer Society, 2013.

[32] Object Management Group BPMN Technical Committee. Business Process Model and Notation, version 2.0, 2013. http://www.omg.org/spec/BPMN/2.0.2/PDF

[33] Maja Pesic, Helen Schonenberg, and Wil M. P. van der Aalst. DECLARE: full support for loosely-structured processes. In *EDOC 2007*, pages 287–300, 2007.

[34] Søren Debois and Tijs Slaats. The analysis of a real life declarative process. In *CIDM 2015*. IEEE, 2015. Accepted for publication.

[35] Davide Sangiorgi and David Walker. *The pi-calculus: a Theory of Mobile Processes*. Cambridge university press, 2003.

[36] M. W. Shields. Concurrent machines. *Computer Journal*, 28(5):449–465, 1985.

[37] Tijs Slaats. *Flexible Process Notations for Cross-organizational Case Management Systems*. PhD thesis, IT University of Copenhagen, January 2015.

[38] Tijs Slaats, Raghava Rao Mukkamala, Thomas T. Hildebrandt, and Morten Marquard. Exformatics declarative case management workflows as DCR graphs. In *BPM '13*, volume 8094 of *LNCS*, pages 339–354. Springer, 2013.

[39] Don Syme, Jack Hu, Luke Hoban, Tao Liu, Dmitry Lomov, James Margetson, Brian McNamara, Joe Pamer, Penny Orwick, Daniel Quirk, et al. The F# 4.0 language specification. Technical report, 2005.

[40] Wil M. P. van der Aalst and Maja Pesic. DecSerFlow: Towards a truly declarative service flow language. In *WS-FM 2006*, volume 4184 of *LNCS*, pages 1–23. Springer, 2006.

6

A Tool for Choreography-Based Analysis of Message-Passing Software

Julien Lange[1], Emilio Tuosto[2] and Nobuko Yoshida[1]

[1]Imperial College London, UK
[2]University of Leicester, UK

Abstract

An appealing characteristic of choreographies is that they provide two complementary views of communicating software: the *global* and the *local* views. Communicating finite-state machines (CFSMs) have been proposed as an expressive formalism to specify local views. Global views have been represented with global graphs, that is graphical choreographies (akin to BPMN and UML) suitable to represent general multiparty session specifications. Global graphs feature expressive constructs such as forking, merging, and joining for representing application-level protocols.

An algorithm for the reconstruction of global graphs from CFSMs has been introduced in [17]; the algorithm ensures that the reconstructed global graph faithfully represents the original CFSMs provided that they satisfy a suitable condition, called generalised multiparty compatibility (GMC). The CFSMs enjoying GMC are guaranteed to interact without deadlocks and other communication errors. After reviewing the basic concepts underlying global graphs, communicating machines and safe communications, we highlight the main features of **ChorGram**, a tool implementing the generalised multiparty compatibility and reconstruction of global graphs of [17]. We overview the architecture of **ChorGram** and present a comprehensive example to illustrate how it is used directly to analyse message-passing software and programs.

6.1 Introduction

Choreographic approaches are becoming popular in the "top-down" development of distributed software. In fact, a choreography-based model features two views of software: the *global view* and the *local view*. The former is a "holistic" specification of the behaviour of the composition of *all* components (and it abstracts away low level details such as asynchrony). The latter view specifies the behaviour of each component in isolation and should be obtained by *projecting* the global behaviour with respect to each component. In this framework, well-formedness of the global view and compliance of the realisation of software with respect to the corresponding projection should guarantee the soundness of the communication of the application.

The recent rise of services, cloud, and micro-services is changing the way software is produced. As a matter of fact, applications are being developed by composing (possibly distributed) independent components which coordinate their execution by exchanging messages. Modern applications offer and rely on public APIs to interact with each other, are deployed on different architectures and devices, and try to accommodate the needs of a vast number of users. The term "API economy" (see e.g., ibm.com/apieconomy) has been coined to refer to such applications. Existing and novel languages, as well as middlewares and programming models, foster this shift in software development. Languages such as Erlang, Elixir, Scala, and Go are paramount examples of this shift and start to be used in a wider range of application domains than the ones they were originally conceived for. For instance, Erlang plays a main role in well-known applications such as WhatsApp [24] and Facebook [21].

The trend described above is dictated by the compelling requirements of openness, scalability, and heterogeneity and caters to new challenges. Firstly, this shift pushes the applicability of top-down software development approaches to their limits. The composition mechanisms required to guarantee the interoperability of applications have to be of an order of magnitude more sophisticated than just the type signature of their APIs, as in traditional software engineering practice. More precisely, in order to attain a correct composition, it is crucial to expose (part of) the communication pattern of components. Hence, developers are responsible to guarantee the correct composition of their components. This is not an easy task. Subtle and hard to fix bugs can be introduced by inappropriate communications.

Our recent work [17] has shown that communication soundness is guaranteed when a set of communicating components enjoys the *generalised multiparty compatibility* property. Moreover, we have defined an algorithm that reconstructs a global view of a system from the composition of its local components. These results enable the realisation of an effective tool-supported approach to the design and analysis of communicating software. In fact, we have developed **ChorGram** [16], a tool supporting the theory of multiparty compatibility and choreography construction, i.e., **ChorGram** implements two fundamentals functionalities: it ensures that a system of CFSMs validates the GMC condition and if so, it returns a choreography which faithfully captures the interactions of the original system.

In this chapter, we introduce **ChorGram** and show how it supports software architects in the design and analysis of software. We first review the theoretical results underlying the tool; Section 6.2 presents our theory only informally and with the help of a simple example. Section 6.3 presents the architecture of the tool, how it integrates with the auxiliary tools it relies upon, and its data flow. Section 6.4 shows an application to a non trivial example. We start from a multiparty compatible application and show how a naive evolution could break its multiparty compatibility. We then use **ChorGram** to analyse and fix the problem. Section 6.5 gives our concluding remarks.

6.2 Overview of the Theory

Here we introduce the key ingredients of our framework which constructs choreographies, i.e., global graphs such as the one in Figure 6.2, from local specifications, i.e., communicating finite-state machines, such as the ones in Figure 6.1.

CFSMs In this framework, we use *communicating finite-state machines* [7] as behavioural specifications of distributed components (i.e., end-point specifications) from which a choreography can be built. CFSMs are a conceptually simple model and are well-established for analysing properties of distributed systems. A system of CFSMs consists of a finite number of automata which communicate with each other through unbounded FIFO channels. There are two channels for each pair of CFSMs in the system, one in each direction. We present the semantics of CFSMs informally through the example below.

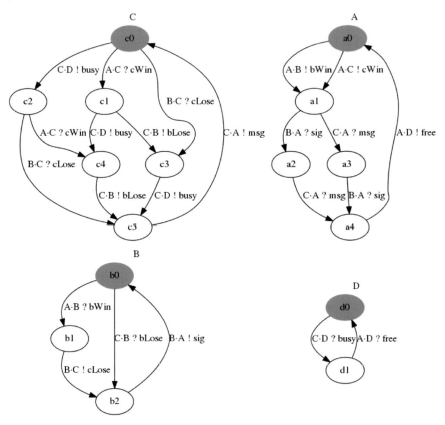

Figure 6.1   Four player game – CFSMs.

Consider the system of four machines in Figure 6.1, whose initial states are highlighted in blue. Each machine has *three* input buffers to receive messages from the other three participants and has access to three output buffers to send messages to other participants. Each transition in a machine is either a send action, e.g., A·B!bWin in machine A or a receive action, e.g., A·B?bWin in machine B. The system realises a protocol of a fictive game where: A*lice* (A) sends either bWin to B*ob* (B) or cWin to C*arol* (C) to decide who wins the game. In the former case, A fires the transition A·B!bWin whereby the message bWin is put in the FIFO buffer AB from A to B, and likewise in the latter case. If B wins (that is the message bWin is on top of the queue AB and B consumes it by taking the transition A·B?bWin), then he sends a notification (cLose) to C to notify her that she has lost. Symmetrically, C notifies B of her victory (bLose). During the game, C notifies D*ave* (D) that she is busy.

After B and C have been notified of the outcome of the game, B sends a signal (sig) to A, while C sends a message (msg) to A. Once the result is sent, A notifies D that C is now free and a new round starts.

Global graph The final product of our framework is the construction of a choreography which is equivalent to the original system of CFSMs. Global graphs [17] were inspired by the generalised global types [10] and BPMN choreography [19]. Given as input the CFSMs from Figure 6.1, our tool generates the global graph in Figure 6.2. The nodes of a global graph are labelled according to their function: a node labelled with ○ indicates the starting point of the interactions; a node labelled with ◎ indicates the termination of the interactions (not used in Figure 6.2); a node labelled with an interaction A → B : msg indicates that participant A sends a message of type msg to B; a node labelled ◈ indicates either a choice, merge, or recursion; a node labelled with ▫ indicates either the start or the end of concurrent interactions. The graphical notation for branch and merge is inspired by process-algebraic notations; the reader familiar with BPMN should note that our ◈ -node corresponds to the ◇ and ◇ gateways in BPMN, while our ▫ -node corresponds to the ◈ gateway in BPMN.

In the global graph of Figure 6.2, the flow of the four player game becomes much clearer. In particular, one can clearly see that either B or C win the game and that, while the results of the game are being announced, C and D are interacting.

Communication soundness properties A (runtime) configuration of a system of CFSMs, is a tuple consisting of the states in which each machine is and the content of each channel.

We say that a machine is in a sending (resp. receiving) state if all its outgoing transitions are send (resp. receive) actions. A state without any outgoing transition is said to be final. A state that is neither final, sending nor receiving is a mixed state.

We say that a configuration is a *deadlock* if all the buffers are empty, there is at least one machine in a receiving state, and all the other machines are either in a receiving state or a final state. A system has the *eventual reception property* [5] if whenever a message has been sent by a participant,

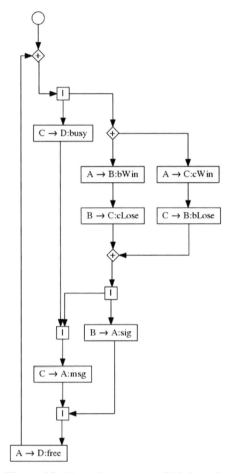

Figure 6.2   Four player game – Global graph.

that message will be eventually received. We say that a system of CFSMs is *communication sound* if none of its reachable configuration is a deadlock and it validates the eventual reception property.

Ensuring communication soundness Our tool checks that the CFSMs validate *generalised multiparty compatibility* (GMC) [17] which guarantees that (*i*) the projections of the generated global graph are equivalent to the original system and (*ii*) the system is *communication sound* (as defined above).

The GMC condition consists of two parts: *representability* and *branching property*. Both parts are checked against the machines and their synchronous executions, i.e., the finite labelled transition system (dubbed TS_0) of the machines executing with the additional constraint that a message can be sent only if its partner is ready to receive it and no other messages are pending in other buffers. For instance, all the synchronous executions of our running example are modelled in the finite labelled transition system in Figure 6.3.

The representability condition essentially requires that for each participant, the projection of TS_0 onto that participant yields an automaton that is bisimilar to the original machine. The branching property condition requires that whenever a branching occurs in TS_0 then either (*i*) the branching commutes, i.e., it corresponds to two independent (concurrent) interactions, or (*ii*) it corresponds to a choice and the following constraints must be met:

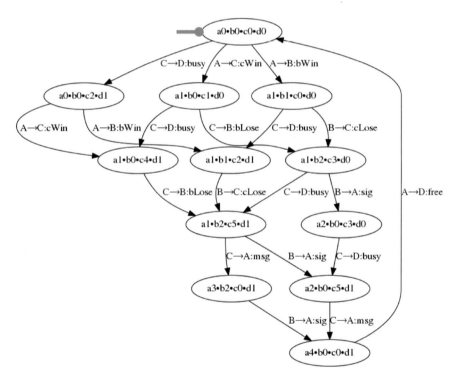

Figure 6.3   Four player game – TS_0.

1. The choice is made by a single participant (the selector).
2. If a participant is not the selector but behaves differently in two branches of the choice, then it must receive different messages in each branch (before the point where its behaviours differ).

Item (1) guarantees that every choice is located at exactly one participant (this is crucial since we are assuming *asynchronous* communications). Item (2) ensures that all the participants involved in the choice are made aware of which branch was chosen by the selector.

Besides guaranteeing communication soundness, our GMC condition ensures that if a system of CFSMs validates it, then we can construct a global graph which is equivalent to the original system, i.e., the global graph contains exactly the same information than the system of CFSMs.

6.3 Architecture

The structure and the work-flow of our tool is illustrated in Figure 6.4. Before commenting on the diagram, we explain its graphical conventions. Dashed arrows represent files used to exchange data; the input files are provided by the user, those of **hkc** are generated by the Haskell module **Representability**. Solid arrows do not represent invocations but rather control/data flow. For instance, the arrow from **TS** represents the fact that the check of the GMC property is made by concurrent threads on the results produced by **TS**.

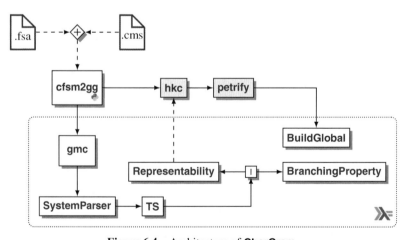

Figure 6.4   Architecture of **ChorGram**.

Data- and control-flow The Python script **cfsm2gg** provides a command line interface to the application and connects it with the external tools **hkc** [6] and **petrify** [8], respectively used to check language equivalence between projections and their corresponding CFSMs and to extract a Petri net from a transition system. The script takes a description of the system in (a file that is in) either of the two formats described in the following paragraph and triggers all the other activities.

The core functionalities are implemented in the Haskell modules (within the dotted box) and are described below.

gmc is the main program; it is invoked by **cfsm2gg**, which passes over the input file (after having set some parameters according to the flags of the invocation). After invoking **SystemParser**, the internal Haskell representation of the system of CFSMs is passed by **gmc** to **TS**, which computes the synchronous transition system $-\mathrm{TS}_0$ (and the bounded one if required with the -b flag of **cfsm2gg**). The synchronous transition system is then checked for generalised multiparty compatibility [17, Definitions 3.4(ii) and 3.5] (but for the language equivalence part [17, Definition 3.4(i)] later checked by invoking **hkc** from **cfsm2gg**). This check is performed in parallel and has the side effect of producing the files to give in input to **hkc**.

cfsm2gg invokes **hkc**, once it has obtained the control back from **gmc**, to check the language equivalence of the original CFSMs with respect to the corresponding projections of the synchronous transition system. Finally, **petrify** is invoked and its output is then transformed by **BuildGlobal** as described in [17] to obtain a global graph (in dot format) of the system. Besides, **cfsm2gg** generates also graphical representation of the communicating machines and the transition systems (again in the dot format).

Input formats The syntax of the input files of gmc can be specified either in the *fsa* (after finite state automata) or *cms* format, the latter being a simple process-algebraic syntax (described below). The format to be used depends on the extension of the file name (.fsa or .cms respectively, and for file names without extensions the default format is *fsa*).

A system consists of a list of automata, each described by specifying an (optional) identifier, its initial state, and its transitions. (Identifiers of CFSMs are strings starting with a letter.) We refer to the example in Figure 6.5 to

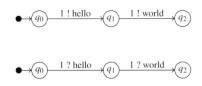

```
.outputs A
.state graph
q0 1 ! hello q1
q1 1 ! world q2
.marking q0
.end

.outputs
.state graph
q0 0 ? hello q1
q1 0 ? world q2
.marking q0
.end
```

The first automaton has an identifier A while for the second no identifier is specified, so the automaton is identified by 1, its position in the file (automata positions start from 0). The lines following each .state graph line yield the transitions followed by the specification of the initial state with the line starting with .marking, and finally with the end of the automaton specification (line starting with .end). Transitions are written as *src m act msg tgt*, where *src* and *tgt* are respectively the source and target state, *m* is the position of the partner CFSM, *act* is the action (! and ? respectively for output and input actions), and *msg* is the message.

Figure 6.5   HelloWorld example – fsa representation.

describe the *fsa* format. Consider the text on the left of Figure 6.5 specifying the (system consisting of) two simple automata depicted on the right.

It is sometimes more convenient to have a more concrete syntax to represent machines. Therefore we define the alternative *cms* format. The idea is that each CFSM of a system is described by a process in the syntax that we now describe.

The *cms* format of a system is a term of the following grammar:

$$S \quad ::= \quad \text{system id of } A_1, \cdots , A_n : A_1 = M_1 \parallel \cdots \parallel A_m = M_m$$

where id is a string used to identify the system, $A_1, \cdots , A_n$ are the names of the machines forming the system, and for each $1 \leqslant i \leqslant m$ (with $m \geqslant n \geqslant 2$) we have a unique defining equation assigning an expression that specifies the behaviour of A_i. We can now model the HelloWorld example of Figure 6.5, as follows:

system helloWorld of A, B : A = $\cdots$ $\parallel$ B = $\cdots$

(where the ellipsis will be defined in a moment). The list of defining equations specify the behaviour M_i of each role A_i, with $1 \leqslant i \leqslant n$, of the system and the behaviour of some auxiliary machines. For each $1 \leqslant i \leqslant m$, the identity A_i cannot appear in the communication actions of the behaviour M_i of the defining equation $A_i = M_i$.

Basically, the behaviour of a machine[1] is specified as a regular expression on an alphabet of actions. We impose some syntactic restrictions to keep out some meaningless terms and define:

$$
\begin{array}{llll@{\qquad}llll}
M & ::= & B{+}M & \text{branching} & B & ::= & pre;\ \text{end} & \text{prefix} \\
pre & ::= & A!\mathsf{m} & \text{output} & & | & pre;\ M & \text{prefix} \\
 & | & A?\mathsf{m} & \text{input} & & | & pre\ \text{do}\ A & \text{iteration}
\end{array}
$$

A machine M is a sum of branches B. A branch is a prefix-guarded behaviour (a machine or end) or it is the invocation to the behaviour of a machine A specified in the set of defining equations of the system. Prefixes yield the possible actions of a participant: in A!m (resp. A?m), the message m is sent to (resp. received from) participant A. The equations for the participants of the helloWorld system are:

$$
A = B!\text{hello};\ B!\text{world};\ \text{end} \qquad \text{and} \qquad B = A?\text{hello};\ A?\text{world};\ \text{end}
$$

Trailing occurrences of end can be omitted, e.g., writing A = B!hello;B!world. Finally, + is right-associative and gives precedence to all the other operators except $\|$, which has the lowest precedence.

6.4 Modelling of an ATM Service

We use a simple scenario to showcase **ChorGram**. We want to design the protocol of a service between an ATM (A), a bank (B), and a customer (C), where, after a successful authentication, the customer C can withdraw cash or check the balance of their account. Such services are enabled only after the ATM has successfully checked the credentials of C. We also require that bank B monitors the usage of the cards of its customers, so that unsuccessful attempts to use it are reported to C (e.g., via an SMS to the customers' mobile).

---

[1] The *cms* format provides a richer and more flexible syntax which we omit here because not used in the examples. The full syntax is described at https://bitbucket.org/emlio_tuosto/chorgram/wiki/Home

6.4.1 ATM Service – Version 1

For the moment, we will assume that the protocol repeats only after a successful withdrawal. Let us start with the description of the bank B:

```
1   B = A ? accessFailed ;
2         C ! failedAttempt
3       +                                    .. Notification of authentication outcome
4       A ? accessGranted ; (
5         A ? checkBalance ;
6         A ! balance
7       +
8       A ? withdraw ; (
9         A ! deny
10        +                                  .. B decides if to allow the withdrawal
11        A ! allow  do B
12      )
13      +
14      A ? quit
15    )
```

The bank B is notified of the outcome of the authentication by the ATM A. If the access fails, B sends a message to the customer C (lines 1–2); otherwise, the bank waits to be told which service has been requested by the customer and acts accordingly (lines 4–14). (The symbol ".." is for single-line comments.)

The specification for the customer C is as follows:

```
1   C = A ! auth ; (
2         A ? authPass ;  (                   .. Services are now enabled
3           A ! checkBalance ;
4             A ? balance
5         +
6           A ! withdraw  do Cw               .. after the request C continues as Cw
7         +
8           A ! quit ;
9             A ? card
10        )
11      +
12      A ? authFail ;
13        A ? card ;
14          B ? failedAttempt
15      )
16   ||
17   Cw = A ? card
18      +
19      A ? money  do C
```

Firstly, C provides the ATM A with their credentials by sending the auth message (line 1). If the authentication fails, the ATM replies with the authFail message; in this case the customer also expects their card back and the message failedAttempt from the bank (line 14). On successful authentication, C can select one of the services offered by the ATM or quit the protocol (lines 3–9). In the latter case, C receives their card and terminates (line 9). To check their balance, C sends the message checkBalance to A and waits for the result (line 3). If C sends A the message withdraw, then C continues to Cw (line 6), namely they expects to receive their cash (in which case the protocol restarts) or their card back.

The most complex participant is the ATM A. It can be specified as follows:

```
1   A = C ? auth ; (
2           C ! authFail ;
3               B ! accessFailed ;
4               C ! card
5           +
6           C ! authPass ;
7               B ! accessGranted ;  (
8                   C ? checkBalance do Ac          .. Ac is specified below
9                   +
10                  C ? withdraw do Aw              .. Aw is specified below
11                  +
12                  C ? quit ;
13                      B ! quit ;
14                      C ! card
15              )
16      )
```

The structure of participant A is very similar to the one of C with the addition of the interactions to liaise with the bank. In case A receives the request for a service from C, it will behave according to Ac (for checking the balance) or to Aw (for withdrawing money). These behaviours are specified below.

```
1   Ac = B ! checkBalance ;  B ? balance ; C ! balance
2   ||
3   Aw = B ! withdraw ;  (
4           B ? deny ;
5               C ! card
6           +
7           B ? allow ;
8               C ! money do A
9       )
```

Auxiliary machine Ac forwards the checkBalance message to B, waits for the balance, and returns it to the customer (line 1). Similarly, auxiliary machine Aw forwards the request for withdrawal to B, and waits for the outcome (lines 3–8). If the withdrawal is denied (line 4), then the card is returned to the customer, otherwise the customer receives the money and the protocol restarts (line 8).

Executing **ChorGram** on the system

```
1   system atm of C, A, B:
2       C = ⋯ ||  Cw = ⋯
3       ||
4       A = ⋯ ||  Ac = ⋯ ||  Aw = ⋯
5       ||
6       B = ⋯
```

we verify that the system is GMC and the resulting global graph is reported in Figure 6.6, where the overall protocol becomes apparent.

6.4.2 ATM Service – Version 2

The previous specification is GMC, but has several drawbacks, the most evident of which is the fact that when the protocol is repeated the customer

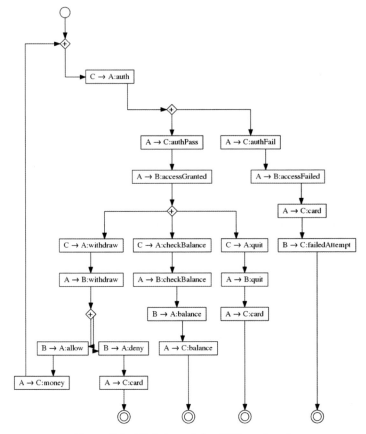

Figure 6.6   Global graph for ATM service v1.

has to re-authenticate. We therefore replace the previous participants C and A with the following ones:

```
1    C = A ! auth ; (
2          A ? authPass do Ca                            .. Now C loops back after authentication
3          +
4          A ? authFail ; A ? card;  B ? failedAttempt
5          )
6    ||
7    Ca = A ! checkBalance;  A ? balance do Cf            .. After successful requests , C decides
8          +                                              .. whether to continue or not to behave as Cf
9          A ! withdraw  ; (
10               A ? card
11               +
12               A ? money  do  Cf
13               )
14   ||
15   Cf = A ! newService  do  Ca
16         +
17         A ! quit;  A ? card
```

```
18    ||
19    A = C ? auth ; (
20            C ! authPass ;  B ! accessGranted  do  Aa
21            +
22            C ! authFail ;  B ! accessFailed  ;C ! card
23          )
24    ||
25    Aa =  C ? checkBalance  do  Ac
26            +
27            C ? withdraw  do  Aw
28    ||
29    Ac =  B ! checkBalance ;  B ? balance  ;C ! balance  do  Af
30    ||
31    Aw =  B ! withdraw ;  (
32            B ? deny ;  C ! card
33            +
34            B ? allow  ;C ! money  do  Af
35          )
36    ||
37    Af =  ( C ? quit ;  C ! card )  +( C ? newService  do  Ac )
```

Now, after successful authentication, the customer C decides which service to invoke, behaving as specified by Ca (lines 7–12). Once the request has been served, the customer executes Cf deciding whether to quit or ask for a new service (lines 15–17). Accordingly, A reacts to service requests as per Aa on lines 25–27 of the above snippet, similarly to the previous version of ATM, but after the completion of each request, A behaves as per Af on line 37 and returns the card to C if a quit message is received or repeats from Aa when a new service is requested.

The verification of the new version of the system with **ChorGram** now highlights some problems as shown by the following output message (slightly manipulated for readability):

```
1     ...
2     gmc:        Branching Property (part ( ii )):  [Rp ([qCf,qAf,qBa],···)]
3                                                    (qCf,qAf,C,A,Tau,newService )
4                                                    (qCf,qAf,C,A,Tau,quit )    No choice awareness
5     ...
```

The above message reports that a reachable configuration where participants C, A, and B respectively are in state qCf, qAf, and qBa is a 'No choice awareness' configuration. This configuration is highlighted in yellow in the synchronous transition system, which is reported in Figure 6.7. Inspecting the synchronous transition system, we note that this configuration leads to deadlocks (the configurations highlighted in orange in Figure 6.7), due to the fact that the participant B is not notified when the quit branch is taken, i.e., B is *not aware* of which branch of the protocol was chosen by C.

Notice that **ChorGram** builds a global graph also when the system violates GMC (not shown for space restrictions). Such a synthesised global graph reflects some of their possible communication sound executions while leaving out traces where communication misbehaviour happen. The global graph of

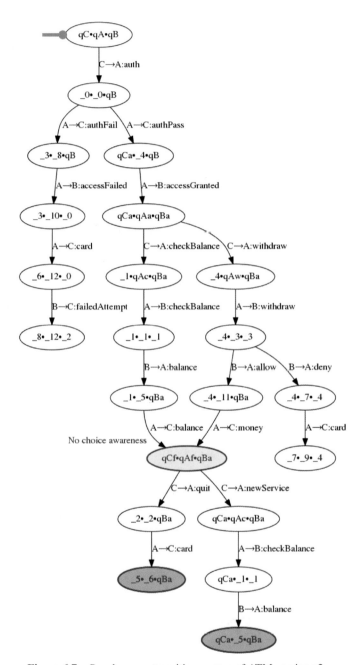

Figure 6.7   Synchronous transition system of ATM service v2.

our second version of the ATM system can also be used to understand what goes wrong in the overall choreography.

6.4.3 ATM Service – Version 3 (fixed)

Besides making the refined specification of Section 6.4.2 GMC, in the next version we also want to let the customer quit the protocol immediately after the authentication. This change makes Cf and Af unnecessary: so, we replace Ca and Cf with the following new versions:

```
1  Ca = A ! checkBalance;  A ? balance  do  Ca
2        +
3        A ! withdraw ; (
4                        A ? card
5                        +
6                        A ? money  do  Ca
7                      )
8        +
9        A ! quit
10  ||
11  Aa = C ? checkBalance  do  Ac
12        +
13        C ? withdraw  do  Aw
14        +
15        C ? quit;
16          B ! quit
```

Note that now A notifies B when the protocol quits (line 15). This modification requires also to modify the bank, which is now:

```
1  B = A ? accessFailed;  C ! failedAttempt      .. The bank tells the customer the attempt failed
2        +
3        A ? accessGranted  do  Ba
4  ||
5  Ba = A ? checkBalance;  A ! balance  do  Ba
6        +
7        A ? withdraw ; (
8            A ! deny
9            +
10           A ! allow  do  Ba
11         )
12        +
13        A ? quit
```

The above changes re-establish GMC, hence communication soundness of the system, as verified by **ChorGram**, which returns the global graph of Figure 6.8.

6.5 Conclusions and Related Work

Conclusions & future work We presented **ChorGram**, a tool supporting the analysis and design of choreography-based development. We have discussed only part of the features of **ChorGram**, those strictly related to

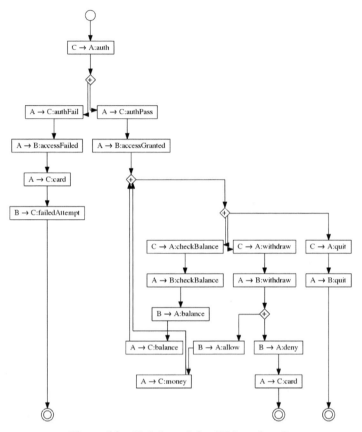

Figure 6.8   Global graph for ATM service v3.

the bottom-up development based on our theory [17], which is itself an extension of previous work on synthesising global types from local specifications [9, 13, 15]. Recently, **ChorGram** has been extended with new functionalities for top-down development. These new functionalities rely on a new semantic framework [11]. We are also planning to plug the "bottom-up" approach advocated here with the classical "top-down" approach [12] as advocated by, e.g., the Scribble specification language [22, 25]. Such an integration would give the flexibility of designing protocols at the global level and obtain the local level automatically, and vice-versa.

As illustrated in Section 6.4, our approach can be used to give feedback to protocol designers. Hence, we are considering integrating **ChorGram** with a framework [20] allowing programmers to obtain real-time feedback wrt. the multiparty compatibility of the system they are designing. Currently, the prototype highlights communication mismatches at the local level and it is sometimes difficult to identify the real cause of such errors [20]. However, it appears that a (possibly partial) global graph can help giving precise feedback to the developer so that they can fix the error(s) easily.

Existing extensions and applications Recent work extends the theory underlying **ChorGram** [17] to communicating timed automata (CTA) [5], i.e., CFSMs which use clocks to constrain when send and receive actions may take place. The authors show that if a system validates some conditions on communication soundness and deadlines, it is possible to construct a choreography with time constraints which is equivalent to the original system of CTAs.

The synthesis of global graphs from local specifications has been applied thus far in two programming languages. A tool to statically detect deadlocks in the Go programming language is available [18]. The tool first extracts CFSMs from each Go-routine in the source code, then feeds them into a slight variation of **ChorGram** (for synchronous semantics) which checks whether the system is multiparty compatible and generates the corresponding global graph (which may be used to track down what may have caused deadlocks). Also, **ChorGram** has been used to model and analyse **genserver** [23], a part of the Erlang OTP standard library widely used in the Erlang community for the development of client/server applications. The analysis highlighted possible coordination errors and was conducted following the pattern showed in Section 6.4. The main difference was that the participants and the corresponding CFSMs had to be extracted from the API documentation of **genserver**.

An interesting use [2, 14] of multiparty compatibility is to support an orchestration mechanism based on the *agreement* of behavioural contracts [4]. Recently this theoretical framework has been used to develop **Diogenes** [1], a middleware supporting designers (and developers) to write *honest* programs [3], namely programs that respect *all* their contracts in *all* their execution contexts. An interesting future work is to integrate **Diogenes** and **ChorGram** in order to adapt components when they are not multiparty compatible. In such cases, (as discussed at the end of Section 6.4.2) **Chor-Gram** synthesises a choreography which, although not faithfully reflecting the behaviour of participants, represents some of their possible communication sound executions. Such a synthesised choreography could then be used to obtain projections that help to attain honesty.

Acknowledgements This work is partially supported by EU FP7 612985 (UPSCALE) and by EPSRC EP/K034413/1, EP/K011715/1, EP/L00058X/1, EP/N027833/1 and EP/N028201/1.

References

[1] Nicola Atzei and Massimo Bartoletti. Developing honest Java programs with Diogenes. In *Formal Techniques for Distributed Objects, Components, and Systems (FORTE)*, pages 52–61, 2016.

[2] Massimo Bartoletti, Julien Lange, Alceste Scalas, and Roberto Zunino. Choreographies in the wild. *Sci. Comput. Program.*, 109:36–60, 2015.

[3] Massimo Bartoletti, Alceste Scalas, Emilio Tuosto, and Roberto Zunino. Honesty by typing. In *FMOODS/FORTE*, volume 7892 of *LNCS*, pages 305–320. Springer, 2013.

[4] Massimo Bartoletti, Emilio Tuosto, and Roberto Zunino. Contract-oriented computing in CO_2. *Scientific Annals in Comp. Sci.*, 22(1):5–60, 2012.

[5] Laura Bocchi, Julien Lange, and Nobuko Yoshida. Meeting deadlines together. In *CONCUR 2015*, pages 283–296, 2015.

[6] Filippo Bonchi and Damien Pous. HKC. http://perso.ens-lyon.fr/damien.pous/hknt/

[7] Daniel Brand and Pitro Zafiropulo. On communicating finite-state machines. *JACM*, 30(2):323–342, 1983.

[8] Jordi Cortadella, Michael Kishinevsky, Alex Kondratyev, Luciano Lavagno, Enric Pastor, and Alexandre Yakovlev. Petrify. http://www.lsi.upc.edu/~jordicf/petrify/

[9] Pierre-Malo Deniélou and Nobuko Yoshida. Multiparty compatibility in communicating automata: Characterisation and synthesis of global session types. In *ICALP 2013*.

[10] Pierre-Malo Deniélou and Nobuko Yoshida. Multiparty session types meet communicating automata. In *ESOP 2012*, pages 194–213, 2012.

[11] Roberto Guanciale and Emilio Tuosto. An Abstract Semantics of the Global View of Choreographies. In *ICE 2016*, pages 67–82, 2016.

[12] Kohei Honda, Nobuko Yoshida, and Marco Carbone. Multiparty asynchronous session types. *J. ACM*, 63(1):9:1–9:67, 2016.

[13] Julien Lange. *On the Synthesis of Choreographies*. PhD thesis, Department of Computer Science, University of Leicester, 2013.

[14] Julien Lange and Alceste Scalas. Choreography synthesis as contract agreement. In *ICE*, volume 131 of *EPTCS*, pages 52–67, 2013.

[15] Julien Lange and Emilio Tuosto. Synthesising choreographies from local session types. In *CONCUR 2012*, pages 225–239, 2012.

[16] Julien Lange and Emilio Tuosto. **ChorGram**: tool support for choreographic development. Available at `https://bitbucket.org/emlio_tuosto/chorgram/wiki/Home`, 2015.

[17] Julien Lange, Emilio Tuosto, and Nobuko Yoshida. From communicating machines to graphical choreographies. In *POPL 2015*, pages 221–232, 2015.

[18] Nicholas Ng and Nobuko Yoshida. Static deadlock detection for concurrent Go by global session graph synthesis. In *CC 2016*, pages 174–184, 2016.

[19] Object Management Group. Business Process Model and Notation. `http://www.bpmn.org`

[20] Roly Perera, Julien Lange, and Simon J. Gay. Multiparty compatibility for concurrent objects. In *PLACES 2016*, pages 73–82, 2016.

[21] Chris Piro. Chat stability and scalability. `https://goo.gl/Z1tpgA`

[22] Scribble. `http://www.scribble.org`

[23] Ramsay Taylor, Emilio Tuosto, Neil Walkinshaw, and John Derrick. Choreography-based analysis of distributed message passing programs. In *PDP 2016*, pages 512–519, 2016.

[24] Paolo D'Incau's blog. `https://goo.gl/eXKng1`, 2013.

[25] Nobuko Yoshida, Raymond Hu, Rumyana Neykova, and Nicholas Ng. The Scribble protocol language. In *TGC 2013*, pages 22–41, 2013.

7

Programming Adaptive Microservice Applications: An AIOCJ Tutorial*

**Saverio Giallorenzo[1], Ivan Lanese[1],
Jacopo Mauro[2] and Maurizio Gabbrielli[1]**

[1]Focus Team, University of Bologna/INRIA, Italy
[2]Department of Informatics, University of Oslo, Norway

Abstract

This tutorial describes AIOCJ, which stands for *Adaptive Interaction Oriented Choreographies in Jolie*, a choreographic language for programming microservice-based applications which can be updated at runtime. The compilation of a single AIOCJ program generates the whole set of distributed microservices that compose the application. Adaptation is performed using adaptation rules. Abstractly, each rule replaces a pre-delimited part of the program with the new code contained in the rule itself. Concretely, at runtime, the application of a rule updates part of the microservices that compose the application so to match the behavior specified by the updated program. Thanks to the properties of choreographies, the adaptive application is free from communication deadlocks and message races even after adaptation.

7.1 Introduction

Today, most applications are distributed, involving multiple participants scattered on the network and interacting by exchanging messages. While still widely used, the standard client-server topology has shown some of its

---

*Supported by the COST Action IC1201 BETTY, by the EU project FP7-644298 *HyVar: Scalable Hybrid Variability for Distributed, Evolving Software Systems*, by the GNCS group of INdAM via project *Logica, Automi e Giochi per Sistemi Auto-adattivi*, and by the EU EIT Digital project *SMAll*.

limitations and peer-to-peer and other interaction patterns are raising in popularity in many contexts, from social networks to business-to-business, from gaming to public services. Programming the intended behavior of such applications requires to understand how the behavior of the single program of one of their nodes combines with the others, to produce the global behavior of the application. In brief, it requires to master the intricacies of concurrency and distribution. There is clearly a tension between the *global* desired behavior of a distributed application and the fact that it is programmed by developing *local* programs. Choreographies [1–5], and more specifically choreographic programming [6], aim at solving this tension by providing to developers a programming language where they directly specify the global behavior. A sample choreography that describes the behavior of an application composed of one client and one seller is:

```
1    product_name@client = getInput( "Insert product name" );
2    quote: client( product_name ) -> seller( sel_product )
```

The execution starts with an action performed by the client: an input request to the local user (line 1). The semicolon at the end of the line is a sequential composition operator, hence the user input should complete before execution proceeds to line 2. Then, a communication between the client and the seller takes place: the client sends a message and the seller receives it. A more detailed description of the choreographic language used in the example above is presented in Section 7.3.

Following the choreographic programming approach, given a choreography, the local programs that implement the global specification are automatically generated by the language compiler, ready for the deployment in the intended locations and machines. For instance, the compilation of the choreography in the example produces the local codes of both the client and the seller. The local code of the client starts with a user interaction, followed by the sending of a message to the seller. The local code of the seller has just one action: the reception of a message from the client.

The choice of a choreographic language also has the advantage of avoiding by construction common errors performed when developing concurrent and distributed applications [7]. Notably, these include communication deadlocks, which may cause the application to block, and message races, which may lead to unexpected behaviors in some executions.

Another advantage of the choreographic approach is that it eases the task of adapting a running distributed application. We recall that nowadays applications are often meant to run for a long time and should adapt to changes of the environment, to updates of requirements, and to the variability of business rules. Adapting distributed applications at runtime, that is without stopping and restarting them, and with limited degradation of the quality of service, is a relevant yet difficult to reach goal. In a choreographic setting, one can simply specify how the global behavior is expected to change. This approach leaves to the compiler and the runtime support the burden of concretely updating the code of each local program. This update should be done avoiding misbehaviors while the adaptation is carried out and ensuring a correct post-adaptation behavior.

This tutorial presents AIOCJ[1], which stands for *Adaptive Interaction Oriented Choreographies in Jolie*, a framework including *i*) a choreographic language, AIOC, for programming microservice-based applications which can be dynamically updated at runtime and *ii*) its runtime environment. The main features of the AIOCJ framework are:

Choreographic approach: the AIOC language allows the programmer to write the behavior of a whole distributed application as a single program;

Runtime adaptability: AIOCJ applications can be updated by writing new pieces of code embodied into AIOC *adaptation rules*. Adaptation rules are dynamically and automatically applied to running AIOCJ applications, providing new features, allowing for new behaviors, and updating underlying business rules.

Microservice architecture: AIOCJ applications are implemented as systems of microservices [8]. Indeed, we found that the microservice architectural style supports the fine-grained distribution and flexibility required by our case. As a consequence, AIOCJ applications can interact using standard application-layer protocols (e.g., SOAP and HTTP) with existing (legacy) software thus also facilitating and supporting the integration of existing systems.

A more technical account of the AIOCJ framework can be found in the literature, describing both the underlying theory [9] and the tool itself [10]. AIOCJ can be downloaded from its website [11], where additional documentation and examples are available.

---

[1]The tutorial refers to version 1.3 of AIOCJ.

7.2 AIOCJ Outline

As described in the Introduction, AIOC is a choreographic language for programming microservice-based applications which can be dynamically updated at runtime. The AIOCJ framework is composed of two parts:

- the AIOCJ Integrated Development Environment (IDE), provided as an Eclipse plugin, that lets developers write both AIOC programs and the adaptation rules that change the behavior of AIOCJ applications at runtime;
- the AIOCJ Runtime Environment (RE), which is used to support the execution and the adaptation of AIOCJ applications.

The AIOCJ IDE (see the screenshot in Figure 7.1) offers standard functionalities such as syntax highlighting and syntax checking. However, the most important functionality of the IDE is the support for code compilation. The target language of the AIOCJ compiler is Jolie [12, 13], the first language natively supporting microservice architectures. A key feature of the Jolie language is its support for a wide range of communication technologies (TCP/IP sockets, Unix domain sockets, Bluetooth) and of protocols (e.g., HTTP, SOAP, JSON-RPC) that makes it extremely useful for system integration. AIOC inherits this ability since it makes the communication capabilities of Jolie available to the AIOC programmer.

Since AIOC is a choreographic language, each AIOC program defines a distributed application. The application is composed of different nodes, each taking a specific *role* in the choreography. Each role has its own local state, and the roles communicate by exchanging messages. The structure of AIOCJ

```
preamble {
  starter: client
  location@client: "socket://client.com:5000"
  location@seller: "socket://seller.com:5050"
  location@bank: "socket://bank.com:6000"
}

aioc {
  continue@client = "y";
  while( continue == "y" )@client{
    product_name@client = getInput( "Insert product name" );
    quote: client( product_name ) -> seller( sel_product );
    price@seller = getInput( "Quote product: " + sel_product );
```

Figure 7.1 The AIOCJ IDE.

applications makes the compilation process of AIOCJ peculiar for two main reasons:

- the compilation of a single AIOC program generates one Jolie microservice for each role involved in the choreography, instead of a unique executable for the whole application;
- the compilation may involve either an AIOC program, or a set of AIOC adaptation rules. In particular, the latter may be compiled even after the compilation, deployment, and launch of the AIOC program. Thus AIOC adaptation rules can be devised and programmed while the application is running, and therefore applied to it at runtime.

Adaptation rules target well-identified parts of AIOC programs. Indeed, an AIOC program may declare some part of its code as adaptable by enclosing it in a scope block. Abstractly, the effect of the application of an AIOC adaptation rule to a given scope is to replace the scope block with new code, contained in the adaptation rule itself. Concretely, when the distributed execution of an AIOC program reaches a scope, the AIOCJ RE checks whether there is any adaptation rule applicable to it. If this is the case, then the running system of microservices adapts so to match the behavior specified by the updated choreography. This adaptation involves coordinating the distribution and execution of the local codes corresponding to the global code in the adaptation rule. If instead no rule applies, the execution proceeds as specified by the code within the scope.

In the rest of this section we describe the architecture of AIOCJ and the workflow that developers, or better DevOps[2], have to follow in order to compile, deploy, and adapt at runtime an AIOCJ application (a more detailed step-by-step description is in Section 7.6). We instead dedicate Sections 7.3 to 7.5 to the description of the AIOC language.

7.2.1 AIOCJ Architecture and Workflow

The AIOCJ runtime environment comprises a few Jolie microservices that support the execution and adaptation of compiled programs. The main microservices of the AIOCJ runtime environment are:

- Adaptation Manager, a microservice in charge of managing the adaptation protocol;
- Adaptation Server, a microservice that contains a set of adaptation rules;

---

[2]DevOps is a portmanteau of "development" and "operations" used to indicate the professional figure involved in the development, deployment, and operation of the application.

- Environment, a microservice used to store values of global properties related to the execution environment. These properties may be used to check whether adaptation rules are applicable or not.

More precisely, a runtime environment includes one Adaptation Manager, zero or more Adaptation Servers, each of them enabling a set of adaptation rules, and, if needed to evaluate the applicability conditions of the rules, one Environment microservice. Adaptation Servers can be added or removed dynamically, thus enabling dynamic changes in the set of rules.

Microservices compiled from AIOC code interact both among themselves, as specified by the choreography, and with the Adaptation Manager, to carry out adaptation. Indeed, when a scope is about to be executed, the Adaptation Manager is invoked to check whether the scope can be executed as it is, or if it must be replaced by the code provided by some adaptation rule, made available by an active Adaptation Server. In fact, when started, an Adaptation Server registers itself at the Adaptation Manager. The Adaptation Manager invokes the registered Adaptation Servers to check whether their adaptation rules are applicable. In order to check applicability, the corresponding Adaptation Server evaluates the *applicability condition* of the rule, possibly interacting with the Environment microservice. The first applicable adaptation rule, if any, is used to replace the code of the scope.

Let us consider an example. Take a simple choreography in AIOC involving two roles, client and seller. Figure 7.2 depicts the process of compilation ① and execution ② of the AIOC. From left to right, we use the IDE to write the AIOC and to compile it into a set of executable Jolie microservices (Client and Seller). To execute the generated application, we first launch the Adaptation Manager and then the two compiled microservices.

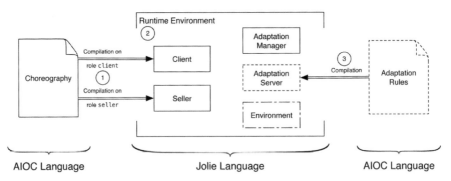

Figure 7.2 The AIOCJ framework — deployment and execution of a choreography.

Now, let us suppose that we want to adapt our application. Assuming that the choreography has at least one `scope`, we only need to write and introduce into the system a new set of adaptation rules. Figure 7.2 depicts the needed steps. From right to left, we write the rules (outlined with dashes) and we compile them using the IDE ③. The compilation of a set of adaptation rules in AIOCJ produces a single Adaptation Server (also outlined with dashes). After the compilation, the generated Adaptation Server is deployed and started, and it registers itself at the Adaptation Manager. If environmental information is needed to evaluate the applicability condition of the rule, then the DevOps has also to deploy the Environment microservice. From now on, until the Adaptation Server is shut down, the rules it contains are active and can be applied to the application. Actual adaptation happens when a `scope` is about to execute, and the applicability condition of the rule for the current `scope` is satisfied. This adaptation is performed automatically and it is completely transparent to the user, except for possible differences in the visible behavior of the new code w.r.t. the original one.

7.3 Choreographic Programming

The main idea of choreographic programming is that a unique program describes the behavior of a whole distributed application. The main construct of such a program are *interactions*, such as:

```
quote: client( product_name ) -> seller ( sel_product )
```

This interaction specifies that role `client` sends a message to role `seller` on operation `quote`. The value of the message is given by the evaluation of expression `product_name` (here just a variable name) in the local state of role `client`. The message will be stored by the `seller` in its local variable `sel_product`. An interaction involves two roles of the choreography, but other choreography constructs involve just one role. For instance, an assignment like `continue@client = "y"`, means that the string `"y"` is assigned to the variable `continue` of role `client`, as specified by the @ operator.

Let us now detail a simple AIOC program implementing a `client`/`seller` interaction featuring a payment via a `bank` (see Listing 7.1). We will use this program as running example throughout the tutorial. Lines 1–6 form the preamble, which specifies some deployment information:

- line 2 declares the `starter` of the choreography, i.e., the first role that needs to be started and the one that coordinates the start of the application by waiting for the other roles to join;

```
1   preamble {
2     starter: client
3     location@client: "socket://client.com:5000"
4     location@seller: "socket://seller.com:5050"
5     location@bank: "socket://bank.com:6000"
6   }
7
8   aioc {
9     continue@client = "y";
10    while( continue == "y" )@client{
11      product_name@client = getInput( "Insert product name" );
12      quote: client( product_name ) -> seller( sel_product );
13      price@seller = getInput( "Quote product: " + sel_product );
14      if ( price > 0 )@seller{
15        quoteResponse: seller( price ) -> client( product_price );
16        accept@client = getInput(
17        "Do you accept to buy the product: " + product name +
18        " at price: " + product_price + "? [y/n]" );
19        if ( accept == "y" )@client{
20          orderPayment: client( product_price ) -> bank( amount );
21          authorisePayment@bank = getInput(
22          "Do you authorise the payment: " + amount + " [y/n]?" );
23          if ( authorisePayment == "y" )@bank{
24            issuePayment: bank( amount ) -> seller( payment );
25            productDelivery: seller() -> client();
26            r@client = show( "Object delivered" )
27          } else {
28            r@client = show( "Payment refused" )
29          }
30        }
31      } else {
32        _r@client = show( "Product " + product_name + " unavailable." )
33      };
34      continue@client = getInput( "Continue shopping? [y/n]" )
35    }
36  }
```

Listing 7.1   Running example: basic choreography.

- lines 3–5 specify how the roles participating to the choreography can be reached. In this case, all the three roles communicate using TCP/IP sockets, as specified by the "socket://" prefix of the URI.

The actual code is introduced by the keyword aioc. After the local assignment at line 9, line 10 introduces a while loop. The @client suffix specifies that the guard is evaluated by the client in its local state. Notice that the decision about whether to enter the loop or not is taken by the client but it impacts also other roles. These roles are notified of the choice by auxiliary communications which are automatically generated. The assignment at line 9 and the while loop starting at line 10 are composed using a semicolon, which represents sequential composition. Line 11 is again an assignment, where built-in function getInput is used to interact with the local user. The function creates a window showing the string in parameter and returns the input of the user. Line 12 is an interaction between the client and the seller. The next interesting construct is at line 14, featuring a conditional. As for while loops, the conditional specifies which role is in charge of evaluating the guard, and other roles are automatically notified of the outcome of the evaluation. Function show (line 26) is a built-in function like getInput, simply showing a message.

Abstracting from the technical details, the choreography specifies that the client asks the quote for a product (line 12), and then decides whether to buy it or not (line 19). In the first case, the client asks the bank to perform the payment (line 20). If the payment is authorized (line 23), then the money is sent to the seller (line 24), which delivers the product to the client (line 25). At the end of the interaction, the client may decide to buy a new product or to stop (line 34).

When writing AIOC programs, beyond the usual syntactic errors, one should pay attention to a semantic error peculiar of choreographic programming. Indeed, a semicolon specifies that the code before the semicolon should be executed before the code after the semicolon. However, since there is no central control, such a constraint can only be enforced if for each pair of statements s and τ such that s is just before the semicolon and τ is just after the semicolon, there is a role occurring in both s and τ. This property is called connectedness [9] and it is needed to enforce the sequentiality of the actions. When connectedness does not hold, AIOCJ IDE alerts the user by showing the error "The sequence is not connected". Instead of asking the programmer to satisfy connectedness, one could extend AIOCJ to automatically insert auxiliary communications to ensure connectedness, similarly to what is done for while loops and conditionals. Such an extension is left as future work.

7.4 Integration with Legacy Software

The example in the previous section shows how one can program a distributed application in AIOCJ. However, such a distributed application is closed: there is no interaction between the application and the outside world, except for basic visual interactions with the users of the application. As we will see below, AIOCJ applications are not necessarily closed systems. Indeed,

AIOCJ provides a strong support to integration with legacy software. We already cited that AIOCJ is based on the microservice technology. As such, it supports interaction with external services via standard application-layer protocols, such as SOAP and HTTP. Such services are seen as functions inside AIOC programs, and can be invoked and used inside expressions.

Let us see how this can be done by refining our running example from Listing 7.1 into the one in Listing 7.2.

```
1   include quoteProduct from "socket://localhost:8000" with SOAP
2   include makePayment from "socket://localhost:8001/IBAN" with HTTP
3
4   preamble {
5     starter: client,
6     location@client: "socket://client.com:5000"
7     location@seller: "socket://seller.com:5050"
8     location@bank: "socket://bank.com:6000"
9   }
10
11  aioc {
12    continue@client = "y";
13    while( continue == "y" )@client{
14      product_name@client = getInput( "Insert product name" );
15      quote: client( product_name ) -> seller( sel_product );
16      price@seller = quoteProduct( sel_product );
17      if ( price > 0 )@seller{
18        quoteResponse: seller( price ) -> client( product_price );
19        accept@client = getInput(
20        "Do you accept to buy the product: " + product_name +
21        " at price: " + product_price + "? [y/n]" );
22        if ( accept == "y" )@client{
23          orderPayment: client( product_price ) -> bank( amount );
24          authorisePayment@bank = makePayment( amount );
25          if ( authorisePayment == "y" )@bank{
26            issuePayment: bank( amount ) -> seller( payment );
27            productDelivery: seller() -> client();
28            r@client = show( "Object delivered" )
29          } else {
30            r@client = show( "Payment refused" )
31          }
32        }
33      } else {
34        _r@client = show( "Product " + product_name + " unavailable." )
35      };
36      continue@client = getInput( "Continue shopping? [y/n]" )
37    }
38  }
```

Listing 7.2   Running example: integration with external services.

In Listing 7.2, lines 1 and 2 declare two external services, `quoteProduct` invoked using SOAP and `makePayment` invoked using HTTP (more precisely, a POST request carrying XML data). Both external services communicate with AIOCJ using TCP/IP sockets. The first service is invoked at line 16 by the `seller` and it is used to check the price of a given product. In principle, such a service can be either publicly available or a private service of the `seller`. Here, we assume that this service gives access to the `seller` IT system, e.g., to the database storing prices of the available products. The second service is invoked at line 24 by the `bank`, and gives access to the `bank` IT system. One can easily imagine to make the example more realistic by adding other external services taking care, e.g., of shipping the product.

We now discuss in more detail how function arguments are encoded for service invocation and how the result is sent back to the caller. In general AIOCJ functions can have an arbitrary number of parameters, separated by commas. The parameters are embedded in a tree structure which is then encoded according to the chosen application-layer data protocol. The tree structure has an empty root with a number of children all named `p` (for parameter) carrying the parameters of the invocation, in the order in which they are specified. The return value instead has basic type (such as string, integer, double) and it is contained in the root of the response message.

For instance, consider a sample function `myFunction`, with three parameters, a string, an integer, and a double. If the data protocol for `myFunction` is SOAP, then the AIOCJ application would send a SOAP message as reported in Listing 7.3. A possible reply to the message above is a SOAP message of the form reported in Listing 7.4.

```
1   <?xml version="1.0" encoding="utf-8" ?>
2   <SOAP-ENV:Envelope
3     xmlns:SOAP-ENV="http://schemas.xmlsoap.org/soap/envelope/"
4     xmlns:xsd="http://www.w3.org/2001/XMLSchema"
5     xmlns:xsi="http://www.w3.org/2001/XMLSchema-instance">
6     <SOAP-ENV:Body>
7       <myFunctionRequest>
8         <p xsi:type="xsd:string">parameter1</p>
9         <p xsi:type="xsd:int">2</p>
10        <p xsi:type="xsd:double">3.14</p>
11      </myFunctionRequest>
12    </SOAP-ENV:Body>
13  </SOAP-ENV:Envelope>
```

Listing 7.3   Function invocation: SOAP message request.

```
1   <?xml version="1.0" encoding="utf-8" ?>
2   <SOAP-ENV:Envelope
3     xmlns:SOAP-ENV="http://schemas.xmlsoap.org/soap/envelope/"
4     xmlns:xsd="http://www.w3.org/2001/XMLSchema"
5     xmlns:xsi="http://www.w3.org/2001/XMLSchema-instance">
6     <SOAP-ENV:Body>
7       <myFunctionResponse xsi:type="xsd:string">
8         responseValue
9       </myFunctionResponse>
10    </SOAP-ENV:Body>
11  </SOAP-ENV:Envelope>
```

Listing 7.4   Function invocation: SOAP message response.

Other application-layer data protocols would produce similar structures. Currently, AIOCJ supports SOAP, HTTP, SODEP (i.e., Jolie's binary data protocol), JSON/RPC, and XML/RPC. As far as the communication medium is concerned, AIOCJ supports other options beyond TCP/IP sockets, namely Bluetooth with URIs of the form `"btl2cap://0050CD00321B:101"` and Unix domain sockets with URIs of the form `"localsocket://var/comm/socket"`. The choice of the communication medium and the choice of the application-layer data protocols are orthogonal.

7.5 Adaptation

We now come to the main feature of AIOCJ, namely the support for adaptation. Adaptation is performed in two stages:

1. when writing the original AIOC program, one should foresee which parts of the code could be adapted in the future, and enclose them into scopes;
2. while the AIOC program is running, one should write adaptation rules to introduce the desired new behavior.

We introduce in Listing 7.5 three scopes to show how adaptation can be enabled in the running example in Listing 7.2.

Scope `transaction-execution` at lines 26–29 encloses the body of the business transaction, with the idea that this can be changed to support integration with a shipper service, or more refined payment protocols. Then, we have two scopes, `success-notification` (lines 30–32) and `failure-notification` (lines 34–36), which are in charge of notifying the `client` of the outcome of the transaction, with the idea that different forms of notification, e.g., through e-mail or SMS, could be implemented in the future. Developers can equip

```
1   include quoteProduct from "socket://localhost:8000" with SOAP
2   include makePayment from "socket://localhost:8001/IBAN" with HTTP
3
4   preamble {
5    starter: client
6    location@client: "socket://client.com:5000"
7    location@seller: "socket://seller.com:5050"
8    location@bank: "socket://bank.com:6000"
9   }
10
11  aioc {
12   continue@client = "y";
13   while( continue == "y" )@client{
14    product_name@client = getInput( "Insert product name" );
15    quote: client( product_name ) -> seller( sel_product );
16    price@seller = quoteProduct( sel_product );
17    if ( price > 0 )@seller{
18     quoteResponse: seller( price ) -> client( product_price );
19     accept@client = getInput(
20     "Do you accept to buy the product: " + product_name +
21     " at price: " + product_price + "? [y/n]" );
22     if ( accept == "y" )@client{
23      orderPayment: client( product_price ) -> bank( amount );
24      authorisePayment@bank = makePayment( amount );
25      if ( authorisePayment == "y" )@bank{
26       scope @seller {
27        issuePayment: bank( amount ) -> seller( payment );
28        productDelivery: seller() -> client()
29       } prop { N.scopename = "transaction-execution" };
30       scope @seller {
31        r@client = show( "Object delivered" )
32       } prop { N.scopename = "success-notification" }
33      } else {
34       scope @seller {
35        r@client = show( "Payment refused" )
36       } prop { N.scopename = "failure-notification" }
37      }
38     }
39    } else {
40     _r@client = show( product_name + " is unavailable." )
41    };
42    continue@client = getInput( "Continue shopping? [y/n]" )
43   }
44  }
```

Listing 7.5 Running example: enabling adaptation.

scopes with properties describing their nature and characteristics. These properties can be used to decide whether a given rule should apply to a given scope or not. In the example, we just use a property scopename to describe each scope. In general, however, many properties can be used. For example, if some `scope` encloses a part of the code which is critical for security reasons, one of its properties could declare the security level of the current code. Such a declaration is under the responsibility of the programmer and it is in no way checked or enforced by the AIOCJ framework.

Note that each `scope` is followed by an annotation @*role* that declares the coordinator of the adaptation procedure of the `scope`. The coordinator is in charge of invoking the Adaptation Manager, which handles the selection of an applicable adaptation rule. The Adaptation Manager can access the internal state of the coordinator to check whether an adaptation rule is applicable or not. The coordinator is also in charge of fetching the local codes compiled from the selected adaptation rule and of distributing them to the other roles.

Remark 1 *We highlight that there is no precise convention on how to place* scopes: *one should try to foresee which parts of the AIOC program are likely to change. As a rule of thumb, parts which are critical for security or performance reasons may be updated to improve the security level or the performance of the application. Parts which instead implement business rules may need to be updated to change the business rules. Finally, parts that manage interactions with external services may need to be updated to match changes in the external services. There is also a trade-off involved in the definition of* scopes. *On the one hand, large* scopes *are rarely useful, since they could be updated only before the beginning of their execution, which can be quite early in the life of the application. On the other hand, small* scopes *may be problematic, since a meaningful update may involve many of them and currently AIOCJ does not provide a way to synchronize when and how* scopes *are updated.*

Now that the application in Listing 7.5 is equipped with scopes, it is ready to be deployed, and offers built-in support for adaptation. While the application is running, a new need may emerge. Assume for instance that the application, meant for trading cheap products, needs to be used also for more expensive ones. In this previously unforeseen setting, the fact that the payment is performed in a single installment and before the shipping of the product may be undesirable for the `client`. One can meet this new need by providing an adaptation rule (see Listing 7.6) where the payment is performed in two installments, each consisting in half of the original `amount`: one sent before and the other after the delivery of the product. This rule targets `scopes` with property `scopename` equal to `transaction-execution` and it applies only

```
1   rule {
2     on { N.scopename == "transaction-execution" and
3           E.split_payment_threshold < price }
4     do {
5       issuePayment: bank( amount / 2 ) -> seller( first_payment );
6       productDelivery: seller() -> client();
7       issuePayment: bank( amount / 2 ) -> seller( second_payment );
8       payment@seller = first_payment + second_payment
9     }
10  }
```

Listing 7.6   Adaptation rule: split payment.

if the `price` of the product is above a `split_payment_threshold` available in the Environment microservice. The idea is that such a threshold may be agreed upon by the `client` and the `seller` or established by some business regulation. We remark that properties of the `scope`, like `N.scopename`, are prefixed by `N` while values provided by the Environment microservice, like `E.split_payment_threshold`, are prefixed by `E`. Names with no prefix refer to variables of the role that coordinates the adaptation of the `scope`, such as `price` in this example.

We note that the above adaptation rule changes the choreography and, as a consequence, the behavior of two of its roles. In general, an adaptation rule can impact an arbitrary number of roles. We also note that the need for adaptation is checked — and adaptation is possibly performed — every time the `scope` is executed. In this example, if the `client` buys many products, some with price above the threshold and some below, the need for adaptation is checked for each item and adaptation is performed only for the ones with a price above the threshold. In essence, purchases of cheap products follow the basic protocol whilst purchases of expensive ones follow the refined protocol introduced by the adaptation rule.

We now consider another need that may emerge. Assume that the `seller` decides to log all its sales, e.g., for tax payment reasons. Again, one may write an adaptation rule (see Listing 7.7) to answer this need. This rule targets the `scope` with property `N.scopename = "success-notification"` (lines 30–32 in Listing 7.5), which was not exactly intended for logging, but can be adapted to do so by taking care of repeating in the new code also the original notification message (line 31 in Listing 7.5, repeated at line 8 in Listing 7.7). The rule exploits both a new role, `logger`, and two external services `log` and `getTime`. External services are declared exactly as in AIOC programs. Note that here we omit the application-layer protocol of both services, hence the default, SOAP, is used.

```
1  rule {
2    include log from "socket://localhost:8002"
3    include getTime from "socket://localhost:8003"
4    newRoles: logger
5    location@logger: "socket://localhost:15000"
6    on { N.scopename == "success-notification" }
7    do {
8      r@client = show( "Object delivered" )
9      |
10     {
11      log: seller( sel_product + " " + payment ) -> logger( entry );
12      time@logger = getTime();
13      log_entry = time + ": " + entry;
14      { r1@logger = log( log_entry ) | r2@logger = show( log_entry ) }
15     }
16   }
17 }
```

Listing 7.7   Adaptation rule: logging.

The additional role is declared using keyword newRoles (line 4). New roles in AIOCJ rules should not be involved in the target AIOC program and take part to the choreography only while the body of the rule executes. As for normal roles, the URI of new roles is declared using the keyword location.

7.6 Deployment and Adaptation Procedure

In this section we describe the steps that DevOps need to follow to deploy the AIOCJ application of Listing 7.5 and to adapt it at runtime. When reporting paths, we use the Unix forward slash notation.

Compiling and Running an AIOC. As already mentioned, AIOCJ IDE runs as an Eclipse plugin. Hence, to create a new AIOC program we create a new project and a new file inside it with .ioc extension. We write the code in Listing 7.5 and we compile it by clicking on the button "Jolie Endpoint Projection" ⏩. The compilation creates three folders in the Eclipse project: epp_aioc, adaptation_manager, and environment.

Within the folder epp_aioc we can find one subfolder for each role in the AIOC program containing all the related code. The main file is named after the role and has the standard Jolie extension .ol. The subfolder needs to be moved in the host corresponding to the location of the role declared in the preamble of the AIOC program. For example, the subfolder client should be moved into the host located at "client.com".

Within the folders `adaptation_manager` and `environment` the main files are, respectively, `main_adaptationManager.ol` and `environment.ol`.

Before starting the compiled AIOC program, we make sure that the external services included in the choreography are running. To run the AIOC program, we first launch the Adaptation Manager with

```
jolie adaptation_manager/main_adaptationManager.ol
```

Then, we run the roles in the choreography, beginning from the `client`, which is declared as the `starter` of the choreography. For instance, the `client` — previously deployed at `"client.com"` — can be launched with

```
jolie client/client.ol
```

At the moment there is no need to run the Environment. As soon as the last role is started, the execution of the AIOCJ application begins.

Adapting a Running AIOC. Adaptation rules are defined using the same Eclipse plugin as AIOC programs. They need to be stored in a new `.ioc` file, either in the same project as the AIOC program or in a new one.

As for AIOC programs, the compilation of a set of adaptation rules is triggered by the "Jolie Endpoint Projection" button $\boxed{\mathcal{F}}$ and produces a folder named `epp_rules`, which corresponds to a unique Adaptation Server. Inside the folder, the main file is `AdaptationServer.ol` within path

```
__adaptation_server/servers/server
```

Also in this case, before starting the Adaptation Server, we make sure that the external services included in the rules are running.

If some adaptation rule has an applicability condition that checks some Environment variables (e.g., variable `E.split_payment_threshold` in Listing 7.6, line 3), the Environment microservice needs to be launched, running the program `environment.ol`. Environment variables can be added and removed both by console interaction or by editing the configuration file `environmentVariables.xml`.

If some adaptation rule needs a new role, the `location` declared for it should be able to interact with the Adaptation Server that contains the rule. To this end, AIOCJ provides a dedicated microservice called Role Supporter, which needs to be deployed in the host corresponding to the target `location`. This is done by moving to the corresponding host the folder

```
role_supporter/ruleN/roleName
```

where N is the sequential number of the rule, from top to bottom, inside the file `.ioc`, and *roleName* is the name of the new role. The folder contains the code of the utility microservice, `RoleSupporter.ol`, and an automatically generated configuration file `config/location.iol`. For instance, the configuration file for the RoleSupporter for role `logger` in the rule in Listing 7.7 (assuming it is the only rule in the `.ioc` file) is

`role_supporter/rule1/logger/config/locations.iol`

If the location of the new role is unspecified, then `"localhost:9059"` is used by default and the corresponding folder is `default_role_supporter`.

Once both the external services and the Role Supporters are running, we can launch the Adaptation Server. When launched, the Adaptation Server registers at the Adaptation Manager and the compiled adaptation rules become enabled. From now on, when a `scope` is reached during execution, the rules in the Adaptation Server are checked for applicability.

Both microservices implementing roles of AIOCJ applications and the ones in AIOCJ RE — namely the Adaptation Manager, the Environment, the Adaptation Servers, and the Role Supporters — can be re-deployed on hosts different from the default ones. This requires to move the corresponding folder, but also to update the configuration files that contain their addresses, including their own configuration file. Notably, no recompilation is needed. We report in Figure 7.3 the dependency graph among the locations of AIOCJ microservices. In the figure, the notation $\boxed{A} \rightarrow \boxed{B}$ means that microservice **A** must know the deployment location of microservice **B**. At the bottom of each box we report the path to the corresponding configuration file for

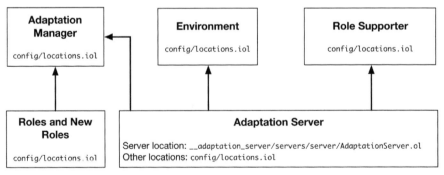

Figure 7.3   Location dependency graph among AIOCJ microservices.

locations, which is `config/locations.iol` except for the deployment location of Adaptation Servers which is directly contained in their own main file.

7.7 Conclusion

In this tutorial we have given a gentle introduction to the AIOCJ framework and to the AIOC language. While both adaptation and choreographies are thoroughly studied in the literature, their combination has not yet been explored in depth. As far as we know, AIOCJ is the only implemented framework in this setting. Theoretical investigations of the interplay between adaptation and multiparty session types [14–16] (which use choreographies as types instead of as a language) have been undertaken. A relevant work considers self-adaptive systems [14]. It uses multiparty session types to monitor that the computation follows the expected patterns, and it performs adaptation by moving from one choreography to the other according to external conditions. However, all possible behaviors are present in the system since the very beginning. Another work studies how to update a system so to preserve the guarantees provided by multiparty session types [15]. Another study, still preliminary, describes multiparty session types that can be updated from both inside and outside the system [16]. None of the three proposals above has been implemented. On the other side, we find two implemented approaches for programming using choreographies, Scribble [4, 17] and Chor [2], but they do not support adaptation. Chor is particularly related to AIOCJ, since they both produce Jolie code and they share part of the codebase. Finally, the main standard in the field of choreographic specifications, WS-CDL [5], does not support adaptation. Moreover, WS-CDL is just a specification language and not an executable one. Further information on choreographies can be found in two surveys. One presents a general description of the theory of choreographies and session types [18]. The other accounts for their use in programming languages [19].

As future work we would like to understand what is needed to make AIOCJ more usable in practice. To this end, we are experimenting by applying AIOCJ to case studies developed for other approaches to adaptation, such as Context-Oriented Programming [20] and distributed [21] and dynamic [22] Aspect-Oriented Programming. Initial results in this direction can be found on the AIOCJ website [11]. Another direction is to provide automated support for the deployment of AIOCJ applications using containerization technologies such as Docker [23].

References

[1] M. Carbone, K. Honda, and N. Yoshida, "Structured communication-centered programming for web services," *ACM Trans. Program. Lang. Syst.*, vol. 34, no. 2, 2012.

[2] M. Carbone and F. Montesi, "Deadlock-Freedom-by-Design: Multiparty Asynchronous Global Programming," in *POPL*, pp. 263–274, ACM, 2013.

[3] I. Lanese, C. Guidi, F. Montesi, and G. Zavattaro, "Bridging the Gap between Interaction- and Process-Oriented Choreographies," in *SEFM*, pp. 323–332, IEEE, 2008.

[4] K. Honda, A. Mukhamedov, G. Brown, T. Chen, and N. Yoshida, "Scribbling interactions with a formal foundation," in *ICDCIT*, vol. 6536 of *LNCS*, pp. 55–75, Springer, 2011.

[5] World Wide Web Consortium, *Web Services Choreography Description Language Version 1.0*, 2005. http://www.w3.org/TR/ws-cdl-10/

[6] F. Montesi, "Kickstarting choreographic programming," in *WS-FM:FASOCC*, vol. 9421 of *LNCS*, pp. 3–10, Springer, 2014.

[7] S. Lu, S. Park, E. Seo, and Y. Zhou, "Learning from mistakes: a comprehensive study on real world concurrency bug characteristics," in *ASPLOS*, pp. 329–339, ACM, 2008.

[8] S. Newman, *Building Microservices*. " O'Reilly Media, Inc.", 2015.

[9] M. Dalla Preda, M. Gabbrielli, S. Giallorenzo, I. Lanese, and J. Mauro, "Dynamic choreographies - safe runtime updates of distributed applications," in *COORDINATION*, vol. 9037 of *LNCS*, pp. 67–82, Springer, 2015.

[10] M. Dalla Preda, S. Giallorenzo, I. Lanese, J. Mauro, and M. Gabbrielli, "AIOCJ: A choreographic framework for safe adaptive distributed applications," in *SLE*, vol. 8706 of *LNCS*, pp. 161–170, Springer, 2014.

[11] "AIOCJ website." http://www.cs.unibo.it/projects/jolie/aiocj.html

[12] "Jolie website." http://www.jolie-lang.org/

[13] F. Montesi, C. Guidi, and G. Zavattaro, "Composing services with JOLIE," in *Proc. of ECOWS'07*, pp. 13–22, IEEE, 2007.

[14] M. Coppo, M. Dezani-Ciancaglini, and B. Venneri, "Self-adaptive multiparty sessions," *Service Oriented Computing and Applications*, vol. 9, no. 3-4, pp. 249–268, 2015.

[15] G. Anderson and J. Rathke, "Dynamic software update for message passing programs," in *APLAS*, vol. 7705 of *LNCS*, pp. 207–222, Springer, 2012.

[16] M. Bravetti *et al.*, "Towards global and local types for adaptation," in *SEFM Workshops*, vol. 8368 of *LNCS*, pp. 3–14, Springer, 2013.

[17] "Scribble website." http://www.jboss.org/scribble

[18] H. Hüttel *et al.*, "Foundations of session types and behavioural contracts," *ACM Computing Surveys*, vol. 49, no. 1, 2016.

[19] D. Ancona *et al.*, "Behavioral types in programming languages," *Foundations and Trends in Programming Languages*, vol. 3, no. 2-3, pp. 95–230, 2016.

[20] R. Hirschfeld, P. Costanza, and O. Nierstrasz, "Context-oriented Programming," *Journal of Object Technology*, vol. 7, no. 3, pp. 125–151, 2008.

[21] R. Pawlak *et al.*, "JAC: an aspect-based distributed dynamic framework," *Software: Practice and Experience*, vol. 34, no. 12, pp. 1119–1148, 2004.

[22] Z. Yang, B. H. C. Cheng, R. E. K. Stirewalt, J. Sowell, S. M. Sadjadi, and P. K. McKinley, "An aspect-oriented approach to dynamic adaptation," in *WOSS*, pp. 85–92, ACM, 2002.

[23] D. Merkel, "Docker: Lightweight linux containers for consistent development and deployment," *Linux J.*, vol. 2014, no. 239, 2014.

8

JaDA – the Java Deadlock Analyzer

Abel Garcia and Cosimo Laneve

Department of Computer Science and Engineering, University of Bologna –
INRIA FOCUS, Mura Anteo Zamboni 7, 40127, Bologna, Italy

Abstract

JaDA is a static deadlock analyzer that targets Java bytecode. The core of
JaDA is a behavioral type system especially designed to record dependencies
between concurrent code. These behavioral types are thereafter analyzed by
means of a fixpoint algorithm that reports potential deadlocks in the original
Java code. We give a practical presentation of JaDA, highlighting the main
connections between the tool and the theory behind it and presenting some of
the features for customising the analysis. We finally assess JaDA against the
current state-of-the-art tools, including a commercial grade one.

8.1 Introduction

In concurrent languages, a *deadlock* is a circular dependency between a set of
threads, each one waiting for an event produced by another thread in the set.
In the Java programming language, deadlocks are usually *resource-related*,
namely they are caused by operations ensuring different threads the exclusive
access to a set of resources. (Java has also so-called *communication-related*
deadlocks, which are common in network based systems. These deadlocks,
which are thoroughly studied in [1,2], are out of the scope of this work.) Java
features threads by means of an ad-hoc class called Thread; this class has two
methods Thread.start() and Thread.join() for spawning and joining
threads. The consistency between threads that share objects is enforced by
synchronized blocks, a linguistic construct that may be defined either for

169

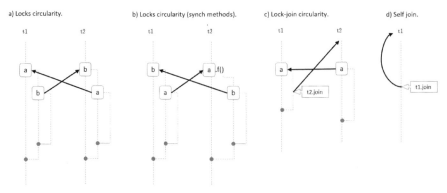

Figure 8.1   Cases of circular dependencies that may lead to deadlocks. (Lock acquisitions are represented with squares, the corresponding release is marked with a circle).

simple code blocks or for method bodies [3, Chapter 17][1]. It turns out that the dependencies defined by synchronized blocks may be circular. These problems are difficult to detect or anticipate, since they may not occur during every execution. Figure 8.1 shows (a timeline representation of) some examples of deadlocked programs. At the time of writing this chapter, the *Oracle Bug Database*[2] reports more than 40 unresolved bugs due to deadlocks, while the *Apache Issue Tracker*[3] reports around 400 unresolved deadlock bugs. Clearly, a deadlock may have catastrophic effects for the overall functionality of a software system.

In this chapter, we present an end-to-end automatic analyzer for detecting potential deadlock bugs of Java programs *at compilation time* – JaDA, the Java Deadlock Analyzer tool. JaDA addresses the compilation target of every Java application – the Java *Virtual Machine Language*, JVML, also called Java bytecode – and extracts abstract models out of it by means of an inference system. These abstract models are successively analyzed.

The decision of addressing JVML instead of Java was motivated by two reasons: Java is too complex and it has no reference semantics. On the contrary, JVML is simple – it has 198 instructions that can be sorted into 7

---

[1]There are also other mechanisms that remain out of the scope of this work, such as, the volatile variables and the higher-level synchronization API defined on package java.util.concurrent.

[2]http://bugs.java.com/

[3]https://issues.apache.org/jira

different groups of similar instructions – and has a reference semantics that is defined by the behavior of the Java *Virtual Machine* (JVM) [3, Chapter 6]. Analyzing JVML has also other relevant advantages: addressing programming languages that are compiled to the same bytecode, such as Scala [4], and the possibility to analyze proprietary software whose sources are not available.

The inference system of JaDA consists of a number of rules that analyze the effects of the instructions on the synchronization process. The types inferred from the bytecode, called *lams* [1, 2, 5, 6], are functional programs that define dependencies between threads. Then JaDA uses a variation of the algorithm defined in [1, 2] for detecting circularities in lams, and reports potential threats as output of the analysis. The tool also exhibits the executions causing deadlocks, by linking the dependencies with the chunk of source code that originated them, thus easing the analysis of false positives.

The current release of JaDA covers most of the JVML, including threads and synchronizations, constructors, arrays, exceptions, static members, interfaces, inheritance, recursive data types. Few synchronization-related features are not covered by the current release, such as wait-notify-notifyAll operations, dynamic class loading and reflection.

The rest of the chapter is organized as follows. Section 8.2 presents a motivating example of a recursive Java program that creates a (statically) unbounded number of threads. This is one of the main achievements so far and the theory overviewed in Section 8.3 will be explained by means of it. Section 8.4 describes the tool in some detail, highlighting implementation issues. Section 8.5 analyses the current limitations of JaDA and Section 8.6 reports an assessment of JaDA with respect to state-of-the-art tools for Java deadlock analysis. Finally we conclude in Section 8.7.

8.2 Example

Figure 8.2 reports the Java class Network and part of its JVML. The main method creates a network of n threads by invoking buildNetwork – say $t_1, \cdots, t_n$ – that are all potentially running in parallel with the caller – say t_0. Every two adjacent threads share an object, which is also created by buildNetwork.

The buildNetwork method will produce a deadlock depending on its actual arguments: it is deadlock-free when it is invoked with two different objects, otherwise it may deadlock (if also n > 0). Therefore, in the

```
class Network{

  public void main(int n){
    Object x = new Object();
    Object y = new Object();

    // deadlock
    buildNetwork(n, x, x);

    // no deadlock
    //buildNetwork(n, x, y);
  }

  public void buildNetwork(int n,
              Object x, Object y){
    if (n==0) {
      takeLocks(x,y) ;
    } else {
      final Object z = new Object() ;

      //anonymous Thread child class
      Thread thr = new Thread(){
        public void run(){
          takeLocks(x,z) ;
      }} ;
      thr.start();
      this.buildNetwork(n-1,z,y) ;
    }
  }

  public void takeLocks(Object x,
                Object y){
    synchronized (x) {
      synchronized (y) { }
    }
  }
}
```

```
public void buildNetwork(int n, Object x, Object y)
 0   iload_1            //n
 1   ifne 13
 4   aload_0            //this
 5   aload_2            //x
 6   aload_3            //y
 7   invokevirtual 24 //takeLocks(x, y):void
10   goto 50
13   new 3
16   dup
17   invokespecial 8  //Object()
20   astore 4          //z
22   new 26
25   dup
26   aload_0            //this
27   aload_2            //x
28   aload 4            //z
30   invokespecial 28 //Network$1(this, x, z)
33   astore 5          //thr
35   aload 5           //thr
37   invokevirtual 31 //start():void
40   aload_0            //this
41   iload_1            //n
42   iconst_1
43   isub
44   aload 4            //z
46   aload_3            //y
47   invokevirtual 36 //buildNetwork(n-1, z, y):void
50   return

public void takeLocks(Object x, Object y)
 0   aload_1;           //x
 1   dup;
 2   astore_3;
 3   monitorenter;      //acquires x
 4   aload_2;           //y
 5   dup;
 6   monitorenter;      //acquires y
 7   monitorexit;       //releases y
 8   aload_3;
 9   monitorexit;       //releases x
16   return;
```

Figure 8.2   Java Network program and corresponding bytecode of methods buildNetwork and takeLocks. Comments in the bytecode give information of the objects used and/or methods invoked in each instruction.

case of Figure 8.2, the program is deadlocked, while it is deadlock free if we comment the instruction buildNetwork(n,x,x) and uncomment buildNetwork(n,x,y).

The problematic issue of Network is that the number of threads is not known statically – n is an argument of main. This is displaycd in the bytecode of buildNetwork in Figure 8.2 by the instructions at addresses 30 and 37 that respectively created a new thread and start it, and by the recursive invocation at instruction 47.

8.3 Overview of JaDA's Theory

JaDA's theory relies on two main techniques: (*i*) an inference type system for extracting abstract models out of JVML instructions, and (*ii*) a fixpoint algorithm for analyzing the models. We overview the two techniques in the following subsections; in the last subsection we discuss the JaDA behavioral types for the buildNetwork example.

8.3.1 The Abstract Behavior of the Network Class

Figure 8.3 details the output of JaDA for the Network class in Figure 8.2. The types have been simplified for readability: the actual JaDA types are more complex and verbose. Some comments (in gray) explain the side effects of invocations, other comments (in yellow) correspond to the lines that are commented in Figure 8.2. The behavior of main begins by calling the constructor of the class Object. Notice that, after such invocation, the structure of x and y is known. Then the type reports the invocation to buildNetwork.

The behavior of takeLocks is the parallel composition of two dependencies corresponding to the acquisition of the locks of x and y. Every dependency is formed by the last held lock and the current element. Notice that every method receives an extra argument corresponding to the last acquired lock at the moment of the invocation, in this case that argument is u.

The behavior of buildNetwork has five states: (*i*) the invocation to takeLocks, (*ii*) the creation and initialization of the object z, (*iii*) the creation and initialization of the thread thr, (*iv*) the spawn of thr, (*v*) and the recursive invocation (in parallel with the spawn of thr). The buildNetwork method also reports one spawned thread as side effect. This may appear contradictory (because buildNetwork spawns n threads). However, in this case JaDA is able to detect that thr is the only thread (from those created) that may be relevant (for the deadlock analysis) in an outer scope. This deduction is done by considering the objects in the record structure of thr.

The constructors of Object and Thread have an empty behavior. On the contrary, the constructor of the class Network$1 is more complex (Network$1 is the name the JVM automatically gives to the anonymous

```
main(this | t,u):T{thr} =
    Object.init(x | t,u) + Object.init(y | t,u) +      //structure of x:x[] and y:y[]
    //deadlock
    buildNetwork(this,_,x,x | t,u)
    //no-deadlock
    //buildNetwork(this,_,x,y | t,u)                    //creates unsync thread: thr

takeLocks(this,x,y | t,u) = t:(u,x) & t:(x,y)

buildNetwork(this,_,x,y | t,u) :T{thr}=
    takeLocks(this,x,y | t,u) +
    Object.init(z | t,u) +                             //z:z[]
    Network$1.init(thr, this, x, z | t, z)+
    //thr:thr[this$0:this[], val$x:x[, val$z: z[]]
    Network$1.run(thr | thr,u1) +
    Network$1.run(thr | thr,u1) & buildNetwork(this,_,z,y | t,u)

Object.init(this | t, u):this[] = 0 //no side effects

Thread.init(this | t, u):this[] = 0 //no side effects

Network$1.init(this, x1, x2, x3 | t, u):this[this$0:x1, val$x:x2, val$z:x3] =
    Thread.init(this | t, u)

Network$1.run(this[this$0:x1,val$x:x2,val$z:x3]|t,u) = takeLocks(x1,x2,x3|t,u)
```

Figure 8.3   BuildNetwork's lams.

Thread child class[4] instantiated inside the method buildNetwork of the class Network). Being defined as an inner class, Network$1 has access to the local variables in the scope in which it has been created, namely the variables x, z and the this reference to the container instance. The JVM addresses this by passing these variables to the constructor of the class and assigning them to internal fields, in this case named valx, valz and this$0. Notice that the behavior of the constructor keeps track of two important things: the invocation to the constructor of the parent class Thread.init and the changes in the carrier object which goes from this to this[this$0:x1, val$x:x2, val$z:x3] where x_i are the formal arguments.

Finally, the behavior of the run method from the class Network$1 contains only the invocation to the takeLocks method. Notice that run method assumes a certain structure from the carrier object.

8.3.2 Behavioral Type Inference

The typing process is done bottom-up, in a compositional way. That is, a type is derived for every JVML instruction; the type of each method is the composition of the types of the instructions it contains. Similarly, the type of a program is the set of type of the methods therein. JaDA types are not standard types, such as integers, booleans, etc. They are models of the abstract behavior of a program, called *behavioral types*, that hold information about concurrency and synchronizations of every execution path.

In particular, the types of instructions retain two key pieces of information in JaDA:

- the *dependencies*, written t:(a,b), to be read as "thread t acquires the lock of object b while it is holding the lock of a", and

- the *method invocations*, written C.m(*args*| t, a), which means that the method m of class C is invoked with arguments *args* (that include the carrier object) in the thread t and while holding the lock of a.

In order to verify the consistency of parallel threads, behavioral types also take into account the (reading and writing) *effects* on objects. The types of the instructions can be composed either sequentially with the + operation, or in parallel with the & operation.

In JaDA, the behaviors of methods are the sequential composition of instructions' types in method's body plus the sum of their effects. The effects

---

[4]https://docs.oracle.com/javase/tutorial/java/javaOO/anonymousclasses.html

also include both threads spawns and thread joins in the method body. It is worth to remark that thread creations and synchronizations in JVML are defined by method invocations of the class Thread; therefore they are typed as method invocations with an ad-hoc management of spawns and joins.

The flow of the inference of behavioral types is described by the chart in Figure 8.4. The algorithm starts with an (empty) *Behavioral Class Table* (BCT), a structure where a behavioral description is associated to every method, and a sorted set of pending methods which initially contains all the methods of the program. The algorithm takes the first element of the set and types it (see below). The resulting effects are compared to the previous state of the BCT: if a change is found, the method is updated and every caller (every method depending on the current one) is added to the pending methods list. The algorithm terminates when the BCT reaches an stable state.

A similar technique is used to type the body of each method. The process is described in the chart shown in Figure 8.5. In this case, the inference process inside a method starts with a queue of pending instructions, which initially contains the first instruction. Each instruction is typed and the instruction *state* is updated (we have defined a set of typing rules in [7]). If the instruction type has been updated then the subsequent instruction(s) need to be typed (again). Notice that there may be several subsequent instructions, for example when the current instruction is a conditional. The state of an instruction contains an abstraction of the operand stack, the local variables, the local heap, the threads created upto that instruction and the chain of acquired locks

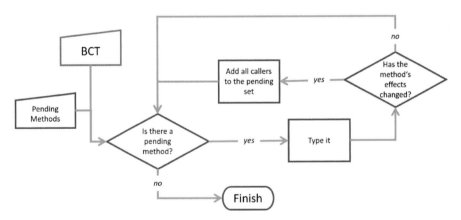

Figure 8.4   Type inference of methods' behaviors in JaDA.

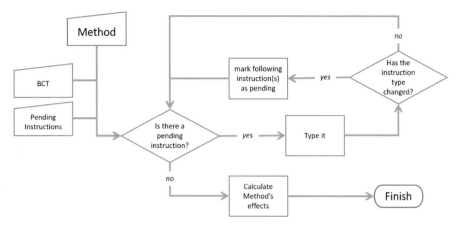

Figure 8.5 Type inference of method's bodies in JaDA.

(this information allows us to define the *lam* [2] of an instruction). Once no state is updated anymore, the type of the corresponding method is computed accordingly.

8.3.3 Analysis of Behavioral Types

The analysis of the inferred types is also performed iteratively. The overall approach is described by the chart in Figure 8.6. The initial step computes the *abstract state* of every method. This state is a sequence of parallel compositions of dependency pairs – function invocations in lams are deleted. The algorithm proceeds instruction by instruction, by *expanding* and *cleaning* its current state. The expansion process unfolds every invocation, the cleaning process removes pairs containing a fresh name (names not belonging to the method arguments or effects). Removing such pairs is crucial for termination because it allows us to keep the set of dependencies finite. In particular, the cleaning is performed by computing the transitive closure of the dependency pairs (this way we recover dependencies that are not direct and involve fresh names) and keeping only those whose elements are not fresh. In case we find a circular dependency formed only by fresh names then a special dependency pair is inserted (and this will ensure the presence of a deadlock). The full details of this algorithm are described in [1, 2].

Once all abstract states have been computed, the algorithm returns the circularities present in the `main` method.

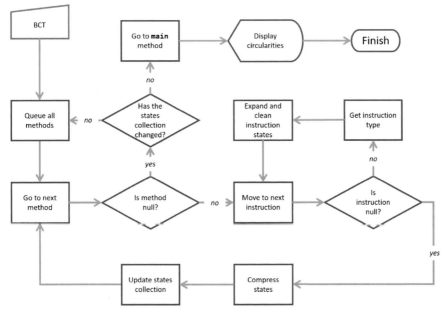

Figure 8.6   JaDA Analysis of behavioral types.

As an example, we apply the algorithm of Figure 8.6 to the `Network` behaviour in Figure 8.3. For simplicity we have excluded the methods with empty behaviour as well as their invocations.

Initially, an empty state is associated to every method. Using this model, we perform the first iteration and we get (we denote a set with [e_1, e_2, $\cdots$] where elements e_i are dependencies $t:(x,y)$; sets of sets are denoted by [[e_1, e_2, $\cdots$], $\cdots$]):

```
main(this | t,u) thr = [
        []    // no states resulting from buildNetwork(this,_,x,y | t,u)
] = []        // expanding and cleaning result empty

takeLocks(this,x,y | t,u) = [
        [ t:(u,x) & t:(x,y) ]
] = [
        [ t:(u,x) & t:(x,y) & t:(u,y)] // t:(u,y) is added by transitivity
]

buildNetwork(this,_,x,y | t,u)  thr = [
        [ t:(u,x) & t:(x,y) & t:(u,y)],  // invocation of takeLocks
        [],                     // invocation of Network$1.run
        [],                     // invocation of Network$1.run and buildNetwork
```

```
] = [
        [ t:(u,x) & t:(x,y) & t:(u,y) ]
]

Network$1.run(this[this$0:x1, val$x:x2, val$z:x3] | t, u) = [
        [ t:(u,x2) & t:(x2,x3) & t:(u,x3) ]        // invocation of takeLocks
] = [
        [ t:(u,x2) & t:(x2,x3) & t:(u,x3) ]
]
```

Since the states of methods is changed (all except main) we perform a second iteration, which gives:

```
main(this | t,u) thr = [
  [ t:(u,x) & t:(x,x) & t:(u,x)] // state of buildNetwork(this,_,x,y | t,u)
] =
[
  []   // the cleaning process removes dependencies that contain fresh names
       // the dependency t:(x,x) is removed because it is a reentrant lock
]

takeLocks(this,x,y | t,u) = [
  [ t:(u,x) & t:(x,y) ]
] = [
  [ t:(u,x) & t:(x,y) & t:(x,y)]   // this is the fixpoint for takeLocks
]

buildNetwork(this,_,x,y | t,u) thr = [
  [ t:(u,x) & t:(x,y) & t:(u,y)],           // invocation of takeLocks
  [ thr:(u,x) & thr:(x,z) & thr:(u,z)],     // invocation of Network$1.run
  [[ thr:(u,x) & thr:(x,z) & thr:(u,z)] & [ t:(u,z) & t:(z,y) & t:(u,y) ]]
                             // invocation of Network$1.run and buildNetwork
] = [
  [ t:(u,x) & t:(x,y) & t:(u,y)],   // this state has not changed
  [ t1:(u,x)],
  [ thr:(u,x) & t_thr:(x,y) & t:(u,y) ] // t_thr:(x,y) is new: it is a
              // dependency between x and y involving the threads t and thr
]

Network$1.run(this[this$0:x1, val$x:x2, val$z:x3] | t, u) = [
  [ t:(u,x2) & t:(x2,x3) & t:(u,x3) ] // invocation of takeLocks(x1,x2,x3 | t,u)
] = [
  [ t:(u,x2) & t:(x2,x3) & t:(u,x3) ] // this is the fixpoint for Network$1.run
]
```

Since buildNetwork is changed, we need a third iteration. The computation of the dependencies of main gives

```
main(this | t,u) thr = [// states resulting from buildNetwork(this,_,x,x | t,u)
        [ t:(u,x) & t:(x,x) & t:(u,x) ],
        [ thr:(u,x) ],
```

```
        [ thr:(u,x) & t_thr:(x,x) & t:(u,y) ]
] =
[
        [t_thr:($,$)]
]
```

In particular, in the states of main, after the transitive closure, contain t_thr: (x,x), which is a circular dependency involving two threads. Instead of writing the dependency in that way (using a fresh name x), we write it as t_thr: ($,$), where x is replaced by a special name $. It is worth to notice that t_thr: ($,$) gives two informations: (*i*) the deadlock is created by threads t and thr, (*ii*) the object name is $, which indicates that the deadlock is produced regardless of the arguments of the invocation. Since t_thr: ($,$) denotes a circularity, the algorithm might stop. Actually, we decided not to stop JaDA at this point, we let it continue in order to collect every circularity.

JaDA output for the Network program is reported in Figure 8.7. In this case, JaDA has been set to analyze only the Network class (see *analysis-extent* in Section 8.4.4). Therefore, it warns about non-analyzed dependencies: the constructors from classes Thread and Object (whose types are considered empty – the actual type of these methods is nevertheless empty). JaDA reports 1 deadlock after the analysis, and outputs its trace. In this

```
[INFO] Analysis started at: 2016/12/08 09:37:55
[INFO]    Creating classpath for the analysis
[INFO]    Checking for classes located under: C:\JaDA\Git\Java-Deadlocks\[PaperExamples]\Network\bin
[INFO]    Analysis will run on the following target methods:
[INFO]        Network.main_(I)V
[INFO]    Calculating method behaviors
[WARNING] Method dependency not analyzed: java/lang/Thread.<init>
[WARNING] Method dependency not analyzed: java/lang/Object.<init>
[INFO]    Calculating method states
[INFO]    Method states calculation completed. Fixpoint process took 4 iterations.
[INFO]
[INFO]    NUMBER OF DEADLOCKS/LIVELOCKS: 1
[INFO]
[INFO]    1)
[INFO]    Deadlock found in method: main

        Thread 204 ($MAIN$) --this is the main thread-- tries to acquire:
main -> buildNetwork:30 -> buildNetwork:21 -> takeLocks:12 -> x (346)
main -> buildNetwork:30 -> buildNetwork:21 -> takeLocks:12 -> y (211)

        Thread 229 (thr) started at Network:15 (buildNetwork) tries to acquire:
main -> buildNetwork:30 -> run:15 -> takeLocks:17 -> x (211)
main -> buildNetwork:30 -> run:15 -> takeLocks:17 -> y (346)

[INFO]
[INFO]
[INFO] Analysis ended at: 2016/12/08 09:37:56. Analyis took 444 ms
```

Figure 8.7   JaDA analysis output for the Network program.

case there are two threads involved in the deadlock: those with id 204 (the one running `main`) and 229. The deadlock is caused by two `monitorenter` instructions on objects 346 and 211, taken in different order by the two threads. The tool outputs the computational traces ending with the two `monitorenter` instructions; the numbers in the traces represent the lines in the source[5].

8.4 The JaDA Tool

In this section we describe the main features of the JaDA tool, as well as, some key implementation details.

8.4.1 Prerequisites

JaDA has been designed to run on bytecode generated by the Java compiler[6] and it assumes that the bytecode has been already checked by the Java Bytecode Verifier (therefore it does not contain either syntactic or semantic errors). JaDA also requires that every dependency is matched by a corresponding bytecode. Although the bytecode is not executed, JaDA computes every necessary information to solve key issues for the analysis, such as the informations about inheritance. The loading of the existing types is done dynamically in a sand-boxed class loader[7] to avoid security risks. The full set of dependencies can be specified in JaDA through a *classpath*-like configuration (see property `class-path` in Section 8.4.4). Finally, JaDA also assumes that the code targeted by the analysis fits with the current limitations of the tool (see Section 8.5).

8.4.2 The Architecture

The JaDA analysis starts by parsing of the bytecode of a program and its dependencies. This is a cumbersome task because of the length and verbosity of the JVML syntax. JaDA relies on the ASM framework [8] for the bytecode extraction and manipulation. (Other third party tools have

---

[5]The line numbers in the output may not accurately match the example in Figure 8.2, because the latter has been slightly reduced for presentation purposes.

[6]We have tested JaDA against the 1.6, 1.7 and 1.8 versions of the Java compiler, and against the 1.8 version of the Eclipse Java Compiler (ECJ).

[7]https://docs.oracle.com/javase/7/docs/api/java/lang/ClassLoader.html

been also designed for manipulating and analyzing the bytecode: the page
https://java-source.net/open-source/bytecode-libraries contains a list of existing tools for this purpose. ASM provides a wide set of
tools for interacting with the bytecode, including code generation and code
analysis. It is also light-weight, open source, very well-documented and up
todate with the latests versions of Java.

Figure 8.8 shows part of the JaDA architecture. In the figure, nodes
are classes while arrows denote inheritance relationships. In the center of
the image, there are the classes of the ASM framework; the other classes
implement the technique so far described.

Values. A basic element of the architecture are the Value objects. JaDA uses
two types of values: RecordTree store the methods' signature in the BCT,
while RecordPtr store the state of local variables and the operand stack.
Updating the Value elements amounts to upgrade every other element of the
JaDA architecture. In the following paragraphs we discuss their functionality.

Frames. The JDAFrame class extends the ASM Frame by defining two important methods: execute and merge. The method execute implements the
typing rules used by JaDA. It relies on an abstract interpreter that executes
symbolically the current instruction with the given stack and local variables
state. The method merge is invoked when the analysis process reiterates
over an already typed frame. This method implements the logics of the *has
changed* condition in the type inference of method bodies, see Section 8.3.2.

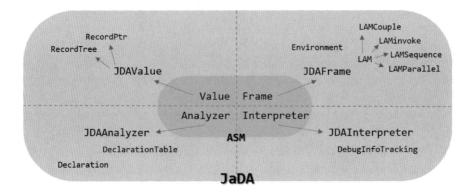

Figure 8.8   JaDA architecture.

```
                          20 ...
                          21 aload_1;        //a
                          22 dup;
                          23 astore 4;
                          25 monitorenter;
for(int i = 0; i < 10; i++){   26 aload_2;        //b
  synchronized (a) {      27 dup;
    synchronized (b) {    28 monitorenter;
    }                     29 monitorexit;
  }                       30 aload 4;
}                         32 monitorexit;
                          40 iinc 3 1;       //i++
                          43 iload_3;        //i
                          44 bipush 10;
                          46 if_icmplt -25;  //LOOP condition
                          48 ...;
```

Figure 8.9 Java while loop with nested synchronizations and the corresponding bytecode.

The decision on whether the subsequent frames must be checked again is taken upon the result of this method. To illustrate this consider the example from Figure 8.9.

When the typing process arrives each instruction for the first time its current Frame changes from the empty frame to the frame containing information about the instruction. Namely this first frame will contain the local variable status, the invocations and the existing locks and threads at each instruction. This changes enforces every frame to calculate its continuation at least one time. The following sequence shows the frames calculation for this chunk of code (only the relevant instructions are included):

```
...
Fr.21:{CurrentThread: main, Locks:{}}
...
Fr.25:{CurrentThread: main, Locks:{a}}
...
Fr.28:{CurrentThread: main, Locks:{a,b}}
Fr.29:{CurrentThread: main, Locks:{a}}
Fr.32:{CurrentThread: main, Locks:{}}
...
Fr.46:{CurrentThread: main, Locks:{}}
Fr.21:{CurrentThread: main, Locks:{}}
Fr.48:{CurrentThread: main, Locks:{}}
...
```

Notice that after calculating the frame 46, there are two possible continuations: 21 and 48. The second time Fr.21 is calculated it produces the same known result, therefore its continuation (Fr.22) is not calculated again. The calculation process continues then sequentially with Fr.48.

Interpreter. The JVM is a *stack machine*, every operation pops a certain number of elements off the stack and pushes on its result. The JDAInterpreter class extends the ASM Interpreter in order to comply with the values representations in JaDA. In particular, JDAInterpreter implements an important feature of our tool, namely the output of the traces potentially causing deadlocks. In fact, it returns the variable names of the objects involved, the stack trace chain and the related line numbers in the original Java code [8].

Analyzer. The ASM default analyzer supports very basic data-flow analysis limited to the scope of a single method. Similarly, JaDA analysis of a method does not go beyond its scope. However JDAAnalyzer extracts the necessary information – the type – that supports the compositional analysis. This part is implemented by the algorithm described in Figure 8.5, which is the building block of our tool. Consequently, JDAAnalyzer analyses the whole program by computing the final state of the BCT according to the algorithm in Figure 8.4. The final step of JDAAnalyzer is the computation of the models of the methods in the BCT as described in the algorithm in Figure 8.6.

8.4.3 The Current JVML Coverage

The theory of JaDA has been studied in [7] where we have defined the typing rules for a number of complex features of JVML, in particular, threads, synchronizations, static constructors, recursive data structures, inheritance and polymorphism, reflection and native methods. For a subset of this language – those featuring threads and synchronizations – we also delivered a correctness proof [9]. In this section we briefly overview our solutions for the main features of JVML that are covered in the current release of JaDA.

Static constructors. Static constructors are problematic because they are not explicitly invoked by the JVM. In fact, those are invoked on-the-fly by the JVM when the first (static) access to the containing class is performed. That is, the code of a static constructor can potentially precede any operation involving a static member of its class. In order to deal with this issue in a sound way, we model every static operation as a non-deterministic choice between the

---

[8]This is possible only when the bytecode has been compiled including debugging information.

type of the operation *per-se* and the typing of the static constructor of the class followed by the original operation. As one can imagine this makes the analysis computationally complex because the number of possible behaviors exponentially increases. The alternative (and the default choice) in JaDA is to assume that static constructors are all executed *before* the main method (see the static-constructor option in Section 8.4.4). This is a safe choice provided that concurrent operations do not occur within static constructors (which is often the case).

Recursive data structures. As discussed in Section 8.3.2, the analysis of JaDA relies on several iterative processes. The termination of these iterations strongly relies on the constraint that the number of object names is always finite. To ensure this finiteness constraint, the recursive objects are abstracted during the inference process. In particular, the inference replaces the field values whose class is already present in the object structure by a generic representative value. These representative values are treated in an ad-hoc way during the analysis of circular dependencies. Namely, they are considered equal to any other object of the same class (that is their identity is not guaranteed). Our assessments indicate that this over-approximation does not jeopardise JaDA's precision when the elements of the recursive structure are pairwise different and threads act in a uniform way on them. On the contrary, the tool may return a number of false positives that is proportional to the dimension of the structure.

Arrays. Since JaDA does not process numerical expressions, it considers array[2] equal to array[3]. Therefore, JaDA manages arrays in a similar way it does for recursive data types. Every element in the array is represented by a unique object and, as for recursive data structures, this may be the cause of over-approximations. For example, JaDA returns a false positive when two different threads in parallel perform a lock operation on different objects of the array.

Inheritance and polymorphism. Inheritance and, in particular, polymorphism are sources of non-determinism. In fact, since it is not possible to resolve the runtime type of an object at static time, we cannot determine in a precise way the instance method being invoked over it. To deal with this issue in a sound way, JaDA substitutes every invocation with the non-deterministic choice among the method implementations in the type hierarchy of the carrier. Enhancing this process to increase the precision of the analysis is currently an ongoing work. In the current release, whenever it is possible to derive the runtime type, we drop the wrong invocations.

Reflection, native methods, alternative concurrency models. In Java, like in many modern programming languages, there is some support for meta-programming, namely the capacity of a program to modify itself during its execution. Java also admits (native) methods and concurrency models that have no bytecode implementation. These methods are treated in an ad-hoc manner by the JVM. In all these cases, since there is not an explicit bytecode implementation, there is no evidence of what will happen at static time. Because of this reason, JaDA by default assumes a void behavior in these situations. Although, users can manually provide the behavior descriptions for methods involving such operations (see custom-types option in Section 8.4.4). This is particularly useful in the case of native methods (which are implemented in C), where users provide a more accurate behavior by analyzing its actual implementation.

8.4.4 Tool Configuration

In order to provide some flexibility, JaDA supports a set of settings to customize the analysis.

<target>: this setting specifies the target file or folder to analyze. It is mandatory. The type of files admitted are: Javaclass files (".class"), Java jar files (".jar") and compressed zip files (".zip"). In the case of folders, the content of the folder is analyzed recursively.

verbose[=<value>]: the value ranges from 1 to 5, the default and more verbose value is 5.

class-path <classpath>: Standard Java classpath description. If the target contains dependencies other than those in the standard library, they must be specified via this option.

target-method <methodName>: fully qualified target method (should be a void method without arguments). It compels JaDA to analyse the specified method. If this option is not set, the analysis chooses the first main method found.

analysis-extent[=<value>]: Indicates the extent of the analysis. Possible values are full: analyzes every dependency including the system and classpath-included libraries; classpath: analyzes every library in the classpath (this is the default value); custom: analyzes the classes specified through the property additional-targets; and self: does not analyze any class but the specified target.

additional-targets <classes>: if analysis-extent is set to custom this property must contain a comma separated list of the fully qualified names of a subset of classes in the *classpath* to include in the analysis. Such a feature is useful for avoiding typing known libraries.

custom-types <file>: a setting file to specify predefined behavioral types.

static-constructors[=<value>]: indicates when the static constructors should be processed, the possibilities are before-all and non-deterministically. The default option is before-all.

8.4.5 Deliverables

JaDA is available in three forms: a demo website [10], a command line tool (see Figure 8.7) and an Eclipse plug-in. All of them share the same core: a prototype implementation of the technique discussed in [7]. At the moment of writing this chapter, the demo website only allows to analyze single-file programs and to use a subset of the options previously described. The command line tool and the Eclipse plug-in are available through direct requests. The Eclipse plug-in output also displays the execution graph causing the deadlock with links to the source code that originates it (see Figure 8.10).

8.5 Current Limitations

The current version of JaDA does not cover a coordination mechanism between thread that is quite usual in Java: the wait-notify-notifyAll operations. There are also other less critical limitations, such as the analysis of native code and reflection operations. However, these features can be covered by manually specifying the behavior of the corresponding methods (see property custom-types in Section 8.4.4).

The methods wait-notify-notifyAll are public and final of the class Object; therefore they are inherited by all classes and cannot be modified.

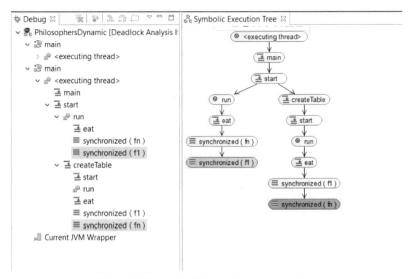

Figure 8.10   JaDA Eclipse plug-in screenshot.

The invocations to `wait`, `notify` and `notifyAll` succeed for threads that already hold the lock of the object a on the stack. In this case, the `wait` instruction moves its own thread to the *wait set* of a and the object is relinquished by performing as many unlock operations as the integer stored in the lock field of a. The instructions `notify` and `notifyAll` respectively wake up one thread and all the threads in the wait set of a. The woken-up threads are re-enabled for thread scheduling, which means competing for acquiring the lock of a again. The winner will lock a as many times it did on a before the `wait`-operation.

Below we briefly describe the solution we are currently investigating for extending JaDA to cover `wait-notify-notifyAll`.

We use two new type of dependency pairs: $t_1 : (a, a^n)$, which means "*the thread t_1 sends a notification on a while holding its lock*, and $t_2 : (a, a^w)$, which means "*the thread t_2 awaits a notification on a while holding its lock*". These two pairs are respectively produced by `notify` and `wait` methods. The problem is that, even if the abstract model retains a term $t_1 : (a, a^n) \& t_2 : (a, a^w)$ expressing that the `wait` and `notify` occur in parallel threads (*notification-wait* matching couple), we cannot conclude that the program is deadlock-free. This because the above term does not convey any information about what operation *has been performed before*. In fact, a wrong ordering might cause the thread t_2 to wait indefinitely. To overcome this problem, we extend JaDA with an additional analysis that detects the *wait* pairs that can potentially remain unsatisfied. This solution is extensively discussed in [7].

8.6 Related Tools and Assessment

JaDA has been assessed with respect to a number of state-of-the-art tools. In particular, in Table 8.1, the tools have been classified according to the type of analysis they perform (see [7] for a discussion about analysis techniques for deadlock detection). We have chosen Chord for static analysis [11], Sherlock for dynamic analysis [12], and GoodLock for hybrid analysis [13]. We have also considered a commercial tool, ThreadSafe [9] [14].

We have analyzed a number of programs that exhibit a variety of sharing patterns. The source of all benchmarks in Table 8.1 is available either at [11, 12] or in the JaDA-deadlocks repository[10].

Since the current release of JaDA does not completely cover JVM, in order to gain preliminary experience, we modified the Java libraries and the multithreaded server programs of RayTracer, MolDyn and MonteCarlo (labelled with "(*)" in the Table 8.1) and implemented them in our system.

Table 8.1 Comparison with different deadlock detection tools. The inner cells show the number of deadlocks detected by each tool. The output labelled "(*)" are related to modified versions of the original programs: see the text

	Static		Hybrid	Dynamic	Commercial
Benchmarks	JaDA	Chord	GoodLock	Sherlock	ThreadSafe
Sor	1	1	7	1	4
RayTracer (*)	0	0	8	2	0
MolDyn (*)	0	0	6	1	0
MonteCarlo (*)	0	0	23	2	0
BuildNetwork	3	0			0
Philosophers2	1	0			1
PhilosophersN	3	0			0
StaticFields	1	1			1
ThreadArrays	1	1			1
ThreadArraysWJoins	1	1			0
ScalaSimpleDeadlock	1				
ScalaPhilosophersN	3				

---

[9]http://www.contemplateltd.com/threadsafe
[10]https://github.com/abelunibo/Java-Deadlocks

This required little programming overhead; in particular, we removed volatile variables, avoided the use of Runnable interfaces for creating threads, and reduced the invocations of native methods involved in I/O operations. Out of the four chosen tools, we were able to install and effectively test only two of them: Chord and ThreadSafe; the results corresponding to GoodLock and Sherlock come from [12] because we were not able to get the sources of the tools and run our new programs (*). We also had problems in testing Chord with some of the examples in the benchmarks, perhaps due to some misconfigurations, that we were not able to solve because Chord has been discontinued.

The first block of programs belongs to a well-known group used as benchmarks for several Java analysis tools. In its current state JaDA only detects 1 deadlock in all of the four analyzed programs from this group. It gives responses that are similar to ThreadSafe and Chord (ThreadSafe appears a bit more imprecise on Sor). The programs in the second block corresponds to examples designed to test our tool against complex deadlock scenarios like the Network program. We notice that both Chord and ThreadSafe fail to detect those kinds of deadlocks. The third group reports the analysis of two examples of Scala programs [4]. These programs have been compiled with the Scala compiler 2.11 whose target is Java bytecode. We remark that, to the best of our knowledge, at the moment of writing this chapter, there is no static deadlock analysis tools for such language (for this reason the entries corresponding to the other tools are empty).

We think that the results in Table 8.1 are encouraging and we hope to deliver more convincing ones as soon as JaDA overcomes its current limitations.

8.7 Conclusions

JaDA is a static deadlock analysis tool that targets JVML. Therefore it supports the analysis of every compiled Java program, as well as, every programs written in languages that are also compiled in JVML, like Scala. The technique underlying JaDA uses a behavioral type system that abstract the main features of the programs with respect to the concurrent operations.

JaDA is designed to run in an automatic fashion, meaning that the inference of the program type and the subsequent analysis could be done unassisted. Nevertheless, user intervention is possible and may enhance the precision of the analysis, for example in presence of native methods.

Even though the tool is still under development, we have been able to asses it by analyzing a set of Java and Scala programs. This contribution also reports a comparison between JaDA's results and those of existing deadlock analysis tools, amongst which is a commercial grade one. The results obtained so far are very promising and we expect to gain more precision as the development continues.

References

[1] N. Kobayashi and C. Laneve, "Deadlock analysis of unbounded process networks," *Information and Computation*, vol. 252, pp. 48–70, 2017.

[2] E. Giachino, N. Kobayashi, and C. Laneve, "Deadlock analysis of unbounded process networks," in *Proceedings of 25th International Conference on Concurrency Theory CONCUR 2014*, vol. 8704 of *Lecture Notes in Computer Science*, pp. 63–77, Springer, 2014.

[3] J. Gosling, W. N. Joy, and G. L. S. Jr., *The Java Language Specification*. Addison-Wesley, 1996.

[4] M. Odersky and al., "An Overview of the Scala Programming Language," Tech. Rep. IC/2004/64, EPFL, Lausanne, Switzerland, 2004.

[5] E. Giachino and C. Laneve, "Deadlock detection in linear recursive programs," in *14th Int. School on Formal Methods for the Design of Computer, Communication, and Software Systems, SFM 2014*, vol. 8483 of *Lecture Notes in Computer Science*, pp. 26–64, Springer, 2014.

[6] E. Giachino, C. Laneve, and M. Lienhardt, "A framework for deadlock detection in core ABS," *Software and Systems Modeling*, vol. 15, no. 4, pp. 1013–1048, 2016.

[7] A. Garcia and C. Laneve, "Deadlock detection of Java Bytecode." A preliminary version is available at http://jada.cs.unibo.it/data/Doc/jada-draft-lncs.pdf, 2016.

[8] E. Bruneton, "Asm 4.0 a java bytecode engineering library." http://download.forge.objectweb.org/asm/asm4-guide.pdf. Last accessed: 2016-12-03.

[9] A. Garcia, "Static analysis of concurrent programs based on behavioral type systems." Available at http://jada.cs.unibo.it/data/Doc/Abel-Garcia-PhD-Thesis-draft.pdf, 2017.

[10] A. Garcia and C. Laneve, "JaDA – the Java Deadlock Analyzer." Available at http://jada.cs.unibo.it, 2016.

[11] M. Naik, C. Park, K. Sen, and D. Gay, "Effective static deadlock detection," in *31st International Conference on Software Engineering (ICSE 2009)*, pp. 386–396, ACM, 2009.

[12] M. Eslamimehr and J. Palsberg, "Sherlock: scalable deadlock detection for concurrent programs," in *Proceedings of the 22nd International Symposium on Foundations of Software Engineering (FSE-22)*, pp. 353–365, ACM, 2014.

[13] S. Bensalem and K. Havelund, "Dynamic deadlock analysis of multi-threaded programs," in *in Hardware and Software Verification and Testing*, vol. 3875 of *Lecture Notes in Computer Science*, pp. 208–223, Springer, 2005.

[14] R. Atkey and D. Sannella, "Threadsafe: Static analysis for java concurrency," *ECEASST*, vol. 72, 2015.

9

Type-Based Analysis
of Linear Communications

Luca Padovani

Dipartimento di Informatica, Università di Torino, Italy

Abstract

This chapter presents a tool called Hypha for the type-based analysis of processes that communicate on linear channels. We describe the specification language used to model the systems under analysis (Section 9.1) followed by the typing rules on which the tool is based in order to verify two properties of systems, *deadlock freedom* and *lock freedom* (Section 9.2). In the final part of the chaper we illustrate the expressiveness and the limitations of the tool discussing a number of examples inspired by representative communication patterns using in parallel computing (Section 9.3) and then discuss closely related work (Section 9.4). The tool can be downloaded from the author's home page, the type system has been described by Padovani [18] and the corresponding reconstruction algorithms by Padovani *et al.* [19, 20].

9.1 Language

The Hypha specification language is a mildly sugared variant of the linear π-calculus [16] whose grammar is shown in Table 9.1. It makes use of booleans, integers, an infinite set $\mathcal{X}$ of names, and comprises *expressions* and *processes*. The syntax shown here is somewhat simplified and tailored to the modeling of the examples discussed in this chapter. Hypha supports other forms that may be useful in the description of protocols with branching points and provide convenient syntactic sugar on top of those given in Table 9.1. The Hypha specification language is appropriate for modeling concurrent processes that exchange messages on private (or *session*) channels [10].

Table 9.1   Syntax of Hypha input language (partial)

Notation				
	b	$\in$	$\{\texttt{true}, \texttt{false}\}$	Booleans
	h, k, m, n	$\in$	$\mathbb{Z}$	Integers
	x, y, z, u, v	$\in$	$\mathbb{X}$	Names
Expression	e	$::=$	b	Boolean constant
		$\mid$	n	Integer constant
		$\mid$	u	Name
		$\mid$	$\cdots$	
Process	P, Q	$::=$	$\{\ \}$	Idle process
		$\mid$	$u!(e_1, \ldots, e_n)$	Output
		$\mid$	$u?(x_1, \ldots, x_n).P$	Input
		$\mid$	$*u?(x_1, \ldots, x_n).P$	Replicated input
		$\mid$	$\texttt{new}\ u_1, \ldots, u_n\ \texttt{in}\ P$	Channel creation
		$\mid$	$\texttt{if}\ e\ \texttt{then}\ P\ \texttt{else}\ Q$	Conditional execution
		$\mid$	$P \mid Q$	Parallel composition
		$\mid$	$\{P\}$	Grouping

For simplicity, in the provided syntax expressions are limited to values and comprise booleans, integers, and names. In the examples we will also make use of a few binary operators (such as +) and relations (such as <). Processes comprise the usual terms of the π-calculus. The term $\{\ \}$ models the idle process that performs no actions. The term $u!(e_1, \ldots, e_n)$ models a process that outputs the tuple $(e_1, \ldots, e_n)$ on the channel u. We omit the parentheses when n is 1 and write, for example, $u!n$ in place of $u!(n)$. We consider two forms of input processes. The term $u?(x_1, \ldots, x_n).P$ models an *ephemeral input process* that waits for *one* message from u, which is supposed to be an n-tuple, and then executes P where x_i is replaced by the i-th component of the tuple. The term $*u?(x_1, \ldots, x_n).P$ models a *persistent input process* (also called *service*) that waits for an arbitrary number of messages. Each time a message is received, a new copy of P (with the variables x_i suitably instantiated) is spawned and the service makes itself available again for further receptions. The term $\texttt{new}\ u_1, \ldots, u_n\ \texttt{in}\ P$ models a process creating new channels $u_1, \ldots, u_n$ with scope P. As usual in the π-calculus, the scope of a channel may broaden as a result of communications (scope extrusion). The terms $\texttt{if}\ e\ \texttt{then}\ P\ \texttt{else}\ Q$ and $P \mid Q$ respectively model conditional and parallel processes P and Q. Finally, $\{P\}$ represents the same process as P and is useful to disambiguate the way processes are

```
*fibo?(n,c).
 if n ≤ 0 then c!1
 else new a,b in { fibo!(n-1,a) | fibo!(n-2,b) | a?(x).b?(y).c!(x+y) }
```

Listing 9.1   Modeling of the recursive Fibonacci function.

grouped. The notions of *free* and *bound names* of a process P, respectively denoted by $\mathsf{fn}(P)$ and $\mathsf{bn}(P)$, are as expected.

Example 1 (recursive Fibonacci function). Listing 9.1 shows the modeling of a service that computes the n-th number in the Fibonacci sequence. The service waits for invocations on channel fibo, each invocation consisting of the number n and a channel c on which the n-th Fibonacci number will be sent. The body of the service closely follows the familiar structure of the recursive definition of the Fibonacci sequence. When $n \leq 0$ the answer is immediately sent over c. When $n > 0$, two new channels a and b are created, the service invokes itself twice to compute the $(n-1)$-th and $(n-2)$-th numbers in the sequence, and then the sum of the two partial results is sent over c. ■

As usual, the operational semantics of the language is defined in terms of a *structural congruence relation*, which identifies terms that are meant to have the same semantics, and a *reduction relation* that defines the proper computation steps. We omit the formal definition of structural congruence, which is essentially the same as in the π-calculus and includes commutativity and associativity laws for parallel composition and the usual laws for shrinking and extending the scope of channels. The second one is the least relation defined by the rules in Table 9.2 and closed by structural congruence and under the following *reduction contexts*:

Reduction context $C \quad ::= \quad [\,] \quad | \quad C \mid P \quad | \quad \mathsf{new}\, u_1, \ldots, u_n\, \mathsf{in}\, C$

The fully-fledged formalization of the language also includes an evaluation relation for compound expressions [18].

To formulate the properties enforced by our typing discipline we introduce a few predicates that describe the pending communications of a process

Table 9.2   Operational semantics of processes

$u!(e_1, \ldots, e_n) \mid u?(x_1, \ldots, x_n).P$	$\longrightarrow$	$P\{e_1/x_1\} \cdots \{e_n/x_n\}$
$u!(e_1, \ldots, e_n) \mid *u?(x_1, \ldots, x_n).P$	$\longrightarrow$	$P\{e_1/x_1\} \cdots \{e_n/x_n\} \mid *u?(x_1, \ldots, x_n).P$
$\mathsf{if}\ \mathsf{true}\ \mathsf{then}\ P\ \mathsf{else}\ Q$	$\longrightarrow$	P
$\mathsf{if}\ \mathsf{false}\ \mathsf{then}\ P\ \mathsf{else}\ Q$	$\longrightarrow$	Q

P with respect to some channel a. We use the obvious extension of bound names we have introduced for processes to reduction contexts:

$$\mathsf{in}(a, P) \stackrel{\mathrm{def}}{\Longleftrightarrow} P = C[a?(x_1, \ldots, x_n).Q] \wedge a \notin \mathsf{bn}(C)$$

$$*\mathsf{in}(a, P) \stackrel{\mathrm{def}}{\Longleftrightarrow} P = C[*a?(x_1, \ldots, x_n).Q] \wedge a \notin \mathsf{bn}(C)$$

$$\mathsf{out}(a, P) \stackrel{\mathrm{def}}{\Longleftrightarrow} P = C[a!(e_1, \ldots, e_n)] \wedge a \notin \mathsf{bn}(C)$$

$$\mathsf{sync}(a, P) \stackrel{\mathrm{def}}{\Longleftrightarrow} (\mathsf{in}(a, P) \vee *\mathsf{in}(a, P)) \wedge \mathsf{out}(a, P)$$

$$\mathsf{wait}(a, P) \stackrel{\mathrm{def}}{\Longleftrightarrow} (\mathsf{in}(a, P) \vee \mathsf{out}(a, P)) \wedge \neg\mathsf{sync}(a, P)$$

In words, $\mathsf{in}(a, P)$ holds if there is a sub-process Q within P that is waiting for a message on channel a. Note that, by definition of reduction context, the input cannot be guarded by other actions. The condition $a \notin \mathsf{bn}(C)$ implies that a is not captured by a binder in C, *i.e.* it occurs free in P. The predicates $\mathsf{out}(a, P)$ and $*\mathsf{in}(a, P)$ are similar, but they regard outputs and persistent inputs, respectively. Therefore, when $\mathsf{in}(a, P)$ holds it means that there is a pending ephemeral input on a and when $\mathsf{out}(a, P)$ holds it means that there is a pending output on a. Then, $\mathsf{sync}(a, P)$ means that there are pending input/output operations on a, but a synchronization on a is *immediately* possible. On the contrary, $\mathsf{wait}(a, P)$ means that there is a pending output or a pending ephemeral input on a, but no immediate synchronization on a is possible. Note the asymmetry in the way pending inputs and outputs trigger the wait predicate. We do not interpret $*\mathsf{in}(a, P)$ as a pending input operation, meaning that we do not require a persistent input process to run infinitely often. At the same time, *any* pending output triggers the wait predicate, even when the output represents a service invocation.

We say that a process P is deadlock free if every residual of P that cannot reduce further contains no pending communications. Formally:

Definition 9.1. P is *deadlock free* if whenever $P \longrightarrow^* \mathsf{new}\, c_1, \ldots, c_n \,\mathsf{in}\, Q \nrightarrow$ we have $\neg\mathsf{wait}(a, Q)$ for every a.

We say that a process P is lock free if every residual Q of P in which there are pending communications can reduce further to a state in which such operations complete. Formally:

Definition 9.2. P is *lock free* if whenever $P \longrightarrow^* \mathsf{new}\, c_1, \ldots, c_n \,\mathsf{in}\, Q$ and $\mathsf{wait}(a, Q)$ there is R such that $Q \longrightarrow^* R$ and $\mathsf{sync}(a, R)$.

In Definitions 9.1 and 9.2, it is important to universally quantifiy over the topmost channel restrictions in a residual of P so that the notion of (dead)lock freedom for P concerns both free and bound channels of P.

It is easy to prove that lock freedom implies deadlock freedom [19]. On the other hand, there exist deadlock-free processes that are not lock free, as shown in the example below.

Example 2 (deadlock-free, locked process). The process

new c in { *forever?(x).forever!x | forever!c | c!42 }

is deadlock free but not lock free. Indeed, the process reduces forever, but no input operation is ever performed on the c channel. As a result, the pending output on c cannot be completed.   ∎

9.2 Type System

In this section we describe a type system that enforces the properties introduced in Section 9.1: well-typed processes are guaranteed to be (dead)lock free. The tool Hypha then implements a type reconstruction algorithm for this type system and finds a typing derivation for a given process, provided there is one. Note that, while the type reconstruction algorithm is complete with respect to the type system, the type system itself is not complete with respect to (dead)lock freedom: there exist (dead)lock free processes that are ill typed according to the type system. In fact, it is undecidable in general to establish whether a π-calculus process is (dead)lock free, hence the type system is necessarily conservative.

Polarities, qualifiers, and types are defined by the following grammar:

$$
\begin{array}{rrcl}
\textbf{Polarity} & p, q & \subseteq & \{?, !\} \\
\textbf{Qualifier} & q & ::= & * \mid \genfrac{}{}{0pt}{}{h}{k} \\
\textbf{Type} & t, s & ::= & \texttt{bool} \mid \texttt{int} \mid \kappa[\bar{t}]^q \mid \alpha \mid \mu\alpha.t
\end{array}
$$

Types comprise the base types `bool` and `int` of boolean and integer values, channel types $\kappa[\bar{t}]^q$, and the usual forms α and $\mu\alpha.t$ for representing recursive types. A channel type $\kappa[\bar{t}]^q$ consists of:

- A *polarity* p specifying the operations allowed on the channel: ∅ means none, $\{?\}$ means input, $\{!\}$ means output, and $\{?, !\}$ means both. We will abbreviate $\{?, !\}$ with # and $\{?\}$ and $\{!\}$ with ? and !, respectively.
- A sequence $t_1, \ldots, t_n$ of types, abbreviated as $\bar{t}$, specifying that each message exchanged over the channel is an n-tuple of values where the i-th value has type t_i.
- A *qualifier* q specifying how many times the channel can or must be used according to its polarity. The qualifier * means that the channel can

be used any number of times. A qualifier of the form $\frac{h}{k}$ means that the channel can only be used once. In this case, h and k are respectively the *level* and the *tickets* associated with the channel: channels with smaller levels must be used *before* channels with greater levels; a channel with k tickets can be sent as a message on another channel at most k times.

We require that, in a recursive type $\mu\alpha.t$, the type variable α can only occur guarded by a channel type constructor. For example, $\mu\alpha.\alpha$ is illegal while $\mu\alpha.?[\alpha]^*$ is allowed. We identify two types modulo renaming of bound type variables and if they have the same infinite unfolding, that is if they denote the same (regular) tree [4]. In particular, we have $\mu\alpha.t = t\{\mu\alpha.t/\alpha\}$.

Qualifiers distinguish *service channels* (with qualifier *) from *linear channels* (with qualifiers of the form h, k). Service channels are used for modeling persistent services, such as fibo in Listing 9.1 and forever in Example 2. Linear channels are used for modeling private communications between pairs of processes. Examples of linear channels are a and b in Listing 9.1 and c in Example 2. The fact that a linear channel can be used for one communication only is not a limitation in practice. Structured private conversations made of arbitrarily many communications can be encoded using a continuation-passing style [5, 12]. We will see several examples of this technique at work in the rest of the chapter. On the other hand, knowing that a channel is linear provides some guarantees on the fact that the channel will not be discarded without being used. This is a necessary (although not sufficient) condition for guaranteeing that communications on linear channels eventually occur.

The level of a linear channel measures the urgency with which the channel must be used: the lower the level is, the sooner the channel must be used. We extend this notion from linear channels to arbitrary types. To compute the level of a type, we define an auxiliary function $|\cdot|$ such that $|t|$ is an element of $\mathbb{Z} \cup \{\perp, \top\}$ where $\perp < n < \top$ for every $n \in \mathbb{Z}$:

$$|t| \stackrel{\text{def}}{=} \begin{cases} \perp & \text{if } t = p[s]^* \text{ and } ? \in p \\ n & \text{if } t = p[s]_m^n \text{ and } p \neq \emptyset \\ \top & \text{otherwise} \end{cases} \tag{9.1}$$

According to this definition, service channels with input capability have the lowest level $\perp$ (first equation). This way we guarantee input receptiveness of services, for the use of a service channel with input capability cannot be postponed by any means. Base values, service channels with output capability, and linear channels with no capabilities have the highest level $\top$ (last equation) because they do not affect (dead)lock freedom in any way. Linear channels with non-empty polarity must be used according to their level (second equation). We say that a (value with) type t is *unlimited* if $|t| = \top$, that it is *linear* if $|t| \in \mathbb{Z}$, that it is *relevant* if $|t| = \perp$.

We define another auxiliary function $\$_k^h$ to *shift* levels and tickets: $\$_k^h t$ is the same as t except when t is a linear channel. In this case, the level/tickets in t are transposed by h and k respectively. Formally:

$$\$_k^h t \stackrel{\text{def}}{=} \begin{cases} p[s]_{m+k}^{n+h} & \text{if } t = p[s]_m^n \text{ and } p \neq \emptyset \\ t & \text{otherwise} \end{cases} \tag{9.2}$$

Note that positive/negative shifting of levels corresponds to decreasing/increasing the urgency with which a value of a given type must be used.

The type system makes use of *type environments* Γ to keep track of the type of names occurring free in processes. A type environment is a finite map from names to types written $u_1 : t_1, \ldots, u_n : t_n$. We write $\text{dom}(\Gamma)$ for the *domain* of Γ, namely the set of names for which there is an association in Γ, and Γ, Γ' for the union of Γ and Γ' when $\text{dom}(\Gamma) \cap \text{dom}(\Gamma') = \emptyset$. We also need a more general way of composing type environments that takes into account the level and tickets of linear channel types and the fact that we can split channel types by distributing different capabilities to different processes. Following [16], we define a partial operator + between types, thus:

$$\begin{aligned} t + t &= t & \text{if } t \text{ is unlimited} \\ p[t]^* + q[t]^* &= (p \cup q)[t]^* \\ p[t]_h^n + q[t]_k^n &= (p \cup q)[t]_{h+k}^n & \text{if } p \cap q = \emptyset \end{aligned} \tag{9.3}$$

Informally, unlimited types combine with themselves without restrictions. The combination of two unlimited/relevant channel types has the union of their polarities. Two linear channel types can be combined only if they have the same level and disjoint polarities, and their combination has the union of their polarities and the sum of their tickets. We extend the partial operator + to type environments:

$$\begin{aligned} \Gamma + \Gamma' &\stackrel{\text{def}}{=} \Gamma, \Gamma' & \text{if } \text{dom}(\Gamma) \cap \text{dom}(\Gamma') = \emptyset \\ (\Gamma, u : t) + (\Gamma', u : s) &\stackrel{\text{def}}{=} (\Gamma + \Gamma'), u : t + s \end{aligned} \tag{9.4}$$

Note that $\Gamma + \Gamma'$ is undefined if there is $u \in \text{dom}(\Gamma) \cap \text{dom}(\Gamma')$ such that $\Gamma(u) + \Gamma'(u)$ is undefined and that $\text{dom}(\Gamma + \Gamma') = \text{dom}(\Gamma) \cup \text{dom}(\Gamma')$. We let $|\Gamma|$ denote the lowest level of the types in the range of Γ, that is

$$|\Gamma| \stackrel{\text{def}}{=} \min\{|\Gamma(u)| \mid u \in \mathsf{dom}(\Gamma)\} \tag{9.5}$$

We say that Γ is unlimited if $|\Gamma| = \top$.

We now turn our attention to the typing rules, which are meant to enforce the following properties of channels:

1. a service channel with input capability must be used by a replicated input process (we refer to this condition as *input receptiveness*);
2. a linear channel cannot be discarded until both its input and output capabilities have been used (we refer to this condition as *linearity*);
3. an operation on a linear channel cannot block channels with lower or equal level (with linearity, this condition guarantees *deadlock freedom*);
4. the use of a linear channel cannot be postponed forever (with linearity and deadlock freedom, this condition guarantees *lock freedom*).

The typing rules allow us to derive judgments of the form $\Gamma \vdash e : t$, stating that e is well typed in the environment Γ and has type t, and of the form $\Gamma \vdash_k P$, stating that P is well typed in the environment Γ. The parameter $k \in \{0, 1\}$ intuitively represents the "cost" for sending a channel over another channel: each output operation consumes k tickets from the channels being sent. The type system is designed in such a way that a well typed, closed process P is guaranteed to be deadlock free if $\emptyset \vdash_0 P$ and lock free if $\emptyset \vdash_1 P$.

Expressions. Because the model has an extremely simple expression language, the corresponding typing rules, shown below, are fairly obvious and extend easily to more complex expressions:

[T-BOOL]   [T-INT]   [T-NAME]
$\emptyset \vdash b : \mathtt{bool}$ $\emptyset \vdash n : \mathtt{int}$ $u : t \vdash u : t$

The important remark concerning these rules is that the type environment used for typing an expression e only contains associations for the free names occurring in e. This makes sure that no linear or relevant resource (namely, channels that must be used at least once) is left unused. Later on we will discuss a structural rule that allows us to discard resources whose use is not necessary in order to enforce (dead)lock freedom.

Idle and grouped processes. Rule [T-IDLE] states that the idle process is well typed in the empty environment only:

[T-IDLE] $\emptyset \vdash_k \{\ \}$

For example, the judgment $a : ![int]_m^n \vdash_k \{\,\}$ is not derivable, because the linear channel a is supposed to be used for one output operation, whereas the process $\{\,\}$ does nothing. This typing rule illustrates a key trait of the type system, making sure that linear channels with pending capabilities are not discarded. If this were not the case, one could write processes like

$$\text{new } a \text{ in } a?(x)$$

which are stuck waiting for messages that will never arrive.

The typing rule for a grouped process is simple and does not enforce any constraint other than typability of the process itself:

$$[\text{T-GROUP}] \qquad \frac{\Gamma \vdash_k P}{\Gamma \vdash_k \{P\}}$$

Outputs. Rule [T-OUT] is used for typing output operations on linear channels:

$$[\text{T-OUT}] \qquad u : ![t]_n^m + v : \$_k^n \, t \vdash_k u!v \qquad 0 < |t|$$

First of all, the channel u being used for the output must indeed have capability $!$. The type of the message v must be t (as specified in the type of the channel u) except that its level is shifted by m and its tickets are shifted by k. The shifting of the level means that the level of t in $![t]_n^m$ is relative to m. This, together with the side condition $0 < |t|$, makes sure that the level of v (the channel being sent as a message) is strictly greater than the level of u. The shifting of the tickets in t accounts for the fact that, by sending v as a message, one ticket from v is consumed. Note that this is necessary only in the judgments for lock freedom ($k = 1$). Below are a few examples:

- The judgment $a : ![?[int]_0^1]_0^2, b : ?[int]_1^3 \vdash_1 a!b$ is derivable because $?[int]_1^3 = \$_1^2 ?[int]_0^1$. Note in particular that the channel to be sent on a must have no tickets, which is in fact what happens to b after 1 ticket is consumed from its type before it travels on a.
- The judgment $a : ![![int]_0^1]_0^0, b : ![int]_0^1 \vdash_1 a!b$ is not derivable because b has no tickets and so it cannot travel on a.
- Let $t = ?[t]_0^0$ and observe that $\#[t]_0^1 = ![t]_0^1 + ?[t]_0^1$. The judgment $a : \#[t]_0^1 \vdash_0 a!a$ is not derivable, despite the message a has the "right" type $?[t]_0^1 = \$_0^1 t$, because the condition $0 < |t| = 0$ is not satisfied. A process like $a!a$ is deadlocked because the occurrence of a that is meant to be used for matching this output operation is the very message sent on a itself.

- The judgment $a : !\,[?[\text{int}]^*]_{0}^{0}, b : ?[\text{int}]^* \vdash_0 a!b$ is not derivable because $0 \not< |?[\text{int}]^*| = \bot$. A service channel with input capability such as b cannot be sent as a message to guarantee input receptiveness.

Rule [T-OUT*] is used for typing outputs on unlimited channels.

$$[\text{T-OUT*}] \qquad u : !\,[t]^* + v : \$_k^n\, t \vdash_k u!v \qquad \bot < |t|$$

There are two key differences between [T-OUT] and [T-OUT*]. First, the condition $\bot < |t|$, where t is the type of v means that only unlimited channels with input capability cannot be communicated. Second, the type of v does not need to match exactly the type t in the channel type of u, but its level can be shifted by an arbitrary amount n. This is the technical realization of polymorphism. In particular, each distinct output on u can shift the type of the message by a different amount, therefore allowing polymorphic recursion. We will often use this feature in the extended examples in the second half of this chapter.

Both [T-OUT] and [T-OUT*] generalize easily to an arbitrary number of message arguments. As an example, the former rule can be generalized as follows:

$$u : ?[t_1, \ldots, t_h]_n^m + v_1 : \$_k^m\, t_1 + \cdots + v_h : \$_k^m\, t_h \vdash_k u!(v_1, \ldots, v_h) \qquad 0 < |t_i|$$

Inputs. Rule [T-IN] is used for typing linear input operations:

$$[\text{T-IN}] \qquad \frac{\Gamma, x : \$_0^n\, t \vdash_k P}{\Gamma + u : ?[t]_m^n \vdash_k u?(x).P} \qquad n < |\Gamma|$$

The channel u must have type $?[t]_m^n$ and the continuation P is typed in an environment where the input polarity of u has been removed and the received message x has been added. The level of the type of x is shifted by n, consistently with the relative interpretation of levels of message types that we have already discussed for [T-OUT]. The tickets of u are irrelevant since u is used for an input operation, not as the content of a message. The condition $n < |\Gamma|$ ensures that the input on u does not block operations on other channels with equal or lower level. In particular, Γ cannot contain service channels with input capability. Below are some typical examples of ill-typed processes that violate this condition:

- $a : ?[\text{int}]_0^1, b : !\,[\text{int}]_0^0 \vdash_k a?(x).b!x$ is not derivable because $1 \not< 0$: the input on a blocks the output on b, but b has lower level than a;

- $a : \#[\mathtt{int}]_0^h \vdash_k a?(x).a!x$ is a degenerate case of the previous example, where the input on a blocks the very output that should synchronize with it. Note that this process is well-typed in the traditional linear π-calculus [16].
- $a : ?[\mathtt{int}]_0^h, c : ?[\mathtt{int}]^* \vdash_k a?(x).{}^*c?(y)$ is not derivable because $|?[\mathtt{int}]^*| = \perp$. To guarantee input receptiveness, we require that replicated inputs cannot be guarded by other operations.

Rule [T-IN*] is used for typing replicated input operations corresponding to persistent services:

$$[\text{T-IN*}] \qquad \frac{\Gamma, x : t \vdash_k P}{\Gamma + u : ?[t]^* \vdash_k {}^*u?(x).P} \qquad \top \leq |\Gamma|$$

This rule differs from [T-IN] in three important ways. First of all, u must be a service channel with input capability. Second, the side condition $\top \leq |\Gamma|$ makes sure that the environment Γ used for typing the body of the service is unlimited. This is because the service can be invoked an arbitrary number of times, hence its body cannot contain linear resources. Third, it may be the case that $u \in \mathrm{dom}(\Gamma)$, because $?[t]^* + ![t]^* = \#[t]^*$ according to (9.3) and $![t]^*$ is unlimited. This means that services may recursively invoke themselves. We use this feature in several examples, including Example 1.

As for [T-OUT] and [T-OUT*], both [T-IN] and [T-IN*] can be easily generalized to handle arbitrary tuples of message arguments.

Conditional and parallel processes. The typing rule for conditional processes is shown below:

$$[\text{T-IF}] \qquad \frac{\Gamma_1 \vdash e : \mathtt{bool} \qquad \Gamma_2 \vdash_k P \qquad \Gamma_2 \vdash_k Q}{\Gamma_1 + \Gamma_2 \vdash_k \mathtt{if}\ e\ \mathtt{then}\ P\ \mathtt{else}\ Q}$$

As usual, the condition must have type $\mathtt{bool}$ and $+$ is used for combining the type environments used in different parts of the process. Note that both branches must be typable using the same type environment, meaning that the linear channels occurring in P and Q must be used in the same order. For example, the judgment

$$a : ?[\mathtt{int}]_0^1, b : ?[\mathtt{int}]_0^2 \vdash_k \mathtt{if}\ e\ \mathtt{then}\ a?(x).b?(y)\ \mathtt{else}\ b?(y).a?(x)$$

is not derivable because the $\mathtt{else}$ branch uses the two linear channels a and b in an order not allowed by their levels.

The typing rule for parallel compositions is shown below:

$$[\text{T-PAR}] \quad \frac{\Gamma_1 \vdash_k P \qquad \Gamma_2 \vdash_k Q}{\Gamma_1 + \Gamma_2 \vdash_k P \mid Q}$$

Because of the definition of the + operator, which combines the types of linear channels only provided that such channels have the same level, the order in which linear channels are used in the branches of the parallel composition must be consistent. For example, the judgment

$$a : \#[\text{int}]_0^1, b : \#[\text{int}]_0^2 \vdash_k a?(x).b!x \mid b?(y).a!y$$

cannot be derived because the second branch violates the side condition of [T-IN] requiring b to have a strictly smaller level than a. Indeed, the whole process is deadlocked.

Channel creation. Restrictions can be used to introduce both linear and service channels. In the former case, the typing rule is

$$[\text{T-NEW}] \quad \frac{\Gamma, a : p[t]_m^n \vdash_k P}{\Gamma \vdash_k \text{new } a \text{ in } P} \qquad p \in \{0, \#\}$$

Note that the rule "guesses" the right level and number of tickets that are necessary for typing P. The polarity is either #, meaning that a must be used for both one input and one output operation in P, or 0, meaning that a is a depleted channel that is not supposed to be used at all in P. The reason why the typing rule accounts for this possibility is purely technical and is necessary to prove that process reductions preserve typing [18].

The typing rule for introducing a service channel is essentially the same:

$$[\text{T-NEW}^*] \quad \frac{\Gamma, a : \#[t]^* \vdash_k P}{\Gamma \vdash_k \text{new } a \text{ in } P}$$

Unlike linear channels, the capability of restricted unlimited channels is always #. Since an unlimited channel a with input capability must be used (*cf.*[T-IN*]), this guarantees that there is always a service waiting for invocations on a. On the other hand, a service channel with output capability does not have to be used (*cf.* (9.1)), therefore imposing that the capability of a is # does not mandate invocations on a.

Unused resources. The following structural rule provides a limited form of weakening whereby it is possible to add unused resources in a type environment, provided that these resources have an unlimited type:

$$[\text{T-WEAK}] \qquad \frac{\Gamma \vdash_k P}{\Gamma + \Gamma' \vdash_k P} \qquad \top \le |\Gamma'|$$

For example, both $a : 0[\texttt{int}]_n^m \vdash_k \{\,\}$ and $x : \texttt{int} \vdash_k \{\,\}$ are derivable, because the type environments only contain resources that impose no usage and therefore can be discarded using [T-WEAK]. On the contrary, neither $a : \,![\texttt{int}]_n^m \vdash_k \{\,\}$ nor $a : ?[\texttt{int}]^* \vdash_k$ are derivable.

The type system refines the one for the linear π-calculus [16], hence all the properties of the linear π-calculus (partial confluence, linear usage of channels, etc.) are still guaranteed. The added value is that the type system also guarantees deadlock/lock freedom.

Theorem 9.3. *The following properties hold:*

1. *If $\emptyset \vdash_0 P$, then P is deadlock free.*
2. *If $\emptyset \vdash_1 P$, then P is lock free.*

Example 3 (recursive Fibonacci function). Let P be the process shown in Listing 9.1. Then it is possible to derive

$$\texttt{fibo} : \#[\texttt{int}, \,![\texttt{int}]_0^0]^* \vdash_k P$$

if and only if $k = 0$. It is not possible to find a derivation for $k = 1$ since the type system cannot establish an upper bound to the time after which the continuation channel c will be used in an invocation $\texttt{fibo!}(n,c)$. In fact, such upper bound depends on n and on the fact that the recursion of the $\texttt{fibo}$ service is well-founded This latter property requires a kind of analysis that eludes the capabilities of the type system. ∎

9.3 Extended Examples

9.3.1 Fibonacci Stream Network

In this section we discuss an alternative modeling of system that computes the sequence of Fibonacci numbers and that is an example of *stream network*, that is a network of communicating processes that exchange infinite

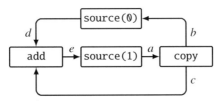

Figure 9.1   Graphical representation of the Fibonacci stream network [8, 22].

sequences (streams) of messages. Figure 9.1 depicts the Fibonacci stream network [8, 22] where the boxes represent processing units and the arrows represent communication channels.

Each source(n) process sends n on the outgoing channel followed by each message received from the incoming channel. The copy process forwards each received message on each of its two outgoing channels. Finally, add sends on the outgoing channel the sum of corresponding messages received from the two incoming channels. Overall, it is easy to see that the stream of messages flowing on channel a corresponds to the Fibonacci sequence $1, 1, 2, 3, 5, \ldots$.

The modeling of the Fibonacci stream network in Hypha's input language is shown in Listing 9.2. There is a service for each of the boxes in Figure 9.1, with source that makes use of an auxiliary service link that acts as a persistent message forwarder. The network itself is created on line 5, where the services are invoked and connected by the channels a through e. The most distinctive aspect of the modeling is the use of continuation passing for the representation of message streams: each channel that connects two combinators is in fact a linear channel (a channel that is meant to be used for one communication only); whenever a message is exchanged on the channel, the payload is paired with a fresh (linear) channel on which the subsequent message will be exchanged. This pattern can be clearly observed in the definitions of link, add, and copy. To improve readability, hereafter we write $\bar{x}$ for a channel name that is meant to represent the continuation of x.

```
{ *link?(x,y).x?(v,x̄).source!(v,x̄,y)
| *source?(n,x,y).new ȳ in { y!(n,ȳ) | link!(x,ȳ) }
| *add?(x,y,z).x?(v,x̄).y?(w,ȳ).new z̄ in { z!(v+w,z̄) | add!(x̄,ȳ,z̄) }
| *copy?(x,y,z).x?(v,x̄).new ȳ,z̄ in { y!(v,ȳ) | z!(v,z̄) | copy!(x̄,ȳ,z̄) }
| source!(1,e,a) | copy!(a,b,c) | source!(0,b,d) | add!(d,c,e) }
```

Listing 9.2   Term representation of the Fibonacci stream network [8, 22].

Hypha infers the following types for the channels used in the system

$$\texttt{source} : \#[\texttt{int},?[\texttt{int},\mu\alpha.?[\texttt{int},\alpha]_2^3]_1^2,\,![\texttt{int},\mu\beta.?[\texttt{int},\beta]_1^3]_0^0]^*$$

$$\texttt{link} : \#[?[\texttt{int},\mu\alpha.?[\texttt{int},\alpha]_2^3]_0^0,\,![\texttt{int},\mu\beta.?[\texttt{int},\beta]_1^3]_1^1]^*$$

$$\texttt{add} : \#[?[\texttt{int},\mu\alpha.?[\texttt{int},\alpha]_1^3]_0^0,\,?[\texttt{int},\mu\beta.?[\texttt{int},\beta]_1^3]_0^1,$$
$$\qquad\quad ![\texttt{int},\mu\gamma.?[\texttt{int},\gamma]_2^3]_0^2]^*$$

$$\texttt{copy} : \#[?[\texttt{int},\mu\alpha.?[\texttt{int},\alpha]_1^3]_0^0,$$
$$\qquad\quad ![\texttt{int},\mu\beta.?[\texttt{int},\beta]_2^3]_0^2,\,![\texttt{int},\mu\gamma.?[\texttt{int},\gamma]_1^3]_0^1]^*$$

$$a,d : \#[\texttt{int},\mu\alpha.?[\texttt{int},\alpha]_1^3]_2^0$$

$$b,e : \#[\texttt{int},\mu\alpha.?[\texttt{int},\alpha]_2^3]_3^2$$

$$c : \#[\texttt{int},\mu\alpha.?[\texttt{int},\alpha]_1^3]_2^1$$

confirming that the system is well typed and therefore lock free. In particular, Theorem 9.3(2) allows us to deduce that every number in the sequence of Fibonacci is computed in finite time. We make some observations concerning the inferred channel types: first, Hypha correctly distinguishes between the channels representing services (such as copy and source) from the linear channels that connect them (such as a and b). Second, the levels associated with linear channels give hints concerning the order of synchronizations in the system. The synchronizations on a and d (with level 0) happen first, followed by that on c (level 1), and then by that on e (level 2). Note however, that the total order on levels does not necessarily reflect the partial order that represents dependencies between channels. For example, b has a strictly greater level than c and yet the synchronizations on these two channels may happen in any order. Concerning the ticket annotations, note that all linear channels require at least 2 tickets because they are used to connect 2 services. For example, a connects source(1) and copy. By contrast, b and e need one more ticket because they are also forwarded by source to link.

9.3.2 Full-Duplex and Half-Duplex Communications

Many parallel algorithms use batteries of processes arranged in a grid that iteratively update array elements and communicate with processes assigned to neighbor elements (Figure 9.2). Processes may communicate according to one out of two modalities: when communication is *full-duplex*, processes simultaneously send messages to each other; when communication is *half-duplex*, only one message travels between two processes at any

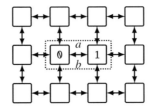

Figure 9.2   Graphical representation of a 4 × 3 bi-dimensional stencil.

moment in time. Correspondingly, we can model the dotted grid fragment in Figure 9.2 as

$$e!\,(0,a,b) \mid f!\,(1,b,a) \tag{9.6}$$

where e and f are service channels defined as either

*full?(n,x,y).new $\bar{x}$ in $\{\ x!\,(n,\bar{x}) \mid y?\,(m,\bar{y}).$full!$(n+m,\bar{x},\bar{y})\ \}$

in case of full-duplex communication or as

*half?(n,x,y).y?$(m,\bar{y})$.new $\bar{x}$ in $\{\ x!\,(n,\bar{x}) \mid$ half!$(n+m,\bar{x},\bar{y})\ \}$

in case of half-duplex communication. In both cases, x is used for sending messages to, and y for receiving messages from, a neighbor process. Each message sent on x carries a payload n as well as a continuation channel $\bar{x}$ used for the communication at the next iteration. Symmetrically, each message received from y contains the neighbor's payload m and a continuation $\bar{y}$. The difference between full and half is that, in the latter case, the sender waits for the message from its neighbor before sending its own.

Overall there are 4 possible configurations of the system (9.6) obtained by instantiating e and f with either full or half. It is easy to see that a configuration is lock free as long as *at least* one of e or f is instantiated with full. Indeed, Hypha infers the types

$$a,b : \#[\text{int}, \mu\alpha.?[\text{int},\alpha]^1_1]^0_2$$

when $e = f =$ full and the types

$$a : \#[\text{int}, \mu\alpha.?[\text{int},\alpha]^2_1]^1_2 \qquad b : \#[\text{int}, \mu\alpha.?[\text{int},\alpha]^2_1]^0_2$$

when $e =$ half and $f =$ full. The case when $e =$ full and $f =$ half is symmetric, while the one when $e = f =$ half is ill typed for deadlock freedom and, therefore, for lock freedom as well.

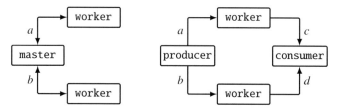

Figure 9.3 Master-worker (left) and producer-consumer (right).

9.3.3 Load Balancing

Figure 9.3 shows two network topologies aimed at taking advantage of parallelism by distributing multiple tasks to independent workers. They differ in that in the *master-worker* topology the same process that produces tasks is also the one that collects the results, whereas in the *producer-consumer* topology (sometimes called "farm") producer and consumer are different processes. The distinction between the two topologies has important consequences at the communication layer since the channels are bi-directional in the former network and uni-directional in the latter.

Listings 9.3 and 9.4 show the modeling of the network topologies in Figure 9.3, both of which are well-typed according to the lock freedom type system implemented in Hypha. For the master-worker network, Hypha infers the types

$$a \; : \; \#[\text{int}, \mu\alpha.\,![\text{int}, ![\text{int}, \alpha]_0^2]_0^1]_2^0$$
$$b \; : \; \#[\text{int}, \mu\alpha.\,![\text{int}, ![\text{int}, \alpha]_1^2]_0^1]_2^1$$

```
{ *master?(n,x,y).
  new x̄,ȳ in { x!(n,x̄) | y!(n+1,ȳ) | x̄?(v,x̄).ȳ?(w,ȳ).master!(n+2,x̄,ȳ) }
| *worker?(n,z).z?(m,z̄).new z̄ in z̄!(m mod n,z̄).worker!(n,z̄)
| master!(0,a,b) | worker!(2,a) | worker!(3,b) }
```

Listing 9.3 Term representation of master-worker (half-duplex channels).

```
{ *producer?(n,x,y).new x̄,ȳ in { x!(n,x̄) | y!(n+1,ȳ) | producer!(n+2,x̄,ȳ) }
| *consumer?(x,y).x?(v,x̄).y?(w,ȳ).{ print!v | print!w | consumer!(x̄,ȳ) }
| *worker?(n,x,y).x?(m,x̄).new ȳ in { y!(m mod n,ȳ) | worker!(n,x̄,ȳ) }
| producer!(0,a,b) | worker!(2,a,c) | worker!(3,b,d) | consumer!(c,d) }
```

Listing 9.4 Term representation of producer-consumer.

which describe a communication protocol whereby the master sends a task (represented as an integer number) to each worker along with a continuation channel. By using this continuation channel, the worker will answer back with the processed task (again represented as an integer number) and another continuation that the master uses for starting another iteration.

For the producer-consumer network Hypha infers the types

$$a \; : \; \#[\text{int}, \mu\alpha.?[\text{int}, \alpha]_1^2]_2^0 \qquad c \; : \; \#[\text{int}, \mu\alpha.?[\text{int}, \alpha]_1^2]_2^1$$

$$b \; : \; \#[\text{int}, \mu\alpha.?[\text{int}, \alpha]_1^2]_2^1 \qquad d \; : \; \#[\text{int}, \mu\alpha.?[\text{int}, \alpha]_1^2]_2^2$$

again confirming that the network is lock free.

9.3.4 Sorting Networks

Figure 9.4 depicts an example of so-called *sorting network*, that is a network of communicating processes whose overall effect is that of sorting an input vector of fixed size, 6 in this case. The network is made of two different kinds of processes: *comparators* (the rectangular boxes) input two values and possibly swap them if the first happens to be larger than the second; *buffers* (the square boxes) simply forward the input value. The input values go through three identical phases; in each stage, the odd-indexed inputs and then the even-indexed inputs are compared to, and possibly swapped with, their successor.

The sorting network in Figure 9.4 is modeled in the linear π-calculus as shown in Listing 9.5. Note the use of auxiliary services odd and even corresponding to the two sub-phases of each phase and linked together by

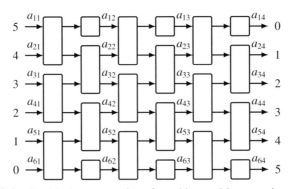

Figure 9.4   Graphical representation of an odd-even 6-input sorting network.

```
1   { *compare?(x,y,l,h).
2       new z in { x?(v).z!v | y?(w).z?(v).if v<w then l!v | h!w else l!w | h!v }
3   | *buffer?(x,y).x?(v).y!v
4   | *provide?(x,n).x!n
5   | *consume?(x,n).x?(v).print!(n,v)
6   | *even?(x1,x2,x3,x4,x5,x6,y1,y2,y3,y4,y5,y6).
7       { compare!(x1,x2,y1,y2) | compare!(x3,x4,y3,y4) | compare!(x5,x6,y5,y6) }
8   | *odd?(x1,x2,x3,x4,x5,x6,y1,y2,y3,y4,y5,y6).
9       { buffer!(x1,y1) | compare!(x2,x3,y2,y3)
10      | buffer!(x6,y6) | compare!(x4,x5,y4,y5) }
11  | *phase?(x1,x2,x3,x4,x5,x6,y1,y2,y3,y4,y5,y6).new z1,z2,z3,z4,z5,z6 in
12      { even!(x1,x2,x3,x4,x5,x6,z1,z2,z3,z4,z5,z6)
13      | odd!(z1,z2,z3,z4,z5,z6,y1,y2,y3,y4,y5,y6) }
14  | phase!(a11,a21,a31,a41,a51,a61,a12,a22,a32,a42,a52,a62)
15  | phase!(a12,a22,a32,a42,a52,a62,a13,a23,a33,a43,a53,a63)
16  | phase!(a13,a23,a33,a43,a53,a63,a14,a24,a34,a44,a54,a64)
17  | provide!(a11,0) | provide!(a21,1) | provide!(a31,2)
18  | provide!(a41,3) | provide!(a51,4) | provide!(a61,5)
19  | consume!(a14,0) | consume!(a24,1) | consume!(a34,2)
20  | consume!(a44,3) | consume!(a54,4) | consume!(a64,5) }
```

Listing 9.5 Term representation of an odd-even 6-input sorting network.

restricted channels z_i. This network is well-typed and Hypha infers the types

$$
\begin{aligned}
\text{comparator} &: \quad \#[?[\text{int}]_0^0, ?[\text{int}]_0^0, ![\text{int}]_0^2, ![\text{int}]_0^2]^* \\
\text{buffer} &: \quad \#[?[\text{int}]_0^0, ![\text{int}]_0^2]^* \\
a_{ij} &: \quad \#[\text{int}]_2^{4(j-1)}
\end{aligned}
$$

confirming that each value sent on a_{i1} is eventually received on some a_{j4}. More specifically, the level 12 assigned with the a_{j4} channels gives an upper bound to the number of synchronizations needed for producing the output.

Comparators input values in parallel (from the channels x and y) and perform an internal synchronization (on a private linear channel z) to join the results of the two receptions and output the results. Alternatively, one could model comparators in such a way that the receive operations on x and y are performed in a fixed order. The choice of a particular modeling affects the levels associated with the input channels, but not the typeability of the network as a whole. This is not the case for buffers: they are operationally irrelevant and are usually omitted in standard presentations of sorting networks. Their use in Listing 9.5 is key for the lock freedom analysis to succeed as they make sure that the levels of the channels connecting one phase to the next one remain aligned.

9.3.5 Ill-typed, Lock-free Process Networks

In general, the problem of verifying whether a π-calculus process is (dead)lock free is undecidable. For this reason, the type system on which Hypha is based is necessarily incomplete, in the sense that there exist processes satisfying Definitions 9.1 and 9.2 which are ill typed according to the type system described in Section 9.2. In this section, we discuss two representative examples of processes that cannot be handled by our type system. In all cases, the inability to find a typing derivation is tightly related to the fact that the type system uses integer numbers for reasoning on the dependencies between linear channels and such numbers measure the (abstract) moment of time at which the synchronization occurs on these channels.

Listing 9.6 shows a process that computes the sequence of prime numbers. The process is modeled after Eratosthenes' sieve. the from process emits the infinite stream of natural numbers starting from 2; the sequence goes through a growing pipeline of `filters`, each filter removing those numbers of the sequence that happen to be a multiple of a given prime number m; if a number n manages to cross the entire pipeline and hits the emitter process output, then it is prime. In this case n is sent on `print` and a new filter removing the multiples of n is inserted at the end of the pipeline. Hypha is able to distinguish linear from service channels and to infer the type of messages exchanged therein, but the process is ill-typed for deadlock freedom even though it is deadlock free. The problem can be traced to the body of `filter`: when the received number n turns out to be a multiple of m, the number is simply discarded and no output is sent on y. So, the recursive invocation of `filter` on line 3 reuses the same output channel y that was received as input. Observe that $\bar{x}$ is received from x, meaning that the level of $\bar{x}$ must necessarily be greater than the level of x, and that the level of $\bar{x}$ must be strictly smaller than the level of y, since the input on performed on $\bar{x}$ at the next iteration of `filter` blocks the possible output on y. Given that the distribution of prime numbers is irregular, there is no upper bound to the number of inputs on x that may be necessary before the next output

```
1   { *from?(n,x).new x̄ in { x!(n,x̄) | from!(n+1,x̄) }
2   | *filter?(m,x,y).x?(n,x̄).{ if n mod m = 0 then filter!(n,x̄,y)
3                          else new ȳ in { y!(n,ȳ) | filter!(m,x̄,ȳ) } }
4   | *output?(x).x?(n,x̄).{ print!n | new y in { filter!(n,x̄,y) | output!y } }
5   | from!(2,a) | output!a }
```

Listing 9.6   Stream Network computing the sequence of prime numbers.

on y is guaranteed to be performed. In general, the type system can handle those cases in which communications occur following a regular pattern that is independent of the content of messages themselves.

The second example we consider is a process stream network (Figure 9.5) that computes the so-called Thue-Morse sequence, that is the sequence of binary digits $011010011001\cdots$ starting with 0 and obtained by appending the boolean complement of the sequence obtained thus far. The term representation of the process network (Listing 9.7) is modeled after its definition in terms of lazy streams [7] and makes use of a set of combinators some of which we have already used for the Fibonacci stream network (Section 9.3.1). The network is lock-free, as witnessed by the fact that the corresponding lazy stream definition can be shown to be productive [7], but also ill typed for deadlock freedom and hence for lock freedom as well. In this network the problematic combinator is zip, which interleaves on the output channel f the digits received from the input channels d and e hence producing messages on f at twice the rate at which they are consumed from d and e. This means that there is no fixed offset between the levels of d and e and that of f that could be dealt with by the typing rule [T-OUT*]. Note that this phenomenon does not manifest in the Fibonacci stream network (Figure 9.1) despite its seemingly similar topology. The key difference is that in the Fibonacci network the add process *combines* the messages received from the two input channels into a single message sent on the output channel. The example is interesting also

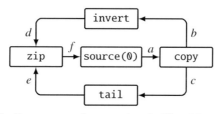

Figure 9.5   Stream network computing the Thue-Morse sequence.

```
1   { *zip?(x,y,z).x?(u,x̄).new z̄ in { z!(u,z̄) | zip!(y,x̄,z̄) }
2   | *invert?(x,y).x?(u,x̄).new ȳ in { y!(1-u,ȳ) | invert!(x̄,ȳ) }
3   | *tail?(x,y).x?(_,x̄).link!(x̄,y)
4   | *link?(x,y).x?(v,x̄).source!(v,x̄,y)
5   | *copy?(x,y,z).x?(v,x̄).new ȳ,z̄ in { y!(v,ȳ) | z!(v,z̄) | copy!(x̄,ȳ,z̄) }
6   | *source?(n,x,y).new ȳ in { y!(n,ȳ) | link!(x,ȳ) }
7   | source!(0,f,a) | copy!(a,b,c) | invert!(b,d) | tail!(c,e) | zip!(d,e,f) }
```

Listing 9.7   Term representation of the stream network computing the Thue-Morse sequence.

because the impossibility to type the term is due to an excess of produced messages rather than the lack thereof, as it was with the sequence of prime numbers.

9.4 Related Work

Binary session type disciplines [12] provide intra-session guarantees of lock freedom but cannot enforce this property in general when multiple sessions are interleaved for types do not carry any information concerning the dependencies between different sessions. Some session type systems [1, 24] are designed in such a way that a plain session type discipline is sufficient to guarantee deadlock freedom. However, only networks with a tree-like communication topology are well typed. Among the examples we have considered, just the recursive Fibonacci (Listing 9.1) and the master-worker (Listing 9.3) fall in this category.

Multiparty session type disciplines [11, 12] extend (dead)lock freedom to sessions involving multiple processes. In these framework a *global type* is used to describe the interactions between participants of a session as opposed to the actions that participants perform on the channel of the session. The global type is given explicitly by the system designer/programmer and a tool is then used to check the consistency of the global type against (a model of) the code that is meant to realize it. This top-down approach is complementary to the one we have pursued in this chapter: Hypha analyzes assemblies of processes knowing nothing about the intended communication topology. In general, global types have been designed for describing delimited interactions within sessions, but they cannot dispense completely from the need of interleaving different sessions, in which case they are unable to prevent (dead)locks. This has led to the study of hybrid approaches [2, 3] that keep track of the order in which different sessions interleave with the purpose of detecting mutual dependencies between sessions that could lead to (dead)locks.

The works most closely related to our own are those where each input/output operation described by a channel/session type is annotated with information that captures the dependencies between different channels/sessions. Such annotations come in the form of integer numbers as in our case, or as abstract events, or as combinations thereof. The original technique and the corresponding analysis tool TyPiCal, which our type system and Hypha are heavily inspired by, were described by Kobayashi [13–15].

These type systems and our own are uncomparable: on the one hand, in Kobayashi's works annotations can be used to reason about dependencies between arbitrary channels, whereas we focus on linear channels only. On the other hand, the form of level polymorphism allowed by rule [T-OUT*] enables the verification of cyclic networks of recursive processes (most of the examples we have examined in this chapter fall in this category) that cannot be successfully handled by Kobayashi's type systems [13–15]. A more recent work [9] improves the precision of the technique, although recursive types (hence recursive communication protocols) are not considered. The annotation-based technique has also been applied directly to binary [17, 23] and multiparty sessions [21].

It has been shown that the approaches imposing a tree-like communication topology [1, 24] are subsumed by those those annotating I/O actions in session types with dependency information [6].

Acknowledgments I'm grateful to the anonymous reviewers whose comments helped me improving both content and presentation of this chapter.

References

[1] Luís Caires, Frank Pfenning, and Bernardo Toninho. Linear logic propositions as session types. *Mathematical Structures in Computer Science*, 26(3):367–423, 2016.

[2] Mario Coppo, Mariangiola Dezani-Ciancaglini, Luca Padovani, and Nobuko Yoshida. Inference of Global Progress Properties for Dynamically Interleaved Multiparty Sessions. In *Proceedings of COORDINATION'13*, LNCS 7890, pages 45–59. Springer, 2013.

[3] Mario Coppo, Mariangiola Dezani-Ciancaglini, Nobuko Yoshida, and Luca Padovani. Global Progress for Dynamically Interleaved Multiparty Sessions. *Mathematical Structures in Computer Science*, 26:238–302, 2016.

[4] Bruno Courcelle. Fundamental properties of infinite trees. *Theoretical Computer Science*, 25:95–169, 1983.

[5] Ornela Dardha, Elena Giachino, and Davide Sangiorgi. Session types revisited. In *Proceedings of PPDP'12*, pages 139–150. ACM, 2012.

[6] Ornela Dardha and Jorge A. Pérez. Comparing deadlock-free session typed processes. In *Proceedings of EXPRESS/SOS'15*, EPTCS 190, pages 1–15, 2015.

[7] Jörg Endrullis, Clemens Grabmayer, Dimitri Hendriks, Ariya Isihara, and Jan Willem Klop. Productivity of stream definitions. *Theoretical Computer Science*, 411(4-5):765–782, 2010.

[8] Marc Geilen and Twan Basten. Kahn process networks and a reactive extension. In *Handbook of Signal Processing Systems*, pages 1041–1081. Springer, 2013.

[9] Elena Giachino, Naoki Kobayashi, and Cosimo Laneve. Deadlock analysis of unbounded process networks. In *Proceedings of CONCUR'14*, LNCS 8704, pages 63–77. Springer, 2014.

[10] Kohei Honda. Types for dyadic interaction. In *Proceedings of CONCUR'93*, LNCS 715, pages 509–523. Springer, 1993.

[11] Kohei Honda, Nobuko Yoshida, and Marco Carbone. Multiparty asynchronous session types. *Journal of the ACM*, 63(1):9, 2016.

[12] Hans Hüttel, Ivan Lanese, Vasco T. Vasconcelos, Luís Caires, Marco Carbone, Pierre-Malo Deniélou, Dimitris Mostrous, Luca Padovani, António Ravara, Emilio Tuosto, Hugo Torres Vieira, and Gianluigi Zavattaro. Foundations of Session Types and Behavioural Contracts. *ACM Computing Surveys*, 49:3:1–3:36, 2016.

[13] Naoki Kobayashi. A type system for lock-free processes. *Information and Computation*, 177(2):122–159, 2002.

[14] Naoki Kobayashi. Type-based information flow analysis for the pi-calculus. *Acta Informatica*, 42(4-5):291–347, 2005.

[15] Naoki Kobayashi. A new type system for deadlock-free processes. In *Proceedings of CONCUR'06*, LNCS 4137, pages 233–247. Springer, 2006.

[16] Naoki Kobayashi, Benjamin C. Pierce, and David N. Turner. Linearity and the pi-calculus. *ACM Transactions on Programming Languages and Systems*, 21(5):914–947, 1999.

[17] Luca Padovani. From Lock Freedom to Progress Using Session Types. In *Proceedings of PLACES'13*, EPTCS 137, pages 3–19, 2013.

[18] Luca Padovani. Deadlock and Lock Freedom in the Linear π-Calculus. In *Proceedings of CSL-LICS'14*, pages 72:1–72:10. ACM, 2014.

[19] Luca Padovani. Type Reconstruction for the Linear π-Calculus with Composite Regular Types. *Logical Methods in Computer Science*, 11:1–45, 2015.

[20] Luca Padovani, Tzu-Chun Chen, and Andrea Tosatto. Type Reconstruction Algorithms for Deadlock-Free and Lock-Free Linear π-Calculi. In *Proceedings of COORDINATION'15*, LNCS 9037, pages 83–98. Springer, 2015.

[21] Luca Padovani, Vasco T. Vasconcelos, and Hugo Torres Vieira. Typing Liveness in Multiparty Communicating Systems. In *Proceedings of COORDINATION'14*, LNCS 8459, pages 147–162. Springer, 2014.

[22] John H. Reppy. *Concurrent Programming in ML*. Cambridge University Press, 1999.

[23] Hugo Torres Vieira and Vasco T. Vasconcelos. Typing progress in communication-centred systems. In *Proceedings of COORDINA-TION'13*, LNCS 7890, pages 236–250. Springer, 2013.

[24] Philip Wadler. Propositions as sessions. *Journal of Functional Programming*, 24(2–3):384–418, 2014.

10

Session Types with Linearity in Haskell

Dominic Orchard[1] and Nobuko Yoshida[2]

[1]University of Kent, UK
[2]Imperial College London, UK

Abstract

Type systems with parametric polymorphism can encode a significant proportion of the information contained in session types. This allows concurrent programming with session-type-like guarantees in languages like ML and Java. However, statically enforcing the linearity properties of session types, in a way that is also natural to program with, is more challenging. Haskell provides various language features that can capture concurrent programming with session types, with full linearity invariants and in a mostly idiomatic style. This chapter overviews various approaches in the literature for session typed programming in Haskell.

As a starting point, we use polymorphic types and simple type-level functions to provide session-typed communication in Haskell without linearity. We then overview and compare the varying approaches to implementing session types with static linearity checks. We conclude with a discussion of the remaining open problems.

The code associated with this chapter can be found at `http://github.com/dorchard/betty-book-haskell-sessions`.

10.1 Introduction

Session types are a kind of behavioural type capturing the communication behaviour of concurrent processes. While there are many variants of session types, they commonly capture the sequence of sends and receives performed over a channel and the types of the messages carried by these interactions. A significant aspect of session types is that they enforce *linear* use of channels:

219

every send must have exactly one receive (no orphan messages), and vice versa (no hanging receives). These properties are often referred to together as *communication safety*. A channel cannot be reused once it has "used up" its capability to perform sends and receives. This aspect of session types makes them hard to implement in languages which do not have built-in notions of linearity and resource consumption in the type system.

The following two example interactions will be used throughout.

Example 1 (Integer equality server and client). Consider a simple server which provides two modes of interaction (services) to clients. If a client chooses the first service, the server can then receive two integers, compare these for equality, send the result back as a boolean, and then return to the start state. The second service tells the server to stop hence it does not return to providing the initial two services.

A potential client requests the first behaviour, sends two integers, receives a boolean, and then requests that the server stop. These server and client behaviours are captured by the following session types, using the notation of Yoshida and Vasconcelos [18], which describe the interaction from the perspective of opposite channel endpoints:

$$Server := \mu\alpha.\&\{\text{eq} : ?\mathbb{Z}.?\mathbb{Z}.!\mathbb{B}.\alpha, \text{ nil} : \textbf{end}\}$$
$$Client := \oplus\{\text{eq} : !\mathbb{Z}.!\mathbb{Z}.?\mathbb{B}. \oplus\{\text{nil} : \textbf{end}\}\}$$

The server has a *recursive* session type, denoted $\mu\alpha.S$ which binds the variable α in scope of a session type S. Session types are typically equi-recursive, such that $\mu\alpha.S \equiv S[\mu\alpha.S/\alpha]$. The operator $\&$ denotes a choice offered between branches, labelled here as eq and nil. In the eq case, two integers are received and a boolean is sent before recursing with α. In the nil case the interaction finishes, denoted by **end**.

The client selects the eq service, denoted by $\oplus$. Two integers are sent and a boolean is received. Then the nil behaviour is selected via $\oplus$, ending the interaction. Session types thus abstract communication over a channel, or equivalently, they describe a channel's *capabilities*.

The two types are *dual*: they describe complementary communication behaviour on opposite end-points of a channel. Duality can be defined inductively as a function on session types:

$$\overline{!\tau.S} = ?\tau.\overline{S} \qquad \overline{\&\{l_i : S_i\}_{i\in I}} = \oplus\{l_i : \overline{S_i}\}_{i\in I} \qquad \overline{\mu\alpha.S} = \mu\alpha.\overline{S[\overline{\alpha}/\alpha]}$$
$$\overline{?\tau.S} = !\tau.\overline{S} \qquad \overline{\oplus\{l_i : S_i\}_{i\in I}} = \&\{l_i : \overline{S_i}\}_{i\in I} \qquad \overline{\textbf{end}} = \textbf{end}$$

Recursion variables come in two flavours: α and their dual $\overline{\alpha}$. The dual of a dualised variable $\overline{(\overline{\alpha})} = \alpha$ is the undualised α. This formulation of duality with recursive types is due to Lindley and Morris [7].

Duality enforces communication safety. If the communication patterns of the server and client do not match then duality does not hold. Duality also encompasses linearity, as any repetition of actions by the server or client leads to non-matching communication behaviour.

Example 2 (Delegating integer equality). Following the expressive power of the π-calculus, session types can also capture *delegation*, where channels are passed over channels. Thus, the types of communicated values τ include session types of communicated channels, written $\langle S \rangle$.

As a permutation on the previous example, we introduce a layer of indirection through delegation. The server, after receiving two integers, now receives a channel over which the resulting boolean should be sent. Dually, the client sends a channel which has the capability of sending a boolean. This is captured by the session types:

$$Server := \mu\alpha.\&\{\text{eq} :?\mathbb{Z}.?\mathbb{Z}.?\langle!\mathbb{B}\rangle.\alpha, \text{nil} : \textbf{end}\}$$
$$Client := \oplus\{\text{eq} :!\mathbb{Z}.!\mathbb{Z}.!\langle!\mathbb{B}\rangle. \oplus\{\text{nil} : \textbf{end}\}\}$$

The server's capability to receive a channel, over which a boolean is sent, is denoted $?\langle!\mathbb{B}\rangle$ whose dual in the client is $!\langle!\mathbb{B}\rangle$: the sending of a channel over which a boolean can be sent.

The reader is referred to the work of Yoshida and Vasconcelos [18] for a full description of a session type theory for the π-calculus on which our more informal presentation is based here.

To unpack the problem of encoding session type linearity in Haskell, we first introduce a relatively simple encoding of session types capturing sequences of send and receive actions on channels and some notion of session duality. However, this approach does not fully enforce linearity (Section 10.2). We then overview the various approaches in the literature for encoding session types in Haskell, focusing on their approach to linearity (Section 10.3). Outstanding problems and open questions in this area are discussed finally in Section 10.4.

Throughout, "Haskell" refers to the variant of Haskell provided by GHC (the Glasgow Haskell Compiler) which provides various type system extensions, the use of which is indicated and explained as required.

10.2 Pre-Session Types in Haskell

Haskell provides a library for message-passing concurrency with channels similar in design to the concurrency primitives of CML [14]. The core primitives have types:

```
newChan   :: IO (Chan a)        writeChan :: Chan a -> a -> IO ()
readChan  :: Chan a -> IO a     forkIO    :: IO () -> IO ThreadId
```

These functions operate within the IO monad for encapsulating side-effectful computations; creating channels (newChan), sending and receiving values on these channels (writeChan and readChan), and forking processes (forkIO) are all effectful. Channels have a single type and are bi-directional. The following program implements Example 1:

```
server c d = do               client c d = do
  x <- readChan c               writeChan c (Just 42)
  case x of                     writeChan c (Just 53)
    Nothing -> return ()        r <- readChan d
    Just x' -> do               putStrLn $ "Result: " ++ show r
      (Just y') <- readChan c   writeChan c Nothing
      writeChan d (x' == y')
      server c d

main = do {c <- newChan; d <- newChan; forkIO (client c d); server c d}
```

The choice between the two services is provided via a Maybe type, where server :: Chan (Maybe Int) -> Chan Bool -> IO (). Two channels are used so that values of different type can be communicated. The channel types ensure *data safety*: communicated values are of the expected type. However, this typing cannot ensure communication safety. For example, the following two alternate clients are well-typed but are communication unsafe:

```
client' c d = do                client'' c d = do
  writeChan c (Just 42)           writeChan c (Just 42)
  writeChan c (Just 53)           readChan c
  writeChan c (Just 53)           r <- readChan d
  r <- readChan d                 putStrLn $ "Result: " ++ show r
  putStrLn $ "Result: " ++ show r writeChan c Nothing
  writeChan c Nothing
```

On the left, an additional message is sent which is left unreceived in the server's channel buffer. On the right, a spurious readChan occurs after the first writeChan leading to a deadlock for the server and client.

```
send c x = do                          fork f = do
    c' <- newChan                          c <- newChan
    writeChan c (Send x c')                c' <- newChan
    return c'                              forkIO (link (c, c'))
                                           forkIO (f c)
recv c = do                                return c'
    (Recv x c') <- readChan c
    return (x, c')                     close c = return ()
```

Figure 10.1 Implementations of the communication-typed combinators where link :: Links => (Chan s, Chan (Dual s)) -> IO ().

A significant proportion of communication safety (mainly the order of interactions) can be enforced with just algebraic data types, polymorphism, and a type-based encoding of duality.

10.2.1 Tracking Send and Receive Actions

Taking inspiration from Gay and Vasconcelos [3], we define the following alternate combinators (with implementations shown in Figure 10.1) and data types:

```
send  :: Chan (Send a t) -> a -> IO (Chan t)
recv  :: Chan (Recv a t) -> IO (a, Chan t)
close :: Chan End -> IO ()

data Send a t = Send a (Chan t)
data Recv a t = Recv a (Chan t)
data End
```

The send combinator takes as parameters a channel which can transfer values of type Send a t and a value x of type a returning a new channel which can transfer the values of type t. This is implemented via the constructor Send, pairing the value x with a new channel c', sending those on the channel c, and returning the new continuation channel c'.

The recv combinator is somewhat dual to this. It takes a channel c on which is received a pair of a value x of type a and channel c' which can transfer values of type t. The pair (x, c') is then returned. The close combinator discards its channel which has only the capability of transferring End values, which are uninhabited (empty data types).

The following implements a non-recursive version of the integer equality server with delegation from Example 2 (for brevity C = Chan):

```
server :: C (Recv Int (Recv Int (Recv (C (Send Bool End)) End))) -> IO ()
server c = do
    (x, c) <- recv c
    (y, c) <- recv c
    (d, c) <- recv c
    d <- send d (x == y)
    close c
    close d
```

The type of the channel c gives a representation of the session type $?\mathbb{Z}.?\mathbb{Z}.?\langle!\mathbb{B}\rangle.$**end** from Example 2. At each step of the program, the channel returned by a send or receive is bound to a variable shadowing the channel variable used *e.g.* (x,c) <- recv c. This programming idiom provides linear channel use.

10.2.2 Partial Safety via a Type-Level Function for Duality

One way to capture duality is via a *type family*. Type families are primitive recursive type functions, with strong syntactic restrictions to enforce termination. We define the (closed) type family Dual:

```
type family Dual s where
          Dual (Send a t) = Recv a (Dual t)
          Dual (Recv a t) = Send a (Dual t)
          Dual End        = End
```

Duality is used to type the fork operation, which spawns a process with a fresh channel, returning a channel of the dual type:

```
fork :: Link s => (Chan s -> IO ()) -> IO (Chan (Dual s))
```

Figure 10.1 shows the implementation which uses a method link of the type class Link to connect sent messages to received messages and vice versa. A client interacting with server above can then be given as:

```
client c = do
    c <- send c 42
    c <- send c 53
    d <- fork (\d' -> do { c <- send c d'; close c })
    (r, d) <- recv d
    putStrLn ("Result: " ++ show r)
    close d

example = do { c' <- fork client; server c' }
```

Thus, the client sends two integers on c then creates a new channel d', which is sent via c before c is closed. On the returned channel d (with dual session type to d'), we receive the result, which is output before closing d. Thus, Chan essentially provides the end-points of a bi-directional channel. The type of client can be given as:[1]

```
client :: (Dual s ~ Recv Bool End, Link s) =>
          Chan (Send Int (Send Int (Send (Chan s) End))) -> IO ()
```

Swapping a send for a recv, or vice versa, means the program will no longer type check. Likewise, sending or receiving a value of the wrong type or at the wrong point in the interaction is also a type error.

10.2.3 Limitations

The approach described so far captures sequences of actions, but cannot enforce exact linear usage of channels; nothing is enforcing the idiom of shadowing each channel variable once it is used. For example, the first few lines of the above example client could be modified to:

```
client c = do
    c <- send c (42 :: Int)
    _ <- send c 53
    c <- send c 53
    ...
```

By discarding the linear variable-shadowing discipline, an extra integer is sent on c in the third line. This is not prevented by the types. While the typing captures the order of interactions, it allows every action to be repeated, and entire session interactions to be repeated. Thus, the session type theory captured above is a kind of Kleene-star-expanded version where sequences of actions in a session type $A_1.\ldots.A_n.$**end** are effectively expanded to allow arbitrary repetition of individual actions and entire interaction sequences: $(A_1^*.\ldots.A_n^*)^*.$**end**.

   We thus need some additional mechanism for enforcing proper linear use of channels, rather than relying on the discipline or morality of a programmer writing against a communication specification. We have also not yet considered branching behaviour or recursion, which are highlighted in the approaches from the literature.

---

[1] A more general type can be inferred, since both Int types can be replaced with arbitrary types of the Num class and Bool with an arbitrary type of the Show class.

10.3 Approaches in the Literature

There are various different approaches in the literature providing session-typed concurrent, communicating programs in Haskell with linearity:

- Neubauer and Thiemann [9] give an encoding of first-order single-channel session types with recursion;

- Using *parameterised monads*, Pucella and Tov [13] provide multiple channels, recursion, and some building blocks for delegation, but require manual manipulation of a session type context;
  (http://hackage.haskell.org/package/simple-sessions)

- Sackman and Eisenbach [15] provide an alternate approach where session types are constructed via a value-level witness;
  (http://hackage.haskell.org/package/sessions)

- Imai et al. [5] extend Pucella-Tov with delegation and a more user-friendly approach to handling multiple channels;
  (http://hackage.haskell.org/package/full-sessions)

- Orchard and Yoshida [11] use an embedding of effect systems into Haskell via graded monads based on a formal encoding of session-typed π-calculus into PCF with an effect system;
  (https://github.com/dorchard/sessions-in-haskell)

- Lindley and Morris [8] provide a *finally tagless* embedding of the GV session-typed functional calculus into Haskell, building on a linear λ-calculus embedding due to Polakow [12].
  (https://github.com/jgbm/GVinHs)

The following table summarises the various implementations' support for desirable session-type implementation features: recursion, delegation, multiple channels (for which we summarise how session contexts are modelled and its members are accessed), idiomatic Haskell code, and whether manual user-given specification of session types is feasible.

	NT04	PT08	SE08	IYA10	OY16	LM16
Recursion	✓	✓ deBruijn	✓labels	✓		Affine
Delegation			✓	✓	✓	✓
Multi-channel		✓	✓	✓	✓	✓
— Contexts		stack	map	list	map	list
— Access		positional	labels	deBruijn	names	member
Idiomatic	✓	✓		✓✓	✓✓	✓
Manual spec	✓	✓	✓ value		✓	✓

We characterise idiomatic Haskell as code which does not require interposing combinators to replace standard syntactic elements of functional languages, *e.g.*, λ-abstraction, application, *let*-binding, recursive bindings, and variables. In the above, for example, PT08 has one tick and IYA10 has two since PT08 must use specialised combinators for handling multiple channel variables whilst IYA10 does not require such combinators, instead using standard Haskell variables.

10.3.1 Note on Recursion and Duality

Early formulations of session types *e.g.* [18], defined duality of recursive types as $\overline{\mu\alpha.S} = \mu\alpha.\overline{S}$. Whilst this duality is suitable for tail-recursive session types, it is inadequate when recursive variables appear in a communicated type [2]. For example, the type $\mu\alpha.!\langle\alpha\rangle$ should have the unfolded dual type of $?\langle\mu\alpha.!\langle\alpha\rangle\rangle$ but under the earlier approach is erroneously $?\langle\mu\alpha.?\langle\alpha\rangle\rangle$. In Section 10.1, duality was defined using dualisable recursion variables, akin to Lindley and Morris [7], which solves this problem. However, all session-type implementations which support delegation and recursion (PT08, IYA10, OY16) implement the erroneous duality. This is an area for implementations to improve upon.

10.3.2 Single Channel; Neubauer and Thiemann [9]

Neubauer and Thiemann provided the first published implementation of session types in Haskell. Their implementation is based on a translation from a simple session-typed calculus that is restricted to one end of a single communication channel. The session type theory is first order (*i.e.*, no channel delegation), but includes alternation and recursive sessions using a representation based on the following data types:

```
data NULL         = NULL        -- the closed session
data EPS          = EPS         -- the empty session
data SEND_MSG m r = SEND_MSG m r -- send message m, then session r
data RECV_MSG m r = RECV_MSG m r -- receive message m, then session r
data ALT l r      = ALT l r      -- alternative session: either l or r
data REC f        = REC (f (REC f)) -- fixed-point of a parametric type
```

Session types are specified by defining a value using the above data constructors which provides a homomorphic type-level representation of the session type. For example, the following value and its type describes a sequence of receiving two integers and sending a bool:

```
simple = RECV_MSG intW (RECV_MSG intW (SEND_MSG boolW EPS))
```

where intW = 0, boolW = False witness the integer and boolean types and simple :: RECV_MSG Int (RECV_MSG Int (SEND_MSG Bool EPS)).

Duality is provided by parameterising such specification values by place-holders for the 'send' and 'receive' actions which can then be applied to SEND_MSG and RECV_MSG in one order or the other to provide the dual specification. For example, the above specification becomes:

```
simple (send :: (forall x y . x -> y -> s x y))
       (recv :: (forall x y . x -> y -> r x y)) =
     recv intW (recv intW (send boolW EPS))
```

This function specialisations to the dual behaviour of the server via (simple RECV_MSG SEND_MSG) and the client (simple SEND_MSG RECV_MSG).

A recursive session type $(\mu\beta.\gamma)$ is represented as a fixed-point, via REC, of a parametric data type representing γ. For Example 1, the body of the server's recursive type &{eq :?$\mathbb{Z}$.?$\mathbb{Z}$.!$\mathbb{B}$.α, nil : **end**} can be represented by the following data type, which also uses ALT:

```
data Exm s r a =
  MkExm (ALT (r Label (r Int (r Int (s Bool a)))) (r Label EPS))
```

where data Label = Eq | Nil. The full specification is constructed as:

```
exampleSpec (send :: (forall x y . x -> y -> s x y))
            (recv :: (forall x y . x -> y -> r x y)) = a0
  where a0 = REC (MkExm (ALT
             (recv Eq (recv intW (recv intW (send boolW a0))))
             (recv Nil EPS)))
```

A computation at one end-point of a channel is represented by the Session data type which is indexed by the session type representation and internally wraps the IO monad. The main communication primitives produce values of Session:

```
class SEND st message nextst | st message -> nextst where
    send     :: message -> Session nextst () -> Session st ()
class RECEIVE st cont | st -> cont where
    receive :: cont -> Session st ()
close    :: Session NULL () -> Session EPS ()
```

The SEND class provides sending values of type message given a continuation session with specification nextst, returning a computation with specification st. The *functional dependency* st message -> nextst enforces that the instantiation of st and message uniquely determines nextst. An instance SEND (SEND_MSG m b) m b specialises send to:

```
send :: m -> Session b () -> Session (SEND_MSG m b) ()
```

The RECEIVE class abstracts receiving, taking a general continuation and returning a computation with communication specified by st. For RECV_MSG and ALT, the receive method is specialised at the types:

```
receive :: (m -> Session x ()) -> Session (RECV_MSG m x) ()
receive :: (RECV s m, RECV s' m') => ALT m m' -> Session (ALT s s') ()
```

with RECV shorthand for RECEIVE. The Example 1 server can be defined:

```
exampleServer socket = do
    (h, _ ,_) <- accept socket
    let session    = receive (ALT (\Eq -> recvNum1) (\Nil -> finish))
        recvNum1    = receive (\x -> recvNum2 x)
        recvNum2 x = receive (\y -> sendEq x y)
        sendEq x y = send (x == y) session
        finish      = close (io $ putStrLn "Fin.")
    str <- hGetContents h
    run session (exampleSpec SEND_MSG RECV_MSG) str h
```

The communication pattern of session (line 3), encoded by its type, must match that of the specification exampleSpec SEND_MSG RECV_MSG as enforced by the run deconstructor which expects a computation of type Session st a and a corresponding specification value of type st. Any deviation from the specification is a static type error. Since computations are wrapped in the indexed Session type, they can only be executed via run and thus are always subject to this linearity check. This contrasts with the simple approach in Section 10.2 where actions on channels produce computations in the (unindexed) IO monad, which allowed arbitrary repetition of actions within the specified behaviour.

10.3.3 Multi-Channel Linearity; Pucella and Tov [13]

Pucella and Tov improve on the previous approach, providing multi-channel session types with recursion and some higher-order support, though not full delegation. Similarly to Neubauer-Thiemann, the basic structure of session

types is represented by several data types: binary type constructors :!: and
:?: for send and receive and Eps for a closed session. Offering and selecting
of choices are represented by binary type constructors :&: and :+:, which
differs to Neubauer-Thiemann who coalesce these dual perspectives into ALT.
Duality is defined as a relation via a type class with a functional dependency
enforcing bijectivity:

```
class Dual r s | r -> s, s -> r
instance Dual r s => Dual (a :!: r) (a :?: s)
instance Dual r s => Dual (a :?: r) (a :!: s)
instance Dual Eps Eps
instance (Dual r1 s1, Dual r2 s2) => Dual (r1 :+: r2) (s1 :&: s2)
instance (Dual r1 s1, Dual r2 s2) => Dual (r1 :&: r2) (s1 :+: s2)
instance Dual r s => Dual (Rec r) (Rec s)
instance Dual (Var v) (Var v)
```

Recursive session types use a De Bruijn encoding where Rec r introduces
a new recursive binder over r and Var n is the De Bruijn index of the nth
binder where n has a unary encoding (*e.g.*, Z, S Z, *etc.*).

Communication is provided by channels Channel c (which we abbreviate
to Chan c) where the type variable c represents the name of the channel. The
session type of a channel c is then a *capability* provided by the data type Cap
c e s which associates session type s to channel c with an environment e of
recursive variables paired with session types.

A *parameterised monad* [1] is used to capture the session types of the
free channels in a computation. Parameterised monads generalise monads to
type constructors indexed by a pair of types akin to pre- and post-conditions.
Its operations are represented via the class:

```
class ParameterisedMonad (m :: k -> k -> * -> *) where
  (>>=) :: m p q a -> (a -> m q r b) -> m p r b
  return :: a -> m p p a
```

The "bind" operation >>= for sequential composition has type indices representing sequential composition of Hoare triples: a computation with post-
condition q can be composed with a computation with pre-condition q.
Relatedly, a pure value of type a can be lifted into a trivial computation which
preserves any pre-condition p in its post-condition.

One of the original examples of parameterised monads is for encoding
first-order single-channel session-typed computations [1]. This is expanded
upon by Pucella and Tov to multi-channels. They provide a parameterised
monad Session, indexed by *stacks* of session type capabilities associated

to channels. Pre-conditions are the channel capabilities at the start of a computation, and post-conditions are the remaining channel capabilities after computation.

Stacks are constructed out of tuples where () is the empty stack. For example, (Chan c e s, (Chan c' e' s', ())) is a stack of two capabilities for channels c and c'. The core communication primitives then manipulate the capability at the top of the stack:

```
send :: Chan c -> a -> Session (Cap c e (a :!: s), x) (Cap c e s, x) ()
recv :: Chan c ->      Session (Cap c e (a :?: s), x) (Cap c e s, x) a
```

For example, sending a value of type a on channel c requires the capability a :!: s at the top of the stack for c in the pre-condition, which becomes s in the post condition. Branching follows a similar scheme.

Recursive behaviour is mediated by combinators which provide the unrolling of a recursive session type (enter) and referencing a bound De-Bruijn-indexed recursive variable via zero and suc:

```
enter :: Chan c -> Session (Cap c e (Rec s), x) (Cap c (s, e) s, x) ()
zero :: Chan c -> Session (Cap c (s,e) (Var Z), x) (Cap c (s,e) s, x) ()
suc  :: Session (Cap t (r, e) (Var (S v)), x) (Cap t e (Var v), x) ()
```

Thus, entering a recursive sessions type adds the body of the type onto the top of De-Bruijn environment stack; zero peeks the session type from the top of the stack and suc pops and decrements the variable. The original paper has a slightly different but equivalent formulation for suc– the above is provided by the online implementation.

Example 1 can then be implemented as follows:

```
server c = do                  client c = do
    enter c                        enter c
    loop                           sel1 c
  where loop = offer c             send c 42
           (do x <- recv c         send c 53
               y <- recv c         x <- recv c
               send c (x == y)     io $ putStrLn $ "Got: " ++ show x
               zero c              zero c
               loop)              sel2 c
           (close c)             close c
```

The types of both can be inferred. For example, the type of server is:

```
server :: Eq a => Chan t -> Session
    (Cap t e (Rec ((a :?: (a :?: (Bool :!: Var Z))) :&: Eps)), x) x ()
```

Dual endpoints of a channel are created by functions `accept` and `request` capturing the notion of *shared channels* [18], called a *rendezvous* here:

```
accept :: Rendezvous r ->
   (forall t. Chan t -> Session (Cap t () r, x) y a) -> Session x y a
request :: Dual r r' => Rendezvous r ->
   (forall t. Chan t -> Session (Cap t () r', x) y a) -> Session x y a
```

Thus, for our example, the server and client processes can be composed by the following code which statically enforces duality through `request`:

```
example = runSession $ do rv <- io newRendezvous
                          forkSession (request rv client)
                          accept rv server
```

With `forkSession :: Session x () () -> Session x () ()` enforcing a closed final state for the forked subcomputation (line 2). Whilst the above code is fairly idiomatic Haskell (modulo the management of recursion variables), the example has only one channel. In the context of multiple channels, the capability of a channel may not be at the top of the session environment stack, thus context manipulating combinators must be used to rearrange the stack:

```
swap :: Session (r, (s, x)) (s, (r, x)) ()
dig  :: Session x x' a -> Session (s, x) (s, x') a
```

where `swap` is akin to exchange and `dig` moves down one place in the stack. Thus, multi-channel code requires the user to understand the type-level stack representation and to manipulate it explicitly. Multi-channel code is therefore non-idiomatic, in the sense that we can't just use Haskell variables on their own.

Example 2 cannot be captured as channels cannot be passed. Pucella and Tov provide a way to send and receive capabilities, however there is no primitive for sending channels along with an associated capability. Imai *et al.* describe a way to build this on top of Pucella and Tov's approach with an existentially quantified channel name, however this is still limited by the lack of a new channel constructor. Instead, channel delegation could be emulated with global shared channels for every delegation but this shifts away from the message-passing paradigm.

In their paper, Pucella and Tov use the `ixdo` notation which copies exactly the style of the `do` notation for monads, but which is desugared by a pre-processor into the operations of the parameterised monad. In modern GHC, this can be replaced with the `RebindableSyntax` extension which desugars the standard `do` notation using any functions in scope named (>>=) and return, regardless of their type. The operations of a parameterised monad can therefore usurp the regular monad operations. Thus, the non-idiomatic pre-processed `ixdo` notation can be replaced with idiomatic `do` notation. The same applies to the work of Sackman and Eisenbach (Section 10.3.4) and Imai *et al.* (Section 10.3.5) who also use parameterised monads. Similarly, GHC's rebindable syntax is reused by Orchard and Yoshida with *graded monads* (Section 10.3.6).

10.3.4 An Alternate Approach; Sackman and Eisenbach [15]

In their unpublished manuscript, Sackman and Eisenbach provide an implementation also using a parameterised monad but with quite a different formulation to Pucella and Tov. The encoding of session environments is instead through type-level finite maps from channel names (as types) to session types. This requires significantly more type-level machinery (implemented mostly using classes with functional dependencies), resulting in much more complicated types than Pucella-Tov. However, they provide a parameterised monad `SessionType` for constructing session-type witnesses at the value level (similarly to Neubauer-Thiemann) which is much easier to read and write than the corresponding type-level representation. Session-based computations are then constructed through another parameterised monad called `SessionChain`.

Sackman-Eisenbach represent session types by type-level lists (via constructors `Cons` and `Nil`) of actions given by parametric data types `Send`, `Recv`, `Select`, `Offer`, `Jump`, and (non parametric) `End` similar to the other representations. For Example 2, the recursive session type of the server can be constructed via value-level terms as:

```
(serverSpec, a) = makeSessionType $ do
  a <- newLabel
  let eq = do {recv intW; recv intW; recvSession (send boolW); jump a}
  a .= offer (eq ~|~ end ~|~ BLNil)
  return a
```

This uses the `SessionType` parameterised monad indexed by `TypeState` types which have further indices managing labels and representing session types. The `makeSessionType` function returns a pair of a value capturing the specification `serverSpec` and the component of the type labelled by a. Labels are used to associate types to channels and for recursive types, where `newLabel` generates a fresh label bound to a. The third line associates to a the expected session behaviour: a choice is offered where `offer` takes a list of behaviours constructed by ~|~ (cons) and BLNil (nil). As in Neubauer-Thiemann, intW and boolW are value witnesses of types. The recursive step is via jump on label a. The type of send illustrates the SessionType parameterised monad:

```
send :: (TyList f, TyList fs) => t -> SessionType
  (TypeState n d u (Cons (lab, f) fs))
  (TypeState n d u (Cons (lab, (Cons (Send (Normal, t)) f)) fs)) ()
```

The final parameter to TypeState provides a type-level list of labelled session types (themselves lists). In the post-condition, the session type f from the head of the list in the pre-condition has Send consed onto the front, parameterised by (Normal, t) indicating the value type t.

The session-type building primitives have computation building counterparts (whose names are prefixed with s, *e.g.* ssend) returning computations in the SessionChain parameterised monad. We elide the details, but show the implementation of the server from Example 2:

```
server = do
  cid <- fork serverChan dual (cons (serverSpec, notDual) nil) client
  c <- createSession serverSpec dual cid
  withChannel c (soffer ((do
      x <- srecv
      y <- srecv
      recvChannel c (\d ->
        withChannel d (do { ssend (x == y); sjump })))
  ~||~ (return ()) ~||~ OfferImplsNil))
```

The session type specification serverSpec is linked to computation to enforce linearity via fork. Above, client refers to the client code which is forked and given a channel whose behaviour is dual to that created locally by createSession, specified by serverSpec. The sjump primitive provides the recursive behaviour but has no target which is implicitly provided by the specification. The withChannel primitive "focuses" the computation on a particular channel such that the communication primitives are not parameterised by channels, similar to Neubauer-Thiemann. This has some advantage over Pucella-Tov, which required manual session-context manipulation, though channel variables still cannot be used directly here. Combined with the complicated type encoding, we therefore characterise this approach as the least idiomatic.

It should be noted that since the appearance of their manuscript, the type checking of functional dependencies in GHC has become more strict (particularly with the additional *Coverage Condition* [16, Def. 7]). At the time of writing, the latest available online implementation of Sackman-Eisenbach fails to type check in multiple places due to the coverage conditions added later to GHC. It is not immediately clear how to remedy this due to their reliance on functional dependencies which do not obey the new coverage condition.

10.3.5 Multi-Channels with Inference; Imai et al. [5]

Imai, Yuen, and Agusa directly extend the Pucella-Tov approach, providing type inference, delegation, and solving the deficiencies with accessing multiple channels. They replace the positional, stack-based approach for multiple channels with a De Bruijn index encoding which is handled implicitly at the type level. For example, send has type

```
send :: (Pickup ss n (Send v a), Update ss n a ss', IsEnded ss F)
    => Channel t n -> v -> Session t ss ss' ()
```

Computations are modelled by the parameterised monad Session as before, but now pre- and post-condition indices ss and ss' are type-level lists of session types, rather than a labelled stack. Whilst these structures are isomorphic, the way session types are accessed within the list representation differs considerably.

A channel Channel t n has a type-level natural number n representing the position of the channel's session type in the list. The constraint Pickup above specifies that at the nth position in ss is the session type Send v a. The constraint Update then states that ss' is the list of session types produced by replacing the nth position in ss with the session type a. The rest of the communication primitives follow a similar scheme to the above, generalising Pucella-Tov primitives to work with the De Bruijn indices instead of just the capability at the top of the stack.

A fresh channel can be created by the following combinator:

```
new :: SList ss l => Session t ss (ss:>Bot) (Channel t l)
```

where l is the length of the list ss as defined by the constraint SList, and thus is a fresh variable for the computation.

Using this library leads to highly idiomatic Haskell code, with no additional combinators required for managing the context of session-typed channels. Both examples can be implemented, with code similar to that shown for Pucella-Tov in Section 10.3.3. The one downside of this approach however is that the types, whilst they can be inferred (which is one of the aims of their work), are complex and difficult to read, let alone write. Relatedly, the type errors can be difficult to understand due to the additional type-level mechanisms for managing the contexts.

10.3.6 Session Types via Effect Types; Orchard and Yoshida [11]

Orchard and Yoshida studied the connection between effect systems and session types. One part of the work showed the encoding of a session-typed π-calculus into a parallel variant of PCF with a general, parameterised effect system. This formal encoding was then combined with an approach for embedding effect systems in Haskell [10] to provide a new implementation of session-typed channels in Haskell. The implementation supports multiple channels in an idiomatic style, delegation, and a restricted form of recursion (affine recursion only).

The embedding of general effect systems in Haskell types is provided by a *graded monad* structure, which generalises monads to type constructors indexed by a type-representation of effect information. This "effect type" has the additional structure of a monoid, encoded using type families. The graded monad structure in Haskell is defined:

```
class Effect (m :: ef -> * -> *) where
    type Unit m :: ef
    type Plus m (f :: ef) (g :: ef) :: ef
    return ::  a -> m (Unit m) a
    (>>=) ::  m f a -> (a -> m g b) -> m (Plus m f g) b
```

Thus a value of type `m f a` denotes a computation with effects described by the type index `f` of kind `ef`. The `return` operation lifts a value to a trivially effectful computation, marked with the type `Unit m`. The "bind" operation (`>>=`) provides the sequential composition of effectful computations, with effect information composed by the type-level binary function `Plus m`. The session type embedding is provided by a graded monad structure for the data type `Process`:

```
data Process (s :: [Map Name Session]) a = Process (IO a)
```

Type indices `s` are finite maps of the form `'[c :-> s, d :-> t, ...]` mapping channel names `c, d` to session types `s, t`. The `Session` kind is given by a data type (representing a standard grammar of session types) promoted by the *data kinds* extension of GHC to the kind-level.

The `Plus` type operation of the `Process` graded monad takes the *union* of two finite maps and sequentially composes the session types of any channels that appear in both of the finite maps. This relies on the closed type family feature of GHC to define type-level functions that can match on their types, *e.g.*, to compare types for equality.

The core send and received primitives then have the following types:

```
send  :: Chan c -> t -> Process '[c :-> t :! End] ()
recv  :: Chan c       -> Process '[c :-> t :? End] t
```

In each, the type-index on Process gives a singleton finite map from the channel name c to the session type. We elide the rest of the combinators. Duality is enforced when a pair of channel endpoints is created by new:

```
new :: (Duality env c) => ((Chan (Ch c), Chan (Op c)) -> Process env t)
    -> Process ((env :\ (Op c)) :\ (Ch c)) t
```

where :\ removes a channel's session type from the environment.

A non-recursive implementation of Example 2 can be defined:

```
server (c :: (Chan (Op "c"))) =    client (c :: (Chan (Ch "c"))) = do
 do l <- recv c                      send c L
    case l of                        subL' c $ do
      L -> subL $ do                   send c 42
             x <- recv c               send c 53
             y <- recv c               new (\(d :: (Chan (Ch "d")), d') ->
             k <- chRecv c               do chSend c d
             k (\d -> send d (x == y))      x <- recv d'
      R -> subR $ subEnd c (return ())     print $ "Got: " ++ show x
```

which are composed by new (\(c, c') -> client c `par` server c').

One advantage of this approach is that most types are easy to write by hand, with a succinct understandable presentation in terms of the finite maps from channel names to session types. Furthermore, the use of multiple channels is idiomatic, using Haskell's normal variables. The major disadvantage of this approach is that the user must give their own explicit type-level names to the channels, *e.g.*, type signatures like Chan (Ch "c") above. For simple examples this is not a burden, but manually managing uniqueness of variables does not scale well.

Furthermore, the approach is brittle due to complex type-level representation and manipulations of finite maps. For example, GHC has difficulty reasoning about the type-level *union* operation (used as Plus) when applied to types involving some polymorphism.

10.3.7 GV in Haskell; Lindley and Morris [8]

GV is a session-typed linear functional calculus, proposed by [17], based on the work of Gay and Vasconcelos [3], and adapted further by Lindley and Morris [6]. The GV presented by Lindley and Morris aims at re-use of standard components, defined as an extension of the linear λ-calculus with session-typed communication primitives. This provides a basis for their Haskell implementation by reusing an embedding of the linear λ-calculus into Haskell due to Polakow [12]. Polakow's embedding provides a "tagless final" encoding of the linear-λ calculus (LLC), meaning that terms of LLC are represented by functions of a type class, whose interpretation/implementation can be varied based on the underlying type. Furthermore, the embedding uses higher-order abstract syntax (HOAS) *i.e.*, binders in LLC are represented by Haskell binders.

To represent the linear types notion of context *consumption*, contexts are split in two with judgments of the form: $\Delta_I \setminus \Delta_O \vdash e : A$ with *input context* Δ_I and output context Δ_O which remains after computing e and thus after some parts of Δ_I have been consumed. Contexts come equipped with the notion of a "hole" (written $\square$) denoting a variable that has been consumed. For example, a linear variable use is typed by $\Delta, x : A, \Delta' \setminus \Delta, \square, \Delta' \vdash x : A$.

The embedding of this linear type system uses natural numbers to represent variables in judgements. Judgements are represented by types `repr` `:: Nat -> [Maybe Nat] -> [Maybe Nat] -> * -> *`. Thus, the LLC term representation is a type indexed by four pieces of information: a natural number denoting a fresh name for a new variable, the input context (a list of `Maybe Nat` where `Just n` is a variable and `Nothing` denotes $\square$), the output context, and the term type.

The core of the embedding for the linear function space fragment, is then given by the LLC class, parameterised by a `repr` type:

```
class LLC (repr :: Nat -> [Maybe Nat] -> [Maybe Nat] -> * -> *) where
  llam :: (LVar repr v a -> repr (S v) (Just v ': i) (Box ': o) b)
         -> repr v i o (a -<> b)
  (^) :: reprv v i h (a -<> b) -> repr v h o a -> prepr v i o b
```

where `LVar` represents linear variables, defined as the type `forall v i o .` `(Consume x i o) => repr v i o` a describing that using a variable leads to its consumption for all input and output contexts `i` and `o`.

The session primitives of GV are added atop the LLC embedding via another tagless final encoding (we elide the primitives for branching):

```
class GV (ch :: * -> *) repr where
  send :: DualS s => repr v i h t -> repr v h o (ch (t <!> s))
                                  -> repr v i o (ch s)
  recv :: DualS s => repr v i o (ch (t <?> s)) -> repr v i o (t * ch s)
  wait :: repr v i o (ch EndIn)                 -> repr v i o One
  fork :: DualS s => repr v i o (ch s -<> ch EndOut)
               -> repr v i o (ch (Dual s))
```

The types involve duality as both a predicate (type constraint) `DualS` and as a type-level function `Dual`.

The approach does not provide recursive sessions so we implement a non-recursive version of Example 1 as:

```
server = llam $ \c ->           client = llam $ \c ->
  recv c 'bind' (llp $ \x c ->    send (const 42) c 'bind' (llam $ \c ->
  recv c 'bind' (llp $ \y c ->    send (const 53) c 'bind' (llam $ \c ->
  send (const (==) $$$ x $$$ y) c))   recv c    'bind' (llp $ \r c ->
example = fork server 'bind' client  wait c    'bind' (llz $ ret r))))
```

This approach cleanly separates the notion of linearity from the channel capabilities of session types. The main downside is that application, λ-abstraction, and composition of terms must be mediated by the combinators of the LLC embedding. Therefore, the approach does not support idiomatic Haskell programming.

10.4 Future Direction and Open Problems

The table at the beginning of Section 10.3 (p. 226) indicates that there is no one implementation that provides all desirable features: a session-typed library for communication-safe concurrency with linearity, delegation, multiple-channels, recursion, idiomatic Haskell code, and the ability to easily give session type specifications by hand. Furthermore, none correctly implements duality with respect to recursion (Section 10.3.1).

So far there appears to be a trade-off between these different features. Pucella and Tov provide an idiomatic system with relatively simple types, but require the manual management of the capability stack. The work of Imai *et al.* provides a highly idiomatic system, but the types are hard to manipulate and understand. Orchard and Yoshida provide types that are easy to write, but at the cost of forcing the user to manually manage fresh channel names. Lindley and Morris handle variables idiomatically, but require additional combinators for application, λ-abstraction and term composition. Sackman and Eisenbach provide session types which are easily specified by-hand with a value witness, but with non-idiomatic code and hard to manipulate types.

One possible solution is to adapt the approach of Orchard and Yoshida with a way to generate fresh channel names at the type-level automatically via a GHC *type checker plugin* (see, *e.g.*, [4]). Alternatively, existential names can be used for fresh names. However, the implementation of type-level finite maps relies on giving an arbitrary ordering to channel names (for the sake of normalisation) which is not possible for existential names. In which case, a type-checker plugin could provide built-in support for finite maps more naturally, rather than using the current (awkward) approach of Orchard and Yoshida.

We have examined the six major session type implementations for Haskell in this chapter. All of them provide static linear checks, leveraging Haskell's flexible type system, but all have some deficiencies; finding a perfectly balanced system remains an open problem.

Acknoweldgements We thank Garrett Morris and the anonymous reviewers for their helpful comments. This work was supported in part by EPSRC grants EP/K011715/1, EP/K034413/1, EP/L00058X/1, EP/M026124/1, and EU project FP7-612985 UpScale.

References

[1] Robert Atkey. Parameterised notions of computation. *Journal of functional programming*, 19(3–4):335–376, 2009.

[2] Giovanni Bernardi, Ornela Dardha, Simon J. Gay, and Dimitrios Kouzapas. On duality relations for session types. In *Trustworthy Global Computing 2014*, pages 51–66, 2014.

[3] Simon J. Gay and Vasco T. Vasconcelos. Linear type theory for asynchronous session types. *Journal of Functional Programming*, 20(01): 19–50, 2010.

[4] Adam Gundry. A typechecker plugin for units of measure: domain-specific constraint solving in GHC Haskell. In *ACM SIGPLAN Notices*, volume 50, pages 11–22. ACM, 2015.

[5] Keigo Imai, Shoji Yuen, and Kiyoshi Agusa. Session Type Inference in Haskell. In *PLACES*, pages 74–91, 2010.

[6] Sam Lindley and J. Garrett Morris. A semantics for propositions as sessions. In *ESOPb*, pages 560–584. Springer, 2015.

[7] Sam Lindley and J. Garrett Morris. Talking Bananas: Structural Recursion for Session Types. In *Proceedings of the 21st ACM SIGPLAN International Conference on Functional Programming*, ICFP 2016, pages 434–447. ACM, 2016.

[8] Sam Lindley and J Garrett Morris. Embedding session types in haskell. In *Proceedings of the 9th International Symposium on Haskell*, pages 133–145. ACM, 2016.

[9] Matthias Neubauer and Peter Thiemann. An Implementation of Session Types. In *PADL*, volume 3057 of *LNCS*, pages 56–70. Springer, 2004.

[10] Dominic Orchard and Tomas Petricek. Embedding effect systems in Haskell. *ACM SIGPLAN Notices*, 49(12):13–24, 2015.

[11] Dominic Orchard and Nobuko Yoshida. Effects as Sessions, Sessions as Effects. *ACM SIGPLAN Notices*, 51(1):568–581, 2016.

[12] Jeff Polakow. Embedding a Full Linear Lambda Calculus in Haskell. *ACM SIGPLAN Notices*, 50(12):177–188, 2016.

[13] Riccardo Pucella and Jesse A. Tov. Haskell Session Types with (Almost) no Class. In *Proc. of Haskell Symposium '08*, pages 25–36. ACM, 2008. ISBN 978-1-60558-064-7.

[14] John H. Reppy. CML: A Higher-Order Concurrent Language. In *PLDI*, pages 293–305, 1991.

[15] Matthew Sackman and Susan Eisenbach. Session Types in Haskell (Updating Message Passing for the 21st Century), 2008. Technical report, Imperial College London.

[16] Martin Sulzmann, Gregory J Duck, Simon Peyton-Jones, and Peter J Stuckey. Understanding functional dependencies via constraint handling rules. *Journal of Functional Programming*, 17(01):83–129, 2007.

[17] Philip Wadler. Propositions as sessions. *Journal of Functional Programming*, 24(2–3):384–418, 2014.

[18] Nobuko Yoshida and Vasco Thudichum Vasconcelos. Language Primi tives and Type Discipline for Structured Communication-Based Programming Revisited: Two Systems for Higher-Order Session Communication. *Electr. Notes Theor. Comput. Sci.*, 171(4):73–93, 2007.

11

An OCaml Implementation of Binary Sessions

Hernán Melgratti[1,2] and Luca Padovani[3]

[1]Departamento de Computación, Universidad de Buenos Aires, Argentina
[2]CONICET-Universidad de Buenos Aires, Instituto de Investigación en Ciencias de la Computación (ICC), Buenos Aires, Argentina
[3]Dipartimento di Informatica, Università di Torino, Italy

Abstract

In this chapter we describe FuSe, a simple OCaml module that implements binary sessions and enables a hybrid form of session type checking without resorting to external tools or extensions of the programming language. The approach combines static and dynamic checks: the former ones are performed at compile time and concern the structure of communication protocols; the latter ones are performed as the program executes and concern the linear usage of session endpoints. We recall the minimum amount of theoretical background for understanding the essential aspects of the approach (Section 11.1) and then describe the API of the OCaml module throughout a series of simple examples (Section 11.2). In the second half of the chapter we detail the implementation of the module (Section 11.3) and discuss a more complex and comprehensive example, also arguing about the effectiveness of the hybrid approach with respect to the early detection of protocol violations (Section 11.4). We conclude with a survey of closely related work (Section 11.5).

The source code of FuSe, which is partially described in this chapter and can be used to compile and run all the examples given therein, can be downloaded from the second author's home page.

11.1 An API for Sessions

We consider the following grammar of types and session types

$$t, s ::= \texttt{bool} \mid \texttt{int} \mid \alpha \mid T \mid [\texttt{l}_i : t_i]_{i \in I} \mid \cdots$$
$$T, S ::= \texttt{end} \mid !t.T \mid ?t.T \mid \&[\texttt{l}_i : T_i]_{i \in I} \mid \oplus[\texttt{l}_i : T_i]_{i \in I} \mid A \mid \overline{A}$$

where types, ranged over by t and s, include basic types, type variables, session types, disjoint sums, and possibly other (unspecified) types. Session types, ranged over by T and S, comprise the usual constructs for denoting depleted session endpoints, input/output operations, branches and choices, as well as possibly dualized session type variables A, B, etc.

The *dual* of a session type T, written $\overline{T}$, is obtained as usual by swapping input and output operations and is defined by the following equations:

$$\overline{\overline{A}} = A \qquad \overline{(?t.T)} = !t.\overline{T} \qquad \overline{\&[\texttt{l}_i : T_i]_{i \in I}} = \oplus[\texttt{l}_i : \overline{T_i}]_{i \in I}$$
$$\overline{\texttt{end}} = \texttt{end} \qquad \overline{(!t.T)} = ?t.\overline{T} \qquad \overline{\oplus[\texttt{l}_i : T_i]_{i \in I}} = \&[\texttt{l}_i : \overline{T_i}]_{i \in I}$$

Following Gay and Vasconcelos [4], our aim is to incorporate binary sessions into a (concurrent) functional language by implementing the API shown in Table 11.1. The `create` function creates a new session and returns a pair with its two peer endpoints with dual session types. The `close` function is used to signal the fact that a session is completed and no more communications are supposed to occur in it. The `send` and `receive` functions are used for sending and receiving a message, respectively: `send` sends a message of type α over an endpoint of type $!\alpha.A$ and returns the same endpoint with its type changed to A to reflect that the communication has occurred; `receive` waits for a message of type α from an endpoint of type $?\alpha.A$ and returns a pair with the message and the same endpoint with its type changed to A. The `branch` and `select` functions deal with sessions that may continue along different paths of interaction, each path being associated with a *label* $\texttt{l}_i$. Intuitively, `select` takes a label $\texttt{l}_k$ and an endpoint of type $\oplus[\texttt{l}_i : A_i]_{i \in I}$

Table 11.1   Application programming interface for binary sessions

```
val create  : unit → A × A̅
val close   : end → unit
val send    : α → !α.A → A
val receive : ?α.A → α × A
val select  : (A̅ₖ → [lᵢ : A̅ᵢ]ᵢ∈ᵢ) → ⊕[lᵢ : Aᵢ]ᵢ∈ᵢ → Aₖ
val branch  : &[lᵢ : Aᵢ]ᵢ∈ᵢ → [lᵢ : Aᵢ]ᵢ∈ᵢ
```

where $k \in I$, sends the label over the endpoint and returns the endpoint with its type changed to A_k, which is the continuation corresponding to the selected label. The most convenient OCaml representation for labels is as functions that *inject* an endpoint (say, of type $\overline{A_k}$) into a disjoint sum $[1_i : \overline{A_i}]_{i \in I}$ where $k \in I$. This explains the type of select's first argument. Dually, receive waits for a label from an endpoint of type $\&[1_i : A_i]_{i \in I}$ and returns the continuation endpoint injected into a disjoint union.

We note a few more differences between the API we implement in this chapter and the one described by Gay and Vasconcelos [4]. First of all, we use parametric polymorphism to give session primitives their most general type. Second, we have a single function create to initiate a new session instead of a pair of accept/request functions to synchronize a service and a client. Our choice is purely a matter of simplicity, the alternative API being realizable on top of the one we present (the API implemented in the FuSe distribution already provides for the accept/request functions, which we will see at work in Section 11.4). Finally, our communication primitives are *synchronous* in that output operations block until the corresponding receive is performed. Again, this choice allows us to provide the simplest implementation of these primitives solely using functions from the standard OCaml library. Asynchronous communication can be implemented by choosing a suitable communication framework.

11.2 A Few Simple Examples

Before looking at the implementation of the communication primitives, we illustrate the API at work on a series of simple examples. In doing so, we assume that the API is defined in a module named Session. The following code implements a client of an "echo" service, a service that waits for a message and bounces it back to the client.

```
let echo_client ep x =
  let ep = Session.send x ep in
  let res, ep = Session.receive ep in
  Session.close ep;
  res
```

The parameter ep has type $!\alpha.?\beta.\text{end}$ and x has type α. The function echo_client starts by sending the message x over the endpoint ep. The construction let rebinds the name ep to the endpoint returned by the primitive send, which now has type $?\beta.\text{end}$. The endpoint is then used for receiving a message of type β from the service. Finally, echo_client closes the session and returns the received message.

The service is implemented by the echo_service function below, which uses the parameter ep of type ?α.!α.end to receive a message x and then to sent it back to the client before closing the session.

```
let echo_service ep =
  let x, ep = Session.receive ep in
  let ep = Session.send x ep in
  Session.close ep
```

There is an interesting asymmetry between (the types of) client and service in that the message x sent by the service is the very same message it receives, whereas the message res received by the client does not necessarily have the same type as the message x it sends. Indeed, there is nothing in echo_client suggesting that x and res are somewhat related. This explains the reason why the session type of the endpoint used by the client (!α.?β.end) is more general than that used by the service (?α.!α.end) aside from the fact that the two session types describe protocols with complementary actions. In particular, !α.?β.end is *not* dual of ?α.!α.end according to the definition of duality given earlier: in order to connect client and service, β must be *unified* with α. The code that connects echo_client and echo_service through a session is shown below:

```
let _ =
  let a, b = Session.create () in
  let _ = Thread.create echo_service a in
  print_endline (echo_client b "Hello, world!")
```

The code creates a new session, whose endpoints are bound to the names a and b. Then, it activates a new thread that applies echo_service to the endpoint a. Finally, it applies echo_client to the remaining endpoint b.

We now wish to generalize the echo service so that a client may decide whether to use the service or to stop the interaction without using it. A service that offers these two choices is illustrated below:

```
let opt_echo_service ep =
  match Session.branch ep with
  | `Msg ep → echo_service ep
  | `End ep → Session.close ep
```

In this case the service uses the branch primitive to wait for a label selected by the client. We use OCaml's polymorphic variant tags (`Msg and

`End in this case) as labels because they do not have to be declared explicitly, unlike data constructors of plain algebraic data types. The initial type of ep is now &[End : end, Msg : ?α. !α. end] and the value returned by branch has type [End : end, Msg : ?α. !α. end]. In the Msg branch the service behaves as before. In the End branch the service closes the session without performing any further communication.

The following function realizes a possible client for opt_echo_service:

```
let opt_echo_client ep opt x =
  if opt then
    let ep = Session.select (fun x → `Msg x) ep
    in echo_client ep x
  else
    let ep = Session.select (fun x → `End x) ep
    in Session.close ep; x
```

This function has type $\oplus$[End : end, Msg : !α.?α.end] $\rightarrow$ bool $\rightarrow$ $\alpha \rightarrow \alpha$ and its behavior depends on the boolean parameter opt: when opt is true, the client selects the label Msg and then follows the same protocol as echo_client; when opt is false, the client selects the label End and then closes the session. Note that we have to η-expand the polymorphic variant tags `Msg and `End so that their type matches that expected by select. When the same label is used several times in the same program, it is convenient to define the η-expansion once, for example as

```
let _Msg x = `Msg x
let _End x = `End x
```

Note also that the messages sent and received now have the same type in the initial type of ep. This is because of the structure of opt_echo_client, which returns either x or the message returned by the service.

A further elaboration of the echo service allows the client to send an arbitrary number of messages before closing the session. In order to describe this protocol we must extend the syntax of session types presented earlier to permit recursive types. In practice, the representation of session types we will choose in Section 11.3 allows us to describe recursive protocols by piggybacking on OCaml's support for equi-recursive types, which is enabled by passing the -rectypes option to the compiler. The implementation of the elaborated echo service is therefore a straightforward recursive function:

```
let rec rec_echo_service ep =
  match Session.branch ep with
```

```
    | `Msg ep → let x, ep = Session.receive ep in
                let ep = Session.send x ep in
                rec_echo_service ep
    | `End ep → Session.close ep
```

Note the recursive call `rec_echo_service` ep in the Msg branch, which allows the server to accept again a choice from the client after replying back to a request. The `rec_echo_service` function now expects an endpoint ep of type $rec\,A\&[End:end, Msg: ?\alpha.!\alpha.A]$ where $rec\,A\,T$ denotes the (equi-recursive) session type T in which occurrences of A stand for the session type itself.

The following client

```
let rec rec_echo_client ep =
  function
  | [] → let ep = Session.select _End ep in
         Session.close ep; []
  | x :: xs → let ep = Session.select _Msg ep in
              let ep = Session.send x ep in
              let y, ep = Session.receive ep in
              y :: rec_echo_client ep xs
```

has type $rec\,A\oplus[End:end, Msg: !\alpha.?\beta.A] \to list\,\alpha \to \beta\,list$ and repeatedly invokes the recursive echo service on each element of a list.

11.3 API Implementation

In order to implement the API presented and used in the previous sections we have to make some choices regarding the OCaml representation of session types and of session endpoints. In doing so we have to take into account the fact that OCaml's type system is not substructural and therefore is unable to statically check that session endpoints are used linearly. In the rest of this section we address these concerns and then detail the implementation of the API in Table 11.1.

Representation of session types. FuSe relies on the encoding of session types proposed by Dardha *et al.* [1] and further refined by Padovani [13]. The basic idea is that a sequence of communications on a session endpoint can be compiled as a sequence of one-shot communications on linear channels

(channels used exactly once) where each exchanged message carries the actual *payload* along with a *continuation*, namely a (fresh) channel on which the subsequent communication takes place.

The image of the encoding thus relies on two types:

- a type $\mathbb{0}$ which is not inhabited, and
- a type $\langle \rho, \sigma \rangle$ which describes channels for receiving messages of type ρ and sending messages of type σ. Both ρ and σ can be instantiated with $\mathbb{0}$ to indicate that no message is respectively received and/or sent.

The correspondence between session types T and types of the form $\langle t, s \rangle$ is given by the map $[\![\cdot]\!]$ defined below

Encoding of session types

$$[\![end]\!] = \langle \mathbb{0}, \mathbb{0} \rangle$$
$$[\![?t.T]\!] = \langle [\![t]\!] \times [\![T]\!], \mathbb{0} \rangle$$
$$[\![!t.T]\!] = \langle \mathbb{0}, [\![t]\!] \times [\![\overline{T}]\!] \rangle$$
$$[\![\&[l_i : T_i]_{i \in I}]\!] = \langle [l_i : [\![T_i]\!]]_{i \in I}, \mathbb{0} \rangle$$
$$[\![\oplus[l_i : T_i]_{i \in I}]\!] = \langle \mathbb{0}, [l_i : [\![\overline{T_i}]\!]]_{i \in I} \rangle$$
$$[\![A]\!] = \langle \rho_A, \sigma_A \rangle$$
$$[\![\overline{A}]\!] = \langle \sigma_A, \rho_A \rangle$$

and extended homomorphically to all types. We assume that for each session type variable A there exist two distinct type variables ρ_A and σ_A that are also different from any other type variable α.

For example, the session type $?\alpha.A$ is encoded as $\langle \alpha \times \langle \rho_A, \sigma_A \rangle, \mathbb{0} \rangle$, which describes a channel for receiving a message of type $\alpha \times \langle \rho_A, \sigma_A \rangle$ consisting of a component of type α (that is the actual payload of the communication) and a component of type $\langle \rho_A, \sigma_A \rangle$ (that is the continuation channel on which the rest of the communication takes place). There is a twist in the encoding of outputs for the session type of the continuation is dualized. The reason for this is that the type associated with the continuation channel in the encoding describes the behavior of the *receiver* of the continuation rather than that of the *sender*. As we will see, this twist provides us with a simple way of expressing duality relations between session types, even when they are (partially) unknown. The encodings of $\oplus[l_i : T_i]_{i \in I}$ and $\&[l_i : T_i]_{i \in I}$ follow the same lines and make use of polymorphic variant types to represent the selected or received choice. As an example, the encoding of $T = \oplus[\mathtt{End} : \mathtt{end}, \mathtt{Msg} : !\alpha.?\beta.\mathtt{end}]$ is computed as follows

$$
\begin{aligned}
[\![T]\!] &= \langle 0, [\mathtt{End} : [\![\mathtt{end}]\!], \mathtt{Msg} : [\![?\alpha.!\beta.\mathtt{end}]\!]] \rangle \\
&= \langle 0, [\mathtt{End} : \langle 0, 0 \rangle, \mathtt{Msg} : \langle \alpha \times [\![!\beta.\mathtt{end}]\!], 0 \rangle] \rangle \\
&= \langle 0, [\mathtt{End} : \langle 0, 0 \rangle, \mathtt{Msg} : \langle \alpha \times \langle 0, \beta \times [\![\mathtt{end}]\!] \rangle, 0 \rangle] \rangle \\
&= \langle 0, [\mathtt{End} : \langle 0, 0 \rangle, \mathtt{Msg} : \langle \alpha \times \langle 0, \beta \times \langle 0, 0 \rangle \rangle, 0 \rangle] \rangle
\end{aligned}
$$

If instead we consider the session type $\overline{T} = \&[\mathtt{End} : \mathtt{end}, \mathtt{Msg} : ?\alpha.!\beta.\mathtt{end}]$, then we derive:

$$
\begin{aligned}
[\![\overline{T}]\!] &= \langle [\mathtt{End} : [\![\mathtt{end}]\!], \mathtt{Msg} : [\![?\alpha.!\beta.\mathtt{end}]\!]], 0 \rangle \\
&= \langle [\mathtt{End} : \langle 0, 0 \rangle, \mathtt{Msg} : \langle \alpha \times [\![!\beta.\mathtt{end}]\!], 0 \rangle], 0 \rangle \\
&= \langle [\mathtt{End} : \langle 0, 0 \rangle, \mathtt{Msg} : \langle \alpha \times \langle 0, \beta \times [\![\mathtt{end}]\!] \rangle, 0 \rangle], 0 \rangle \\
&= \langle [\mathtt{End} : \langle 0, 0 \rangle, \mathtt{Msg} : \langle \alpha \times \langle 0, \beta \times \langle 0, 0 \rangle \rangle, 0 \rangle], 0 \rangle
\end{aligned}
$$

Remarkably we observe that the encoding of $\overline{T}$ can be obtained from that of T by swapping the two components of the resulting channel types. This is a general property:

Theorem 1 *If $[\![T]\!] = \langle t, s \rangle$, then $[\![\overline{T}]\!] = \langle s, t \rangle$.*

An equivalent way of expressing this result is the following: if $[\![T]\!] = \langle t_1, t_2 \rangle$ and $[\![S]\!] = \langle s_1, s_2 \rangle$, then

$$
T = \overline{S} \iff [\![T]\!] = [\![\overline{S}]\!] \iff t_1 = s_2 \wedge t_2 = s_1
$$

meaning that the chosen encoding allows us to reduce session type duality to type equality. This property holds also for unknown or partially known session types. In particular, $[\![A]\!] = \langle \rho_A, \sigma_A \rangle$ and $[\![\overline{A}]\!] = \langle \sigma_A, \rho_A \rangle$.

We end the discussion of session type representation with two remarks. First, although the representation of session types chosen in FuSe is based on the continuation-passing encoding of sessions into the linear π-calculus [1], we will implement the communication primitives in FuSe so that only the payload (or the labels) are actually exchanged. Therefore, the semantics of FuSe communication primitives is consistent with that given in [4] and the components corresponding to continuations in the above types are solely used to relate the types of session endpoints as these are passed to, and returned from, FuSe communication primitives. Second, the OCaml type system is not substructural and there is no way to qualify types of the form $\langle t, s \rangle$ as linear, which is a fundamental requirement for the type safety of the API. We will overcome this limitation by means of a mechanism that detects linearity violations at runtime. Similar mechanisms have been proposed by Tov and Pucella [20] and by Hu and Yoshida [5].

Table 11.2 OCaml interface of the API for binary sessions

```
module Session : sig
  type 0
  type (ρ,σ) st (* OCaml syntax for ⟨ρ,σ⟩ *)
  val create  : unit → (ρ,σ) st × (σ,ρ) st
  val close   : (0,0) st → unit
  val send    : α → (0,(α × (σ,ρ) st)) st → (ρ,σ) st
  val receive : ((α × (ρ,σ) st),0) st → α × (ρ,σ) st
  val select  : ((σ,ρ) st → α) → (0,[>] as α) st → (ρ,σ) st
  val branch  : ([>] as α,0) st → α
end
```

Having chosen the representation of session types, we can see in Table 11.2 the OCaml interface of the module that implements the binary session API. In OCaml syntax, the type ⟨*t, s*⟩ is written (*t, s*) st. There is a direct correspondence between the signatures of the functions in Table 11.2 and those shown in Table 11.1 so we only make a couple of remarks. First, we extensively use Theorem 1 whenever we need to refer to a session type and its dual. This can be seen in the signatures of create, send and select where both (ρ,σ) st and (σ,ρ) st occur. Second, in the types of select and branch the syntax [>] as α means that α can only be instantiated with a polymorphic variant type. Without this constraint the signatures of select and branch would be too general and the API unsafe: it would be possible to receive a label sent with select, or to branch over a message sent with send. Note that the constraint imposed by [>] as α extends to every occurrence of α in the same signature.

By comparing Tables 11.1 and 11.2 it is clear that the encoding makes session types difficult to read. This problem becomes more severe as the protocols become more involved. The distribution of FuSe includes an auxiliary tool, called rosetta, that implements the inverse of the encoding to pretty print encoded session types into their familiar notation. The tool can be useful not only for documentation purposes but also to decipher the likely obscure type error messages issued by OCaml. Hereafter, when presenting session types inferred by OCaml, we will often show them as pretty printed by rosetta for better clarity.

Representation of session endpoints. Session primitives can be easily implemented on top of any framework providing channel-based communications. FuSe is based on the Event module of OCaml's standard library, which provides communication primitives in the style of Concurrent ML [16] and the abstract type *t* Event.channel for representing channels carrying messages of type *t*. It is convenient to wrap the Event module so as to implement *unsafe communication channels*, thus:

```
module UnsafeChannel : sig
  type t
  val create    : unit → t
  val send      : α → t → unit
  val receive   : t → α
end = struct
  type t          = unit Event.channel
  let create      = Event.new_channel
  let send x u    = Event.sync
                    (Event.send u (Obj.magic x))
  let receive u = Obj.magic
                    (Event.sync (Event.receive u))
end
```

We just need three operations on unsafe channels, create, send and receive. The first one creates a new unsafe channel, which is simply an Event channel for exchanging messages of type unit. The choice of unit over any other OCaml type is immaterial: the messages exchanged over a session can be of different types, hence the type parameter we choose here is meaningless because we will perform unsafe cast at each communication. These casts cannot interfere with the internals of the Event module because *t* Event.channel is parametric on the type *t* of messages and therefore the operations in Event cannot make any assumption on their content. The implementation of send and receive on unsafe channels is a straightforward adaptation of the corresponding primitives of the Event module. Observe that, consistently with the communication API of Concurrent ML, Event.send and Event.receive do not perform communications themselves. Rather, they create *communication events* which occur only when they are synchronized through the primitive Event.sync. The Obj.magic function from the standard OCaml library has type $\alpha \to \beta$ and performs the necessary unsafe casts.

We now have all the ingredients for giving the concrete representation of (encoded) session types. This representation is kept private to the FuSe module so that the user can only manipulate session endpoint through the provided API:

```
type (α,β) st = { chan : UnsafeChannel.t;
                  mutable valid : bool }
```

A session type is represented as a record with two fields: the chan field is a reference to the unsafe channel on which messages are exchanged; the mutable valid field is a boolean flag that indicates whether the endpoint can be safely used or not. Every operation that uses the endpoint first checks whether the endpoint is valid. If this is the case, the valid flag of the endpoint is reset to false so that any subsequent attempt to reuse the same endpoint can be detected. Otherwise, an InvalidEndpoint exception is raised. It is convenient to encapsulate this functionality in an auxiliary function use, which is private to the module and whose implementation is shown below:

```
let use u = if u.valid then u.valid ← false
            else raise InvalidEndpoint
```

In principle, checking that the valid field is true and resetting it to false should be performed atomically, to account for the possibility that several threads are attempting to use the same endpoint simultaneously. In practice, since OCaml's scheduler is not preemptive and use allocates no memory, the execution of use is guaranteed to be performed atomically in OCaml's runtime environment. Different programming languages might require a more robust handling of the validity flag [13].

Whenever an operation on a session endpoint completes and the session endpoint is returned, its valid flag should be set to true again. Doing so on the existing record, though, would be unsafe. Instead, a new record referring to the very same unsafe channel must be created. Again it is convenient to provide this functionality as a private, auxiliary function fresh:

```
let fresh u = { u with valid = true }
```

Implementation of communication primitives. A new session is initiatied by creating a new unsafe channel ch and returning the two peer endpoints of the session, which both refer to the same channel. The valid flag of each peer is set to true, indicating that it can be safely used:

```
let create () = let ch = UnsafeChannel.create ()
                in { chan = ch; valid = true },
                   { chan = ch; valid = true }
```

The implementation of close simply invalidates the endpoint. OCaml's garbage collector takes care of any further finalization that may be necessary to clean up the corresponding unsafe channel:

```
let close = use
```

The send operation starts by checking that the endpoint is valid and, in this case, invalidates it. Then, the message x is transmitted over the underlying unsafe channel and a refreshed version of the endpoint is returned. The receive operation is analogous, except that it returns a pair containing the message received from the underlying unsafe channel and the refreshed endpoint:

```
let send x u =
    use u; UnsafeChannel.send x u.chan; fresh u
let receive u =
    use u; (UnsafeChannel.receive u.chan, fresh u)
```

The select operation is behaviorally equivalent to send, since its purpose is to transmit the selected label (which is its first argument) over the channel. On the other hand the branch operation injects the refreshed session endpoint with the function received from the channel:

```
let select = send
let branch u =
    use u; UnsafeChannel.receive u.chan (fresh u)
```

We conclude this section showing the type inferred by OCaml for the rec_echo_client defined in Section 11.1:

```
val rec_echo_client :
    (0,[> `End of (0,0) st
        | `Msg of (β × (0,γ × (0,α) st) st,0) st]
      as α) st → β list → γ list
```

As expected, the type is rather difficult to understand. Part of this difficulty is a consequence of the fact that the type expression *t* as α, which is used in OCaml also to denote a recursive type, is placed in a position such that *t* does not correspond to the encoding of a session type. It is only by unfolding this recursive type that one recovers an image of the encoding function. The same signature pretty printed by rosetta becomes

```
val rec_echo_client :
    rec X.⊕[ End: end | Msg: !α.?β.X ] →
    α list → β list
```

whose interpretation is straightforward.

11.4 Extended Example: The Bookshop

In this section we develop a FuSe version of a known example from the literature [4], where mother and child order books from an online bookshop. The purpose of the programming exercise is threefold. First, we see a usage instance of the accept and request primitives provided by FuSe for establishing sessions over service channels. Second, we discuss a nontrivial example in which the session types automatically inferred by OCaml are at the same time more general and more precise than those given by Gay and Vasconcelos [4]. This is made possible thanks to the support for parametric polymorphism and subtyping in session types that FuSe inherits for free from OCaml's type system. Finally, we use the example to argue about the effectiveness of the FuSe implementation of binary sessions in detecting protocol violations, considering that FuSe combines both static and dynamic checks.

Service channels in FuSe are provided by the module Service, whose signature is shown below.

```
module Service : sig
   type α t
   val create   : unit → α t
   val accept   : (ρ,σ) st t → (ρ,σ) st
   val request  : (ρ,σ) st t → (σ,ρ) st
   val spawn    : ((ρ,σ) st → unit) → (ρ,σ) st t
end
```

The type A Service.t describe a service channel that allows initiation of sessions of type $\overline{A}$. A session is created when two threads invoke accept and request over the same service channel. In this case, accept returns a session endpoint of type A and request returns its peer of type $\overline{A}$.

The bookshop is modeled as a function that waits for session initiations on the service channel showAccess and invokes shopLoop at each connection:

```
let shop shopAccess =
   shopLoop (Service.accept shopAccess) []
```

A session initiated with the bookshop is handled by the function shopLoop, which operates over the established session endpoint s and the current list order of books in the shopping cart. The shopLoop function is recursive and repeatedly offers the possibility of adding a new book by selecting the Add label. When Checkout is selected instead, the bookshop waits for a credit card number and an address and sends back an estimated delivery date computed by an unspecified `deliveryOn` function.

```
let rec shopLoop s order =
  match Session.branch s with
  | `Add s →
      let book, s = Session.receive s in
      shopLoop s (book :: order)
  | `CheckOut s →
      let card, s = Session.receive s in
      let address, s = Session.receive s in
      let s = Session.send (deliveryOn order) s in
      Session.close s
```

The type inferred by OCaml for `shopLoop` is

```
val shopLoop :
  rec X.&[ Add: ?α.X | CheckOut: ?β.?γ.!day.end ]
  → α list → unit
```

which is structurally the same given by Gay and Vasconcelos [4], except for the type variables α, β and γ. Indeed, the body of `shopLoop` does not use the received values `book`, `card` and `address` and therefore their type remains generic.

We now model a mother process placing an order for two books, one chosen by her and another selected by her son. In principle, the mother could let the son select his own book by delegating the session with the bookshop to him. However, the mother wants to be sure that her son will buy just one book that is suitable for his age. To enforce these constraints, the mother sends her son a voucher, that is a function providing a controlled interface with the bookshop. Overall, the mother is modeled thus:

```
let mother card addr shopAccess sonAccess book =
  let c = Service.request shopAccess in
  let c = Session.select _Add c in
  let c = Session.send book c in
  let s = Service.request sonAccess in
  let s = Session.send (voucher card addr c) s in
  Session.close s
```

where the parameters `card`, `addr` and `book` stand for information about payment, delivery address and mother's book. In addition, `shopAccess` and sonAccess are the service channels for connecting with the bookshop and the son, respectively. The mother establishes a session c with the bookshop and adds book to the shopping cart. Afterwards, she initiates another session s for sending the voucher to her son. The voucher is modeled by the function:

```
let voucher card address c book =
  let c =
    if isChildrensBook book then
      let c = Session.select _Add c in
      Session.send book c
    else c
  in
  let c = Session.select _CheckOut c in
  let c = Session.send card c in
  let c = Session.send address c in
  let day, c = Session.receive c in
  Session.close c
```

where book is chosen by the son. If book is appropriate – something that is checked by the unspecified function isChildrensBook – the book is added to the shopping cart. Then, the order is completed and the connection with the bookshop closed.

For voucher and mother OCaml infers the following types:

```
val voucher :  α → β →
  rec X.⊕[ Add: !γ.X | CheckOut: !α.!β.?δ.end ]
  → γ → unit
val mother :  α → β →
  &[ Add: ?γ.rec X.&[ Add: ?δ.X
               | CheckOut: ?α.?β.!ε.end ]
  ] Service.t → ?(δ → unit).end Service.t → γ
  → unit
```

In contrast to the type of mother given by Gay and Vasconcelos [4], the type inferred by OCaml makes it clear that mother always adds *at least one book* to the shopping cart. The connection between mother and shopLoop is still possible because the protocol followed by mother is more deterministic than – or a supertype of [3] – the one she is supposed to follow.

To finish the exercise we model the son as the following function:

```
let son sonAccess book =
  let s = Service.accept sonAccess in
```

```
let f, s = Session.receive s in
f book;
Session.close s
```

where sonAccess is the service channel used for accepting requests from his mother and book is the book he wishes to purchase. Note that the mother sends a function (obtained as the partial application of voucher) which is saturated by the son who provides the chosen book.

Overall, the code for connecting the three peers is shown below:

```
let _ =
    let mCard = "0123 4567 7654 3210" in
    let mAddr = "17 session type rd" in
    let mBook = "Life of Ada Lovelace" in
    let sBook = "1984" in
    let shopAccess = Service.create () in
    let sonAccess = Service.create () in
    let _ = Thread.create shop shopAccess in
    let _ = Thread.create (son sonAccess) sBook in
    mother mCard mAddr shopAccess sonAccess mBook
```

It is not possible to qualify session endpoints as linear resources in OCaml. This means that there are well-typed programs that, by using session endpoints non-linearly, cause communication errors and/or protocol violations. In the rest of this section we use the example developed so far to do some considerations concerning the effectiveness of the library in detecting programming errors involving session endpoints. In particular we argue that, despite the lack of linear qualification of session endpoints, OCaml's type system is still capable of detecting a fair number of linearity violations. In the worst case, those violations that escape OCaml's type checker are at least detected at runtime with the mechanism we have put in place in Section 11.3.

We can simulate a linearity violation by replacing the session endpoint bound by a let with _. For example, we can replace line 10 in the body of voucher with

10   `let _ = Session.send address c in`

so that the very same endpoint c is used both for this send and also for the subsequent receive. This linearity violation is detected by OCaml because the type of a session endpoint used for an output is incompatible (*i.e.*, not unifiable) with that of an endpoint used for an input. Now suppose that we replace line 8 in the same function with

```
8    let _ = Session.select _CheckOut c in
```

so that the same endpoint c is used for both a select and the subsequent send. Even if select and send are both output operations, the type of messages resulting from the encoding of a plain message output has a topmost × constructor which is incompatible with the polymorphic variant type resulting from the encoding of a label selection. Therefore, also this linearity violation is detected by OCaml's type checker. In general, any linearity violation arising from the use of different communication primitives is detected by OCaml. Consider then line 9, and suppose that we replace it with

```
9    let _ = Session.send card c in
```

so that the same endpoint c is used for sending both card and address. In this case the session endpoint is used for performing two plain outputs and the sent messages have compabile (*i.e.*, unifiable) types. Therefore, taken in isolation, the voucher function would be well typed. In the context of the whole program, however, OCaml detects a type error also in this case. The point is that the faulty version of voucher now implements a different protocol than before. In particular, it appears as if voucher sends just one message after selecting CheckOut and before receiving the estimated delivery date. On the contrary, the protocol of the bookshop as implemented by shopLoop still expects to receive two messages before the delivery date is sent back to the client. Therefore, the protocols of the bookshop and the one inferred by the combination of mother and voucher are no longer dual to each other and the session request performed by mother to the bookshop is ill typed. For this problem to go undetected, there must be *another* linearity violation in the body of shopLoop, in the place that corresponds exactly to the point where the same violation occurs in voucher.

A simpler example of linearity violation that goes undetected by OCaml's type checker can be obtained by duplicating the f book application in the body of the son function. This modification might correspond either to a genuine programming error or to a malicious attempt of son to purchase more than one book. The reason why this duplication results into a linearity violation is that the closure corresponding to f contains the session endpoint c from mother, so applying f twice results in two uses of the same c. This error is detected by the type system of Gay and Vasconcelos [4] where the function f has a linear arrow type. In FuSe, the program compiles correctly, but the second application of f triggers the runtime mechanism that detects linearity violations causing the InvalidEndpoint exception to be raised.

11.5 Related Work

Several libraries of binary sessions have been proposed for different functional programming languages. Most libraries for Haskell [6, 12, 15, 17] use a monad that encapsulates the endpoints of open sessions. Besides being a necessity dictated by the laziness of the language, the monad prevents programmers from accessing session endpoints directly thus guaranteeing that endpoint linearity is not violated. The monad also tracks the evolution of the type of session endpoints automatically, not requiring the programmer to rebind explicitly the same endpoint over and over again. However, the monad has a cost in terms of either expressiveness, usability, or portability: the monad defined by Neubauer and Thiemann [12] supports communication on a single channel only and is therefore incapable of expressing session interleaving or delegation. Pucella and Tov [15] propose a monad that stores a stack of endpoints (or, better, of their capabilities) allowing for session interleaving and delegation to some extent. The price for this generality is that the programmer has to write explicit monadic actions to reach the channel/capability to be used within the monad; also for this reason delegation is severely limited. Imai *et al.* [6] show how to avoid writing such explicit actions relying on a form of type-level computations. Lindley and Morris [9] describe another Haskell embedding of session types that provides first-class channels. Linearity is enforced statically using higher-order abstract syntax.

A different approach is taken in Alms [19, 21], a general-purpose programming language whose type system supports parametric polymorphism, abstract and algebraic data types, and built-in affine types as well. Tov [19] illustrates how to build a library of binary sessions on top of these features. Because Alms' type system is substructural, affine usage of session endpoints is guaranteed statically by the fact that session types are qualified as affine. Further embeddings of session types in other experimental and domain-specific languages with substructural type systems have been described by Mazurak and Zdancewic [10], Lindley and Morris [8], and Morris [11].

Scalas and Yoshida [18] propose a library of binary session for Scala that is very related to our approach. As in FuSe, Scalas and Yoshida use a runtime mechanism to compensate for the lack of affine/linear types in Scala and work with the encoded representation of session types given by Dardha *et al.* [1]. A notable difference is that Scala type system is nominal, so that encoded session types are represented by Scala (case) classes which must be either provided by the programmer or generated from the protocol. This means that the protocol cannot be inferred automatically from the code and that the subtyping relation between session types is constrained by the (fixed) subclassing relation between the classes that represent them.

The main source of inspiration for the representation of session types in FuSe originates from the continuation-passing encoding of binary sessions [1] and partially studied also in some earlier works [2, 7]. Our representation of encoded session types allows session type duality to be expressed solely in terms of type equality, whereas the representation chosen by Dardha *et al.* [1] requires a residual albeit simple notion of duality for the topmost channel type capability. Another difference is that we consider the encoding at the type level only, not requiring the explicit exchange of continuation channels for the implementation of communication primitives. For these reasons, the soundness of the encoding [1] cannot be used directly to argument about the soundness of FuSe's typing discipline. Padovani [13] formalizes FuSe's approach to binary sessions along with the necessary conditions under which the program does not raise exceptions. The same paper also illustrates a simple monadic API built on top of the primitives in Table 11.1 and investigates the overhead of the various approaches to linearity.

In addition to the features described in this chapter, FuSe supports sequential composition of session types. This feature makes it possible to describe with greater precision protocols whose set of (finite) traces is *context-free* as opposed to regular [14,22]. As discussed by Thiemann and Vasconcelos [22], these protocols arise naturally in the serialization of structured data types. Currently, FuSe provides the first and only implementation of context-free session type checking and inference.

Acknowledgments We thank the anonymous reviewers for their careful reading of the chapter and suggestions of improvements and related work.

References

[1] Ornela Dardha, Elena Giachino, and Davide Sangiorgi. Session types revisited. In *Proceedings of PPDP'12*, pages 139–150. ACM, 2012.

[2] Romain Demangeon and Kohei Honda. Full abstraction in a subtyped pi-calculus with linear types. In *Proceedings of CONCUR'11*, LNCS 6901, pages 280–296. Springer, 2011.

[3] Simon Gay and Malcolm Hole. Subtyping for Session Types in the π-calculus. *Acta Informatica*, 42(2–3):191–225, 2005.

[4] Simon J. Gay and Vasco Thudichum Vasconcelos. Linear type theory for asynchronous session types. *Journal of Functional Programming*, 20(1):19–50, 2010.

[5] Raymond Hu and Nobuko Yoshida. Hybrid Session Verification through Endpoint API Generation. In *Proceedings of FASE'16*, LNCS 9633, pages 401–418. Springer, 2016.

[6] Keigo Imai, Shoji Yuen, and Kiyoshi Agusa. Session Type Inference in Haskell. In *Proceedings of PLACES'10*, EPTCS 69, pages 74–91, 2010.

[7] Naoki Kobayashi. Type systems for concurrent programs. In *10th Anniversary Colloquium of UNU/IIST*, LNCS 2757, pages 439–453. Springer, 2002. Extended version available at http://www.kb.ecei. tohoku.ac.jp/ koba/papers/tutorial-type-extended.pdf

[8] Sam Lindley and J. Garrett Morris. Lightweight Functional Session Types. Unpublished manuscript available at http://homepages.inf. ed.ac.uk/slindley/papers/fst-draft-february2015.pdf, 2015.

[9] Sam Lindley and J. Garrett Morris. Embedding session types in haskell. In *Proceedings of Haskell'16*, Haskell 2016, pages 133–145, New York, NY, USA, 2016. ACM.

[10] Karl Mazurak and Steve Zdancewic. Lolliproc: to concurrency from classical linear logic via curry-howard and control. In *Proceeding of ICFP'10*, pages 39–50. ACM, 2010.

[11] J. Garrett Morris. The best of both worlds: linear functional programming without compromise. In *Proceedings of ICFP'16*, pages 448–461. ACM, 2016.

[12] Matthias Neubauer and Peter Thiemann. An implementation of session types. In *Proceedings of PADL'04*, LNCS 3057, pages 56–70. Springer, 2004.

[13] Luca Padovani. A Simple Library Implementation of Binary Sessions. *Journal of Functional Programming*, 27, 2017.

[14] Luca Padovani. Context-Free Session Type Inference. In *Proceedings of the 26th European Symposium on Programming (ESOP'17)*, LNCS. Springer, 2017.

[15] Riccardo Pucella and Jesse A. Tov. Haskell session types with (almost) no class. In *Proceedings of HASKELL'08*, pages 25–36. ACM, 2008.

[16] John H. Reppy. *Concurrent Programming in ML*. Cambridge University Press, 1999.

[17] Matthew Sackman and Susan Eisenbach. Session Types in Haskell: Updating Message Passing for the 21st Century. Technical report, Imperial College London, 2008. Available at `http://pubs.doc.ic.ac.uk/session-types-in-haskell/`

[18] Alceste Scalas and Nobuko Yoshida. Lightweight Session Programming in Scala. In *Proceedings of ECOOP'16*, LIPIcs 56, pages 21:1–21:28. Schloss Dagstuhl, 2016.

[19] Jesse A. Tov. *Practical Programming with Substructural Types*. PhD thesis, Northeastern University, 2012.

[20] Jesse A. Tov and Riccardo Pucella. Stateful Contracts for Affine Types. In *Proceedings of ESOP'10*, LNCS 6012, pages 550–569. Springer, 2010.

[21] Jesse A. Tov and Riccardo Pucella. Practical affine types. In *Proceedings of POPL'11*, pages 447–458. ACM, 2011.

[22] Vasco T. Vasconcelos and Peter Thiemann. Context-free session types. In *Proceedings of ICFP'16*, pages 462–475. ACM, 2016.

12

Lightweight Functional Session Types

Sam Lindley and J. Garrett Morris

University of Edinburgh, Edinburgh, UK

Abstract

Row types provide an account of extensibility that combines well with parametric polymorphism and type inference. We discuss the integration of row types and session types in a concurrent functional programming language, and how row types can be used to describe extensibility in session-typed communication.

12.1 Introduction

In prior work, we have developed a core linear λ-calculus with session types called GV [13]. GV is inspired by a functional language with session types developed by Gay and Vasconcelos [7], which we term LAST (for Linear Asynchronous Session Types), and by the propositions-as-types correspondence between session types and linear logic first introduced by Caires and Pfenning [4] and later adapted to the classical setting by Wadler [23]. We have given direct proofs of deadlock freedom, determinism, and termination for GV. We have also given semantics-preserving translations between GV and Wadler's process calculus CP, showing a strong connection between GV's small-step operational semantics and cut elimination in classical linear logic.

In this article, we demonstrate that we can build practical languages based on the primitives and properties of GV. We introduce a language, FST, that extends GV with polymorphism, row types, and subkinding, integrating linear and unlimited data types. FST, while more expressive, is still deadlock-free, deterministic, and terminating. We consider several extensions of FST. Recursion and recursive session types support the definition of long-running services and repeated behavior. Adding recursion and recursive session types results in a system that is no longer terminating, but is still deadlock free and deterministic. Access points support a more flexible mechanism for session initiation. Adding access points results in a system that is not deadlock-free, deterministic, or terminating, but that still satisfies subject reduction and a weak form of progress.

Outline. The article proceeds as follows. Section 12.2 presents some examples illustrating FST and its extensions. Section 12.3 gives a formal account of FST, a linear variant of System F, incorporating polymorphism, row-typing, subkinding, and session types.

Section 12.4 explores extensions of FST with recursion, recursive types, and access points, and demonstrates the expressivity of access points with encodings of state cells, nondeterministic choice, and recursion.

Section 12.5 describes a practical implementation of FST in Links, a functional language for web programming, and discusses our adaptation of the existing Links syntax and type inference mechanisms to support linearity and session types.

Section 12.6 concludes.

In this version of the article, we focus on the FST type system, and omit the formal semantics and statements of correctness. An extended version including the formal semantics and correctness proofs is available online [15].

12.2 A First Look

Before giving a formal account of the syntax and type system of FST, we present some simple examples of programming in FST. We use a desktop calculator as a running example. Despite its simplicity, it will motivate the features of FST.

A One-Shot Calculator Server. We begin with a process that implements a calculator server. We specify it as a function of one channel, c, on which it will communicate with a user of the calculator.

$$
\begin{aligned}
\text{calc } c = \textbf{offer } c \ \{ \text{Add } c \to\ & \textbf{let } \langle x,c \rangle = \textbf{receive } c \textbf{ in} \\
& \textbf{let } \langle y,c \rangle = \textbf{receive } c \textbf{ in} \\
& \textbf{send } \langle x+y,c \rangle \\
\text{Neg } c \to\ & \textbf{let } \langle x,c \rangle = \textbf{receive } c \textbf{ in} \\
& \textbf{send } \langle -x,c \rangle \}
\end{aligned}
$$

On receiving a channel c, the function calc offers a choice of two behaviors, labeled Add and Neg on c. In the Add case, it then expects to read two values from c and send their sum along c. The Neg case is similar. The session type of channel c encodes these interactions, so the type of calc is

$$\mathsf{calc} : \&\{\mathsf{Add} : ?\mathsf{Int}.?\mathsf{Int}.!\mathsf{Int}.\mathsf{End}, \mathsf{Neg} : ?\mathsf{Int}.!\mathsf{Int}.\mathsf{End}\} \to \mathsf{End}$$

where the session type $!T.S$ denotes sending a value of type T followed by behavior S, $?T.S$ denotes reading a value of type T followed by behavior S, and $\&\{\ell : S, \dots, \ell_n : S_n\}$ denotes offering an n-ary choice, with the behavior of the ith branch given by S_i.

Next, we consider a client for the calculator server:

$$\mathsf{user}_1 \; c = \mathbf{let} \; c = \mathbf{select} \; \mathsf{Add} \; c \; \mathbf{in} \; \mathbf{let} \; \langle x, c \rangle = \mathbf{receive} \; (\mathbf{send} \; \langle 19, \mathbf{send} \; \langle 23, c \rangle \rangle) \; \mathbf{in} \; x$$

Like calc, the user_1 function is passed the channel on which it communicates with the calculator. It begins by selecting the Add behavior, which is compatible with the choice offered by calc. Its subsequent behavior is unsurprising. We could give the channel a type dual to that provided by the calculator:

$$\mathsf{user}_1 : \oplus\{\mathsf{Add} : !\mathsf{Int}.!\mathsf{Int}.?\mathsf{Int}.\mathsf{End}, \mathsf{Neg} : !\mathsf{Int}.?\mathsf{Int}.\mathsf{End}\} \to \mathsf{Int}$$

However, this type overspecifies the behavior of user_1 as the Neg branch is unused in the definition of user_1. In FST, we can use row polymorphism to abstract over the irrelevant labels in a choice, as follows:

$$\mathsf{user}_1 : \forall \rho. \oplus \{\mathsf{Add} : !\mathsf{Int}.!\mathsf{Int}.?\mathsf{Int}.\mathsf{End}; \rho\} \to \mathsf{Int}$$

This type specifies that the argument to user_1 may be instantiated to any session type that offers a choice of Add with a suitable behavior along with arbitrary other choices. FST includes explicit type abstractions and type annotations on bound variables; we omit both in the examples in order to improve readability. Our concrete implementation of FST in Links, is able to reconstruct omitted types and type abstractions using a fairly standard Hindley-Milner-style type inference algorithm.

We can plug the calculator server and the user together as follows

$$\mathbf{let} \; c = \mathbf{fork} \; \mathsf{calc} \; \mathbf{in} \; \mathsf{user}_1 \; c$$

yielding the number 42. The **fork** primitive creates a new child process and a channel through which it can communicate with its parent process.

Recursive Session Types. The one-shot calculator server allows only one operation to be performed before the communication is exhausted. If we add support for recursive session types, then we can define a calculator that allows an arbitrary number of operations to be performed. In order to make the example more interesting, we define a calculator server with a memory.

$$\text{calc}_{\text{rec}} : \text{Int} \to (\textbf{rec } \sigma.\&\{\text{Add} : ?\text{Int}.?\text{Int}.!\text{Int}.\sigma,$$
$$\text{Neg} : ?\text{Int}.!\text{Int}.\sigma,$$
$$\text{M}^+ : ?\text{Int}.\sigma,$$
$$\text{MR} : !\text{Int}.\sigma$$
$$\text{Stop} : \text{End}\}) \to \text{End}$$
$$\text{calc}_{\text{rec}} \; m \; c = \textbf{offer } c \; \{\text{Add } c \; \to \textbf{let } \langle x, c \rangle = \textbf{receive } c \textbf{ in}$$
$$\textbf{let } \langle y, c \rangle = \textbf{receive } c \textbf{ in}$$
$$\text{calc}_{\text{rec}} \; m \; (\textbf{send } \langle x + y, c \rangle)$$
$$\text{Neg } c \; \to \textbf{let } \langle x, c \rangle = \textbf{receive } c \textbf{ in}$$
$$\text{calc}_{\text{rec}} \; m \; (\textbf{send } \langle -x, c \rangle)$$
$$\text{M}^+ \; c \; \to \textbf{let } \langle x, c \rangle = \textbf{receive } c \textbf{ in calc}_{\text{rec}} \; (m + x) \; c$$
$$\text{MR } c \; \to \textbf{let } c = \textbf{send } \langle m, c \rangle \textbf{ in calc}_{\text{rec}} \; m \; c\}$$
$$\text{Stop } c \to c\}$$

The idea is that selecting M^+ adds a number to that currently stored in memory and MR reads the current value of the memory. A user must now explicitly select Stop in order to terminate communication with the server.

$$\text{user}_2 : \forall \rho \rho'. \oplus \{\text{Add} : !\text{Int}.!\text{Int}.?\text{Int}.\oplus\{\text{Stop} : \text{End}; \rho\}; \rho'\} \to \text{Int}$$
$$\text{user}_2 \; c = \textbf{let } \langle x, c \rangle = \textbf{receive } (\textbf{send } \langle 19, \textbf{send } \langle 23, \textbf{select } \text{Add } c \rangle \rangle) \textbf{ in}$$
$$\textbf{select } \text{Stop } c; x$$

With the row variables instantiated appropriately, we can plug user$_2$ and the recursive calculator together

$$\textbf{let } c = \textbf{fork } \text{calc}_{\text{rec}} \; 0 \textbf{ in } \text{user}_2 \; c$$

again yielding 42.

The examples we have seen so far could be implemented using subtyping instead of row polymorphism. We now consider a function that cannot be implemented with subtyping. Suppose we wish to abstract over the memory add operation. We define a function that can be used to communicate with any calculator server that supports M^+ and arbitrary other operations.

$$\text{mAdd} : \forall \rho.\text{Int} \to (\textbf{rec } \sigma. \oplus \{\text{M}^+ : !\text{Int}.\sigma; \rho\}) \to (\textbf{rec } \sigma. \oplus \{\text{M}^+ : !\text{Int}.\sigma; \rho\})$$
$$\text{mAdd} \; n \; c = \textbf{send } \langle n, \textbf{select } \text{M}^+ \; c \rangle$$

The key feature of this function is that the row variable ρ appears both contravariantly (inside the second argument) and covariantly (inside the return type) in the type of mAdd. Thus, in a system with subtyping but without row typing, one would have to explicitly instantiate ρ, ruling out an extensible calculator server implementation. Let us use mAdd to define a client that invokes multiple calculator operations.

user₃ :
$\forall \rho \rho' \rho''.$
$\oplus\{M^+ : !Int.\oplus\{M^+ : !Int.\oplus\{MR : ?Int.\oplus\{Stop : End; \rho\}; \rho'\}; \rho''\}\} \rightarrow Int$
user₃ $c =$ **let** $c =$ **select** MR (mAdd 19 (mAdd 23 c)) **in**
      let $\langle x, c \rangle =$ **receive** c **in**
      select Stop $c; x$

We can plug user₃ and the recursive calculator together as before

$$\textbf{let } c = \textbf{fork } \text{calc}_{\textbf{rec}} \; 0 \textbf{ in } \text{user}_3 \; c$$

again yielding 42.

Access Points. A key limitation of the examples we have seen so far is that they allow only one user to connect to a calculator server at a time. Access points provide a more flexible mechanism for session initiation than the **fork** primitive. Intuitively, we can think of access points as providing a matchmaking service for processes. Processes may either accept or request connections at a given access point; accepting and requesting processes are paired non-deterministically. We now adapt our calculator server to synchronize on an access point instead of a fixed channel:

calc_AP : $\forall \alpha.$Int $\rightarrow$ AP (&{Add : ?Int.?Int.!Int.End,
                              Neg : ?Int.!Int.End,
                              M^+ : ?Int.End,
                              MR : !Int.End}) $\rightarrow \alpha$
calc_AP m $a =$ **let** $c =$ **accept** a **in**
          offer c {
              Add $c \rightarrow$**let** $\langle x, c \rangle =$ **receive** c **in**
                      let $\langle y, c \rangle =$ **receive** c **in**
                      let $c =$ **send** $\langle x + y, c \rangle$ **in** calc_AP m a
              Neg $c \rightarrow$**let** $\langle x, c \rangle =$ **receive** c **in**
                      let $c =$ **send** $\langle -x, c \rangle$ **in** calc_AP m a
              M^+ $c \rightarrow$ **let** $\langle x, c \rangle =$ **receive** c **in** calc_AP $(m + x)$ a
              MR $c \rightarrow$ **let** $c =$ **send** $\langle m, c \rangle$ **in** calc_AP m a}

Unlike calc_rec, this calculator server never stops; rather, it will persist until the access point is no longer accessible by any client code, at which point it may be garbage collected. As calc_rec never returns, it is polymorphic in its return type. In general, an access point a has type AP S for some session type S. The expression **accept** a returns an end point of type S and **request** a returns an end point of type $\bar{S}$.

We can connect our original user to calc_{AP}. We use the **new** operator to create a fresh access point and the **spawn** operator to create child threads (without any shared channels).

$$\textbf{let } a = \textbf{new in spawn } (\lambda \langle \rangle . \text{calc}_{\text{AP}} \; 0 \; a); \text{user}_1 \; (\textbf{request } a)$$

The result of evaluation is again 42. More interestingly, we can connect multiple clients to the same server concurrently.

> **let** $a = $ **new in**
> **let** mAdd $n \; a = $ **send** $\langle n, \textbf{select } \textsf{M}^+ \; (\textbf{request } a) \rangle$ **in**
> **let** mRecall $a = $ **let** $\langle x, c \rangle = $ **receive** $(\textbf{select } \textsf{M}^+ \; (\textbf{request } a))$ **in**
> **spawn** $(\lambda \langle \rangle . \text{calc}_{\text{AP}} \; 0 \; a)$;
> **spawn** $(\lambda \langle \rangle . \text{mAdd } 19 \; (\textbf{request } a))$;
> **spawn** $(\lambda \langle \rangle . \text{mAdd } 23 \; (\textbf{request } a))$;
> mRecall a

The result of evaluating this code is non-deterministic. Depending on the scheduler it may yield 0, 19, 23, or 42.

12.3 The Core Language

The calculus we present in this section, FST (F with Session Types), is a call-by-value linear variant of System F with subkinding, row types, and session types. It combines a variant of GV, our session-typed linear λ-calculus [13], with the row typing and subkinding of our previous core language for Links [11], and the similar approach to subkinding for linearity of Mazurak et al's lightweight linear types [17].

As our focus is programming with session types rather than their logical connections, we make some simplifications compared to our earlier work [13]. Specifically, we have a single unlimited self-dual type of closed channels, and we omit the operation for linking channels together.

12.3.1 Syntax

To avoid duplication and keep the concurrent semantics of FST simple, we strive to implement as much as possible in the functional core of FST, and limit the session typing constructs to the essentials. The only session type constructors are for output, input, and closed channels, and no special typing rules are needed for the primitives, which are specified as constants. Other features such as choice and selection can be straightforwardly encoded using features of the functional core.

Types. The syntax of types and kinds is given in Figure 12.1. The function type $A \to^Y B$ takes an argument of type A and returns a value of type B and has linearity Y. (We write $A \to B$ as an abbreviation for $A \to^\bullet B$.) The record type $\langle R \rangle$ has fields given by the labels of row R. The variant type $[R]$ admits tagged values given by the labels of row R. The polymorphic type $\forall \alpha^{K(Y,Z)}.A$ is parameterized over the type variable α of kind $K(Y,Z)$.

The input type $?A.S$ receives an input of type A and proceeds as the session type S. Dually, the output type $!A.S$ sends an output of type A and proceeds as the session type S. The type End terminates a session; it is its own dual. We let σ range over session type variables and the dual of session type variable σ is $\overline{\sigma}$.

Row Types. Records and variants are defined in terms of row types. Intuitively, a row type represents a mapping from labels to ordinary types. In fact, rows also track absent labels, which are, for instance, needed to type polymorphic record extension (a record can only be extended with labels that are not already present). A row type includes a list of distinct labels, each of which is annotated with a presence type. The presence type indicates whether the label is present with type A ($\mathsf{Pre}(A)$), absent (Abs), or polymorphic in its presence (θ).

Row types are either *closed* or *open*. A closed row type ends in $\cdot$. An open row type ends in a *row variable* ρ or its dual $\overline{\rho}$; the latter are only meaningful for session-kinded rows. The mapping from labels to ordinary types represented by a closed row type is defined only on the labels that are

Ordinary Types	$A,B ::= A \to^Y B$	Labels	ℓ
	$\mid \langle R \rangle \mid [R]$	Label Sets	$\mathscr{L} ::= \{\ell_1, \ldots, \ell_k\}$
	$\mid \forall \alpha^{K(Y,Z)}.A \mid \alpha \mid \overline{\alpha}$	Kinds	$J ::= K(Y,Z)$
	$\mid S$	Primary Kinds	$K ::= \mathsf{Type}$
Session Types	$S ::= !A.S \mid ?A.S$		$\mid \mathsf{Row}\,\mathscr{L}$
	$\mid \mathsf{End} \mid \sigma \mid \overline{\sigma}$		$\mid \mathsf{Presence}$
Row Types	$R ::= \cdot \mid \ell : P; R \mid \rho \mid \overline{\rho}$	Linearity	$Y ::= \bullet \mid \circ$
Presence Types	$P ::= \mathsf{Abs} \mid \mathsf{Pre}(A) \mid \theta \mid \overline{\theta}$	Restriction	$Z ::= \pi \mid \star$
Types	$T ::= A \mid R \mid P$	Type Variables	$\alpha, \sigma, \rho, \theta$

Figure 12.1 Syntax of types and kinds.

explicitly listed in the row type, and cannot be extended. In contrast, the row variable in an open row type can be instantiated in order to extend the row type with additional labels. As usual, we identify rows up to reordering of labels.

$$\ell_1 : P_1; \ell_2 : P_2; R = \ell_2 : P_2; \ell_1 : P_1; R$$

Furthermore, absent labels in closed rows are redundant:

$$\ell : \mathsf{Abs}; \ell_1 : P_1, \ldots; \ell_n : P_n; \cdot = \ell_1 : P_1, \ldots; \ell_n : P_n; \cdot$$

Duality. The syntactic duality operation on type variables extends to a semantic duality operation on session types and is lifted homomorphically to session row types, and session presence types:

$$
\begin{aligned}
\overline{?A.S} &= {!}A.\overline{S} \\
\overline{{!}A.S} &= ?A.\overline{S} \\
\overline{\mathsf{End}} &= \mathsf{End} \\
\overline{\overline{\alpha}} &= \alpha
\end{aligned}
\qquad
\begin{aligned}
\overline{\cdot} &= \cdot \\
\overline{\ell : P; R} &= \ell : \overline{P}; \overline{R} \\
\overline{\overline{\rho}} &= \rho
\end{aligned}
\qquad
\begin{aligned}
\overline{\mathsf{Abs}} &= \mathsf{Abs} \\
\overline{\mathsf{Pre}(S)} &= \mathsf{Pre}(\overline{S}) \\
\overline{\overline{\theta}} &= \theta
\end{aligned}
$$

Kinds. Types are classified by kinds. Ordinary types have kind Type. Row types R have kind $\mathsf{Row}_{\mathscr{L}}$ where $\mathscr{L}$ is a set of labels not allowed in R. Presence types have kind Presence.

The three primary kinds are refined with a simple subkinding discipline, similar to the system described in our previous work on Links [11] and the system of Mazurak et al. on lightweight linear types [17]. A primary kind K is parameterized by a *linearity* Y and a *restriction* Z. The linearity can be either unlimited ($\bullet$) or linear ($\circ$). The restriction can be session typed (π) or unconstrained ($\star$). The interpretation of these parameters on row and presence kinds is pointwise on the ordinary types contained within the row or presence types inhabiting those kinds. For instance, the kind $\mathsf{Row}_{\mathscr{L}}(\circ, \pi)$ is inhabited by row types of linear session type and the kind $\mathsf{Presence}(\bullet, \star)$ by presence types of unlimited unconstrained ordinary types.

By convention we use α for ordinary type variables or for type variables of unspecified kind, ρ for type variables of row kind, and θ for type variables of presence kind. We sometimes omit the primary kind, either inferring it from context or assuming a default of Type. For instance, we write $\alpha^{\bullet, \star}$ instead of $\alpha^{\mathsf{Type}(\bullet, \star)}$.

Subkinding. The two sources of subkinding are the linearity and restriction parameters.

$$\frac{}{\vdash \bullet \leq \circ} \qquad \frac{}{\vdash \pi \leq \star} \qquad \frac{\vdash Y \leq Y' \qquad \vdash Z \leq Z'}{\vdash K(Y,Z) \leq K(Y',Z')}$$

Our notion of linearity corresponds to usage, not alias freedom. Thus, any unlimited type can be used linearly, but not vice versa.

Kind and Type Environments.

$$
\begin{array}{lll}
\text{Kind Environments} & \Delta ::= \cdot \mid \Delta, \alpha : K(Y,Z) \\
\text{Type Environments} & \Gamma ::= \cdot \mid \Gamma, x : A
\end{array}
$$

Kind environments map type variables to kinds. Type environments map term variables to types.

Terms. The syntax of terms and values is given in Figure 12.2. We let x range over term variables and c range over constants. Lambda abstractions $\lambda^Y x^A.M$ are annotated with linearity Y. Type abstractions $\Lambda\alpha^J.V$ are annotated with kind J. Note that the body of a type abstraction is restricted to be a syntactic value in the spirit of the ML value restriction (in order to avoid problems with polymorphic linearity and with polymorphic session types). Records are introduced with the unit record $\langle\rangle$ and record extension $\langle\ell = M; N\rangle$ constructs. They are eliminated with the binding forms **let** $\langle\rangle \leftarrow M$ **in** N and **let** $\langle\ell = x; y\rangle \leftarrow M$ **in** N, the latter of which binds the value labeled by ℓ to x and the remainder of the record to y. Conventional projections $M.\ell$

$$
\begin{array}{lll}
\text{Terms} & L,M,N ::= & x \mid c \\
& & \mid \ \lambda^Y x^A.M \mid L\,M \\
& & \mid \ \Lambda\alpha^J.V \mid M\,T \\
& & \mid \ \langle\rangle \mid \langle\ell = M; N\rangle \\
& & \mid \ \textbf{let}\ \langle\rangle \leftarrow M\ \textbf{in}\ N \\
& & \mid \ \textbf{let}\ \langle\ell = x; y\rangle \leftarrow M\ \textbf{in}\ N \\
& & \mid \ (\ell\,M)^R \mid \textbf{case}\ L\ \{\ell\,x \to M; y \to N\} \\
& & \mid \ \textbf{case}_\perp L \\
\text{Values} & V,W \ \ ::= & x \\
& & \mid \ \lambda^Y x^A.M \\
& & \mid \ \Lambda\alpha^{K(Y,Z)}.V \\
& & \mid \ \langle\rangle \mid \langle\ell = V; W\rangle \\
& & \mid \ (\ell\,V)^R \\
\text{Constants}\ c & ::= & \textbf{send} \mid \textbf{receive} \mid \textbf{fork}
\end{array}
$$

Figure 12.2 Syntax of terms and values.

are definable using this form, but note that because projection discards the remainder of the record, its applicability to records with linear components is limited. Variants are introduced with the injection $\ell\ M$ and eliminated with **case** $L\ \{\ell\ x \to M; y \to N\}$. Hypothetical empty variants are eliminated with **case**$_\perp L$.

Concurrency. The concurrency features of FST are provided by special constants. The term **send** $\langle V, W \rangle$ sends V along channel W, returning the updated channel. The term **receive** W receives a value along channel W, and returns a pair of the value and the updated channel. The term **fork** $(\lambda x.M)$ returns one end of a channel and forks a new process M in which x is bound to the other end of the channel.

Notation. We use the following abbreviations:

$$\textbf{let } x = M \textbf{ in } N \stackrel{\text{def}}{=} (\lambda x.N)\ M \qquad\qquad \langle A_1, \ldots, A_k \rangle \stackrel{\text{def}}{=} \langle 1 : A_1; \ldots; k : A_k; \cdot \rangle$$
$$M; N \stackrel{\text{def}}{=} \textbf{let } x = M \textbf{ in } N, \ x \text{ fresh} \qquad\qquad \overrightarrow{\ell} \stackrel{\text{def}}{=} \ell_1, \ldots, \ell_k$$
$$\ell : A \stackrel{\text{def}}{=} \ell : \mathsf{Pre}(A) \qquad\qquad \overrightarrow{\ell : P} \stackrel{\text{def}}{=} \ell_1 : P_1, \ldots, \ell_k : P_k$$

We interpret n-ary record and case extension at the type and term levels in the standard way. For instance

$$\langle \overrightarrow{\ell : P}; R \rangle \stackrel{\text{def}}{=} \langle \ell_1 : P_1; \langle \ldots; \langle \ell_n : P_n; R \rangle \ldots \rangle \rangle$$

and

$$\textbf{case } L\ \{\cdot\} \stackrel{\text{def}}{=} \textbf{case}_\perp L$$
$$\textbf{case } L\ \{z \to N\} \stackrel{\text{def}}{=} \textbf{let } z = L \textbf{ in } N$$
$$\textbf{case } L\ \{\ell\ x \to N; \chi\} \stackrel{\text{def}}{=} \textbf{case } L\ \{\ell\ x \to N; z \to \textbf{case } z\ \{\chi\}\}$$

where we let χ range over sequences of cases:

$$\chi ::= \cdot \mid z \to N \mid \ell\ x \to N; \chi$$

We write $\mathsf{fv}(M)$ for the free variables of M. We write $\mathsf{ftv}(T)$ for the free type variables of a type T and $\mathsf{ftv}(\Gamma)$ for the free type variables of type environment Γ. We write $\mathsf{dom}(\Gamma)$ for the domain of type environment Γ.

12.3.2 Typing and Kinding Judgments

The kinding rules are given in Figure 12.3. The kinding judgment $\Delta \vdash A : K(Y, Z)$ states that in kind environment Δ, the type A has kind $K(Y, Z)$. Type variables in the kind environment are well-kinded. The rules for forming

$$\boxed{\Delta \vdash T : K(Y,Z)}$$

FUNCTION
$$\frac{\Delta \vdash A : \mathsf{Type}(Y,\star) \qquad \Delta \vdash B : \mathsf{Type}(Y',\star)}{\Delta \vdash A \rightarrow^{Y''} B : \mathsf{Type}(Y'',\star)}$$

FORALL
$$\frac{\Delta, \alpha : K(\bullet,Z) \vdash A : \mathsf{Type}(Y,\star)}{\Delta \vdash \forall \alpha^{K(Y',Z)}.A : \mathsf{Type}(Y,\star)}$$

RECORD
$$\frac{\Delta \vdash R : \mathsf{Row}_{\emptyset}(Y,\star)}{\Delta \vdash \langle R \rangle : \mathsf{Type}(Y,\star)}$$

VARIANT
$$\frac{\Delta \vdash R : \mathsf{Row}_{\emptyset}(Y,\star)}{\Delta \vdash [R] : \mathsf{Type}(Y,\star)}$$

INPUT
$$\frac{\Delta \vdash A : \mathsf{Type}(Y,\star) \qquad \Delta \vdash S : \mathsf{Type}(Y',\pi)}{\Delta \vdash ?A.S : \mathsf{Type}(\circ,\pi)}$$

OUTPUT
$$\frac{\Delta \vdash A : \mathsf{Type}(Y,\star) \qquad \Delta \vdash S : \mathsf{Type}(Y',\pi)}{\Delta \vdash !A.S : \mathsf{Type}(\circ,\pi)}$$

END
$$\frac{}{\Delta \vdash \mathsf{End} : \mathsf{Type}(\bullet,\pi)}$$

EMPTYROW
$$\frac{}{\Delta \vdash \cdot : \mathsf{Row}_{\mathscr{L}}(Y,Z)}$$

EXTENDROW
$$\frac{\Delta \vdash P : \mathsf{Presence}(Y,Z) \qquad \Delta \vdash R : \mathsf{Row}_{\mathscr{L} \uplus \{\ell\}}(Y,Z)}{\Delta \vdash (\ell : P; R) : \mathsf{Row}_{\mathscr{L}}(Y,Z)}$$

ABSENT
$$\frac{}{\Delta \vdash \mathsf{Abs} : \mathsf{Presence}(Y,Z)}$$

PRESENT
$$\frac{\Delta \vdash A : \mathsf{Type}(Y,Z)}{\Delta \vdash \mathsf{Pre}(A) : \mathsf{Presence}(Y,Z)}$$

TYVAR
$$\frac{\alpha : K(Y,Z) \in \Delta}{\Delta \vdash \alpha : K(Y,Z)}$$

DUALTYVAR
$$\frac{\alpha : K(Y,\pi) \in \Delta}{\Delta \vdash \overline{\alpha} : K(Y,\pi)}$$

UPCAST
$$\frac{\vdash J \leq J' \qquad \Delta \vdash T : J}{\Delta \vdash T : J'}$$

Figure 12.3 Kinding rules.

function, record, variant, universally quantified, and presence types follow the syntactic structure of types. Because of the subkinding relation, a record is linear if any of its fields are linear, and similarly for variants. Recall that $\mathsf{Row}_{\mathscr{L}}$ is the kind of row types whose labels cannot appear in $\mathscr{L}$. (To be clear, this constraint applies equally to absent and present labels; it is a constraint on the form of *row types*. In contrast, $\ell : \mathsf{Abs}$ in a row type is a constraint on *terms*.) An empty row has kind $\mathsf{Row}_{\mathscr{L}}(Y,Z)$ for any label set $\mathscr{L}$, linearity Y, and restriction Z. The use of disjoint union in the EXTENDROW rule ensures that row types have distinct labels. A row type can only be used to build a record or variant if it has kind $\mathsf{Row}_{\emptyset}$; this constraint ensures that any absent labels in an open row type are mentioned explicitly.

In Figure 12.4 we define two auxiliary judgments that for use in the typing rules. The linearity judgment $\Delta \vdash \Gamma : Y$ is the pointwise extension of the kinding judgment restricted to the linearity component of the kind. It

$$\boxed{\Delta \vdash \Gamma : Y}$$

$$
\text{L-EMPTY} \qquad\qquad
\begin{array}{c}
\text{L-EXTEND} \\
\dfrac{\Delta \vdash \Gamma : Y \qquad \Delta \vdash A : K(Y,Z)}{\Delta \vdash (\Gamma, x : A) : Y}
\end{array}
$$

$$
\dfrac{}{\Delta \vdash \cdot : Y}
$$

$$\boxed{\Delta \vdash \Gamma = \Gamma_1 + \Gamma_2}$$

$$
\text{C-EMPTY} \qquad\qquad
\begin{array}{c}
\text{C-}\bullet \\
\dfrac{\Delta \vdash A : \mathsf{Type}(\bullet, \star) \qquad \Delta \vdash \Gamma = \Gamma_1 + \Gamma_2}{\Delta \vdash \Gamma, x : A = (\Gamma_1, x : A) + (\Gamma_2, x : A)}
\end{array}
$$

$$
\dfrac{}{\Delta \vdash \cdot = \cdot + \cdot}
$$

$$
\begin{array}{c}
\text{C-}\circ\text{-LEFT} \\
\dfrac{\Delta \vdash A : \mathsf{Type}(\circ, \star) \qquad \Delta \vdash \Gamma = \Gamma_1 + \Gamma_2}{\Delta \vdash \Gamma, x : A = (\Gamma_1, x : A) + \Gamma_2}
\end{array}
\qquad
\begin{array}{c}
\text{C-}\circ\text{-RIGHT} \\
\dfrac{\Delta \vdash A : \mathsf{Type}(\circ, \star) \qquad \Delta \vdash \Gamma = \Gamma_1 + \Gamma_2}{\Delta \vdash \Gamma, x : A = \Gamma_1 + (\Gamma_2, x : A)}
\end{array}
$$

Figure 12.4 Linearity of contexts and context splitting.

states that in kind environment Δ, each type in environment Γ has linearity Y. The type environment splitting judgment $\Delta \vdash \Gamma = \Gamma_1 + \Gamma_2$ states that in kind environment Δ, the type environment Γ can be split into type environments Γ_1 and Γ_2. Contraction of unlimited types is built into this judgment.

The typing rules are given in Figure 12.5. The typing judgment $\Delta; \Gamma \vdash M : A$ states that in kind environment Δ and type environment Γ, the term M has type A. We assume that Γ and A are well-kinded with respect to Δ. If Δ and Γ are empty (that is, M is a closed term), then we will often omit them, writing $\vdash M : A$ for $\cdot; \cdot \vdash M : A$.

We assume a signature Σ mapping constants to their types. The definition of Σ on the basic concurrency primitives is given in Figure 12.6.

The EXTEND rule is strict in the sense that it requires a label to be absent from a record before the record can be extended with the label. The CASE rule refines the type of the value being matched so that in the type of the variable bound by the default branch, the non-matched label is absent.

Selection and Choice. Traditional accounts of session types include types for selection and choice. Following our previous work [13], inspired by Kobayashi [8], we encode selection and choice using variant types.

$$
\begin{aligned}
\oplus\{R\} &\stackrel{\text{def}}{=} ![\overline{R}].\mathsf{End} \\
\&\{R\} &\stackrel{\text{def}}{=} ?[\overline{R}].\mathsf{End} \\
\textbf{select } \ell\, M &\stackrel{\text{def}}{=} \textbf{fork } (\lambda x.\textbf{send } \langle \ell\, x, M \rangle) \\
\textbf{offer } L\, \{\chi\} &\stackrel{\text{def}}{=} \textbf{let } \langle x, z \rangle = \textbf{receive } L \textbf{ in case } x\, \{\chi\}
\end{aligned}
$$

$$\boxed{\Delta;\Gamma \vdash M : A}$$

VAR
$$\frac{\Delta \vdash \Gamma : \bullet}{\Delta;\Gamma,x:A \vdash x : A}$$

CONST
$$\frac{\Sigma(c) = A}{\Delta;\cdot \vdash c : A}$$

LINLAM
$$\frac{\Delta;\Gamma,x:A \vdash M : B}{\Delta;\Gamma \vdash \lambda^{\circ}x^A.M : A \to^{\circ} B}$$

UNLLAM
$$\frac{\Delta \vdash \Gamma : \bullet \qquad \Delta;\Gamma,x:A \vdash M : B}{\Delta;\Gamma \vdash \lambda^{\bullet}x^A.M : A \to^{\bullet} B}$$

APP
$$\frac{\Delta;\Gamma_1 \vdash L : A \to^{Y} B \qquad \Delta;\Gamma_2 \vdash M : A}{\Delta;\Gamma_1 + \Gamma_2 \vdash L\,M : B}$$

POLYLAM
$$\frac{\Delta,\alpha :: K(\bullet,Z);\Gamma \vdash V : A \qquad \alpha \notin \mathrm{ftv}(\Gamma)}{\Delta;\Gamma \vdash \Lambda\alpha^{K(Y,Z)}.V : \forall\alpha^{K(Y,Z)}.A}$$

POLYAPP
$$\frac{\Delta;\Gamma \vdash M : \forall\alpha^{K(Y,Z)}.A \qquad \Delta \vdash T :: K(Y,Z)}{\Delta;\Gamma \vdash M\,T : A[\alpha := T]}$$

UNIT
$$\frac{\Delta \vdash \Gamma : \bullet}{\Delta;\Gamma \vdash \langle\rangle : \langle\rangle}$$

LETUNIT
$$\frac{\Delta;\Gamma_1 \vdash M : \langle\rangle \qquad \Delta;\Gamma_2 \vdash N : B}{\Delta;\Gamma_1 + \Gamma_2 \vdash \mathbf{let}\ \langle\rangle \leftarrow M\ \mathbf{in}\ N : B}$$

CASEZERO
$$\frac{\Delta;\Gamma \vdash L : [\,]}{\Delta;\Gamma \vdash \mathbf{case}_{\perp} L : B}$$

EXTEND
$$\frac{\Delta;\Gamma_1 \vdash M : A \qquad \Delta;\Gamma_2 \vdash N : \langle\ell : \mathsf{Abs};R\rangle}{\Delta;\Gamma_1 + \Gamma_2 \vdash \langle\ell = M;N\rangle : \langle\ell : \mathsf{Pre}(A);R\rangle}$$

LETRECORD
$$\frac{\Delta;\Gamma_1 \vdash M : \langle\ell : \mathsf{Pre}(A);R\rangle \qquad \Delta;\Gamma_2,x:A,y:\langle R\rangle \vdash N : B}{\Delta;\Gamma_1 + \Gamma_2 \vdash \mathbf{let}\ \langle\ell = x;y\rangle \leftarrow M\ \mathbf{in}\ N : B}$$

INJECT
$$\frac{\Delta;\Gamma \vdash M : A}{\Delta;\Gamma \vdash (\ell\,M)^R : [\ell : \mathsf{Pre}(A);R]}$$

CASE
$$\frac{\Delta;\Gamma_1 \vdash L : [\ell : \mathsf{Pre}(A);R] \qquad \Delta;\Gamma_2,x:A \vdash M : B \qquad \Delta;\Gamma_2,y:[\ell : \mathsf{Abs};R] \vdash N : B}{\Delta;\Gamma_1 + \Gamma_2 \vdash \mathbf{case}\ L\ \{\ell\,x \to M;y \to N\} : B}$$

Figure 12.5 Typing rules.

$$\Sigma(\mathbf{send}) = \forall\alpha^{\circ,\star}.\forall\sigma^{\circ,\pi}.\langle\alpha, !\alpha.\sigma\rangle \to^{\bullet} \sigma$$
$$\Sigma(\mathbf{receive}) = \forall\alpha^{\circ,\star}.\forall\sigma^{\circ,\pi}.?\alpha.\sigma \to^{\bullet} \langle\alpha,\sigma\rangle$$
$$\Sigma(\mathbf{fork}) = \forall\sigma^{\circ,\pi}.\forall\alpha^{\bullet,\star}.(\sigma \to^{\circ} \alpha) \to^{\bullet} \overline{\sigma}$$

Figure 12.6 Type schemas for constants.

The encoding of **select** uses **fork** in order to generate a fresh channel of the continuation type. In the implementation of Links we support selection and choice in the source language. This is primarily for programming convenience. One might imagine desugaring these using the rules above, and then potentially rediscovering them in the back-end for performance reasons.

Semantics. In the extended version of this article [15] we give an asynchronous small-step operational semantics for FST. Following Gay and Vasconcelos [7], whose calculus we call LAST (for Linear Asynchronous Session Types), we factor the semantics into functional and concurrent reduction relations, and introduce explicit buffers to provide asynchrony. For the functional fragment of the language, we give a standard left-to-right call-by-value semantics. The semantics of the concurrent portion of the language is given by a reduction relation on configurations of process and buffers. This semantics differs from our previous work on GV [13] in that is relies on explicit buffers, allowing asynchrony between the sending and receiving of a message, and it uses standard β-reduction instead of weak explicit substitutions [10]. FST, like GV but unlike LAST, is deadlock-free, deterministic, and terminating.

12.4 Extensions

FST can be straightforwardly extended with additional features.

If we add a fixed point constant, then we lose termination, but deadlock freedom and determinism continue to hold. Another standard extension supported by Links is recursive types. While care is needed in defining the dual of a recursive session type, the treatment is otherwise quite standard. Negative recursive types allow a fixed point combinator to be defined, so again we lose termination, but deadlock freedom and determinism continue to hold.

The price we pay for the strong properties we obtain is that our model of concurrency is rather weak. For instance, it gives us no way of implementing a server with any notion of shared state. Drawing on LAST (and previous work on session-typed π-calculi), Links supports *access points*, which provide a much more expressive model of concurrency at the cost of introducing deadlock. Nevertheless, it is often possible to locally restrict code to a deadlock-free subset of Links.

12.4.1 Recursion

The grammar of session types we have presented so far is rather limited; for example, it cannot express repeated behavior. As illustrated in Section 12.2, we can use recursive session types to define a calculator that supports multiple calculations. In order to support this kind of example, we can straightforwardly extend FST with equi-recursive types. We add a kinding rule for recursive types and identify each recursive type with its unrolling.

REC

$$\frac{\Delta, \alpha : \mathsf{Type}(Y,Z) \vdash A : \mathsf{Type}(Y,Z)}{\Delta \vdash \mathbf{rec}\ \alpha^{Y,Z}.A : \mathsf{Type}(Y,Z)} \qquad \mathbf{rec}\ \alpha^{Y,Z}.A = A[\mathbf{rec}\ \alpha^{Y,Z}.A/\alpha]$$

It is well-known [2, 3] that recursive types complicate the definition of duality, particularly when the recursion variable appears as a carried type (that is, as A in ?$A.S$ or !$A.S$). For example, consider the simple recursive session type $\mathbf{rec}\ \sigma^{\circ,\pi}.?\sigma.\sigma$. The dual of this type is not $\mathbf{rec}\ \sigma^{\circ,\pi}.!\sigma.\sigma$, as one would obtain by taking the dual of the body of the recursive type directly, but is $\mathbf{rec}\ \sigma^{\circ,\pi}.!\overline{\sigma}.\sigma$ instead.

Bernardi and Hennessy [2] point out that even existing definitions that correctly handle the above instance of recursion variables appearing inside a carried type often fail for other examples. The underlying difficulty arises from attempting to define duality in a setting in which the duality operator may not be applied to atomic type variables. Bernardi and Hennessy show that is is possible to give a correct definition in such a setting, but we prefer the more compositional definition that arises naturally when one admits duals of atomic type variables [16] (something that we want anyway as our calculus is polymorphic).

$$\overline{\mathbf{rec}\ \sigma^{X,\pi}.S} = \mathbf{rec}\ \sigma^{X,\pi}.\overline{(S[\overline{\sigma}/\sigma])}$$

Having added recursive types, one can of course encode a fixed point combinator. Alternatively, we can add a fixed point constant to FST, even without recursive types:

$$\Sigma(\mathbf{fix}) = \forall\alpha^{\bullet,\star}.\forall\beta^{\bullet,\star}.((\alpha \to^{\bullet} \beta) \to^{\bullet} (\alpha \to^{\bullet} \beta)) \to^{\bullet} (\alpha \to^{\bullet} \beta)$$

Of course, these extensions allows us to write nonterminating programs, but it is straightforward to show that subject reduction, progress, deadlock freedom, and determinism continue to hold.

12.4.2 Access Points

In order to extend FST with access points, we replace the constant **fork** with four new constants:

$$\begin{aligned}
\Sigma(\mathbf{spawn}) &= \forall\alpha^{\bullet,\star}.(\langle\rangle \to^{\circ} \alpha) \to^{\bullet} \langle\rangle \\
\Sigma(\mathbf{new}) &= \forall\sigma^{\circ,\pi}.\langle\rangle \to^{\bullet} \mathsf{AP}\ \sigma \\
\Sigma(\mathbf{accept}) &= \forall\sigma^{\circ,\pi}.\mathsf{AP}\ \sigma \to^{\bullet} \sigma \\
\Sigma(\mathbf{request}) &= \forall\sigma^{\circ,\pi}.\mathsf{AP}\ \sigma \to^{\bullet} \overline{\sigma}
\end{aligned}$$

A process M is spawned with **spawn** M, where M is a thunk that returns an arbitrary unlimited value; we can define **spawn** in terms of **fork** and vice versa:

$$\textbf{spawn } M \stackrel{\text{def}}{=} (\lambda x^{\text{End}}.\langle\rangle)(\textbf{fork } (\lambda x^{\text{End}}.M \langle\rangle))$$

$$\textbf{fork } M \stackrel{\text{def}}{=} \textbf{let } z = \textbf{new } \langle\rangle \textbf{ in spawn } (\lambda x.M (\textbf{accept } z)); \textbf{request } z$$

Session-typed channels are created through access points. A fresh access point of type AP S is created with **new**. Given an access point L of type AP S we can create a new server channel (**accept** L), of session type S, or client channel (**request** L), of session type $\overline{S}$. Processes can accept and request an arbitrary number of times on any given access point. Access points are synchronous in the sense that each **accept** will block until it is paired up with a corresponding **request** and vice-versa.

Adding access points exposes the difference between asynchronous and synchronous semantics. Here is an example of a term that reduces to a value under an asynchronous semantics, but deadlocks under a synchronous semantics.

$$\textbf{let } z = \textbf{new } \langle\rangle \textbf{ in}$$
$$\textbf{let } z' = \textbf{new } \langle\rangle \textbf{ in}$$
$$\textbf{spawn } (\lambda\langle\rangle.\textbf{let } x = \textbf{accept } z \textbf{ in}$$
$$\qquad\qquad \textbf{let } y = \textbf{accept } z' \textbf{ in send } \langle 0,x\rangle; \textbf{let } \langle v,y\rangle = \textbf{receive } y \textbf{ in } v);$$
$$\textbf{let } x = \textbf{request } z' \textbf{ in}$$
$$\textbf{let } y = \textbf{request } z \textbf{ in send } \langle 0,x\rangle; \textbf{let } \langle v,y\rangle = \textbf{receive } y \textbf{ in } v$$

Under an asynchronous semantics, both sends happen followed by both receives, and the term reduces to the value 0. Under a synchronous semantics both sends are blocked and the term is deadlocked.

Shared State. With access points we can implement shared state cells.

State $A = \text{AP } (!A.\text{End})$

newCell : $\forall \alpha^{\bullet,\star}.\langle\rangle \to \text{State } \alpha$
newCell $v = \textbf{let } x = \textbf{new } \langle\rangle \textbf{ in spawn } (\lambda\langle\rangle.\textbf{send } \langle v, \textbf{accept } x\rangle); x$

put : $\forall \alpha^{\bullet,\star}.\text{State } \alpha \to \alpha \to \langle\rangle$
put $x\, v = \textbf{let } \langle \_,\_\rangle = \textbf{receive } (\textbf{request } x) \textbf{ in spawn } (\lambda\langle\rangle.\textbf{send } \langle v, \textbf{accept } x\rangle); \langle\rangle$

get : $\forall \alpha^{\bullet,\star}.\text{State } \alpha \to \alpha$
get $x = \textbf{let } \langle v,\_\rangle = \textbf{receive } (\textbf{request } x) \textbf{ in spawn } (\lambda\langle\rangle.\textbf{send } \langle v, \textbf{accept } x\rangle); v$

Nondeterminism. We can straightforwardly encode nondeterministic choice by using an access point to generate a nondeterministic boolean value. Suppose that we have $\Delta; \Gamma \vdash M : T$ and $\Delta; \Gamma \vdash N : T$. The following term will nondeterministically choose between terms M and N:

> **let** $z = $ **new** $\langle\rangle$ **in**
> **spawn** $(\lambda\langle\rangle.\textbf{send } \langle\textsf{True}, \textbf{accept } z\rangle)$;
> **spawn** $(\lambda\langle\rangle.\textbf{send } \langle\textsf{False}, \textbf{accept } z\rangle)$;
> **let** $\langle x, _\rangle = $ **receive** (**request** z) **in**
> **case** x $\{\textsf{True} \rightarrow M; \textsf{False} \rightarrow N\}$

One process is left waiting on **accept** z. However, as z cannot escape, this process can be safely garbage collected.

Recursion. Recursion can in fact be encoded using access points. We have already seen that access points are expressive enough to simulate higher-order state. We can now use Landin's knot (back-patching) [9] to implement recursion. For instance, the following term loops forever:

> **let** $x = \textsf{newCell}_{\langle\rangle\rightarrow\langle\rangle}\ (\lambda\langle\rangle.\langle\rangle)$ **in** put $\langle x, \lambda\langle\rangle.\textsf{get } x\ \langle\rangle\rangle$; get $x\ \langle\rangle$

12.5 Links with Session Types

Version 0.6 of the Links web programming language includes an extension based on FST. It is available online from the Links website:

`http://links-lang.org/`

Links is a functional programming language for the web. From a single source program, Links generates code to run on all three tiers of a web application: the browser, the server, and the database. Links is a call-by-value language with support for ML-style type inference (extended with support for first-class polymorphism similar to that of provided by the impredicative polymorphism extension of GHC [22]). It incorporates a row-type system that is used for records, variants, and effects, and provides equi-recursive types. Subkinding is used to distinguish base types from other types. This is important for enforcing the constraint that generated SQL queries must return a list of records whose fields are of base type [11].

In order to keep the presentation uniform and self-contained we use the concrete syntax of FST throughout rather than that of Links. However, all of the examples presented in this article can be written directly in Links with essentially the same abstract syntax, modulo the fact that Links uses Hindley-Milner style type inference.

12.5.1 Design Choices

Before implementing session types for Links we considered a number of design choices. Linearity is central to our description of session types. Most existing functional languages (including vanilla Links) do not provide native support for linear types. We considered three broad approaches:

1. encode linearity using existing features of the programming language (as in Pucella and Tov's Haskell encoding of session types [19] or our Haskell encoding of session types [14])
2. stratify the language so that the linear fragment of the language is separated out from the host language (as in Toninho et al's work [20])
3. bake linearity into the type system of the whole language (as in LAST [7])

The appeal of the first approach is that it does not require any new language features, assuming the starting point is a language with a sufficiently rich type system—for example, one that is able to conveniently encode parameterized monads [1], or parameterized higher-order abstract syntax [5]. The second approach is somewhere in between. It allows a linear language to be embedded in an existing host language without disrupting the host language. The third approach requires linearity to pervade the whole type system, but opens up interesting possibilities for code reuse, for instance through polymorphism over linearity [24] or through subkinding [17].

Given that we are in the business of developing our own programming language, we decided to pursue the third option. We wanted to include the full expressivity of our language in the linear fragment, so we did not see a significant benefit in stratification, and we wanted to explore possibilities for code-reuse offered by baking linearity into the type system. We were also presented with another choice regarding how to accommodate code reuse. Given that Links already supported subkinding [11] we elected to adopt the linear subkinding approach of Mazurak et al. [17].

An advantage of the LAST (and FST) approach to session typing is that channels are first class and hence support compositional programming. This is in contrast to the parameterized monad approach and approaches based on process calculi, in which channels are just names. For example, in FST with recursive types we can define broadcasting a value to a whole list of channels:

$$\text{broadcast} : \forall \alpha^{\bullet,\star} \sigma^{\circ,\pi}.\alpha \to \text{LinList}\ (!\alpha.\sigma) \to \text{LinList}\ \sigma$$
$$\text{broadcast}\ v\ xs = \text{linMap}\ (\lambda x.\textbf{send}\ \langle v,x \rangle)\ xs$$

where LinList A is a linear list data type and linMap is the map operation over linear lists:

$$\text{LinList } A = \mathbf{rec}\ \alpha^{\circ,\star}.[\text{Nil}; \text{Cons} : \langle A, \alpha \rangle]$$

$$\text{linMap} : \forall \alpha^{\circ,\star} \beta^{(\circ,\star)}.(\alpha \to \beta) \to \text{LinList } \alpha \to \text{LinList } \beta$$

$$\text{linMap } f\ xs = \mathbf{case}\ xs\ \{\text{Nil} \qquad\qquad \to \text{Nil}$$
$$\qquad\qquad\qquad\qquad \text{Cons } \langle x, xs \rangle \to \text{Cons } \langle f\ x, \text{linMap } f\ xs \rangle\}$$

An attendant drawback to having first-class channels is that one must explicitly rebind channels after each operation. This is in contrast to the parameterized monad approach and approaches based on process calculi, which implicitly rebind channels after each communication. In order to mitigate the need to explicitly rebind channels, we introduce process calculus style syntactic sugar inspired by previous work on the correspondence between classical linear logic and functional sessions [12, 13, 23]. To ease the job of writing a parser, we explicitly delimit process calculus style syntactic sugar with special brackets $\lhd - \rhd$.

$$\lhd x(y).Q \rhd \overset{\text{def}}{=} \mathbf{let}\ \langle x, y \rangle = \mathbf{receive}\ x\ \mathbf{in}\ \lhd Q \rhd$$

$$\lhd x[M].Q \rhd \overset{\text{def}}{=} \mathbf{let}\ x = \mathbf{send}\langle M, x \rangle\ \mathbf{in}\ \lhd Q \rhd$$

$$\lhd \ell\ x.Q \rhd \overset{\text{def}}{=} \mathbf{let}\ x = \mathbf{select}\ \ell\ x\ \mathbf{in}\ \lhd Q \rhd$$

$$\lhd \mathbf{offer}\ x\ \{\ell_i \to Q_i\}_i \rhd \overset{\text{def}}{=} \mathbf{offer}\ x\ \{\ell_i(x) \to \lhd Q_i \rhd\}_i$$

$$\lhd \{M\} \rhd \overset{\text{def}}{=} M$$

We let Q range over process calculus style terms. The desugaring of input, output, selection, and branching is direct. The $\{-\}$ brackets allow values to be returned from the tail of a process calculus expression. As an example, we can more concisely rewrite the one-shot calculator server of Section 12.2 as follows:

$$\text{sugarCalc} = \lambda c.\lhd \mathbf{offer}\ c\ \{\text{Add} \to c(x).c(y).c[x+y].\{\langle\rangle\}$$
$$\text{Neg} \to c(x).c[-x].\{\langle\rangle\}\} \rhd$$

In general, the syntactic sugar allows us to take advantage of a process-calculus style for communication-heavy sequences of code, but switch back to a functional style for compositional programming.

12.5.2 Type Reconstruction

Vanilla Links provides type inference, as in many other typed functional languages. However, as a consequence of the typing of application, the types of higher-order functions in FST are not uniquely determined by their uses.

As an example, consider the application operator in FST, implemented by the following term:

$$\Lambda \alpha_1^{\bullet,\star}, \alpha_2^{\bullet,\star}.\lambda^{Y_1} f^{\alpha_1 \to^{Y_2} \alpha_2}.\lambda^{Y_3} x^{\alpha_1}.f \, x$$

This term is well-typed for arbitrary choices of Y_1 and Y_2, and any choice of Y_3 more constraining than Y_2, giving six distinct well-typed instantiations in all.

There are several ways we might hope to restore complete type inference, but they each come with significant additional complexity. We could introduce bounded quantification over linearities, combining the approaches of Tov and Pucella [21] and Walker [24]; in addition to introducing new forms of quantification, the implications of the resulting system for type inference have not been studied. Another approach was recently proposed by Morris [18]. His approach captures all the variations of the term above in a single term, and provides complete type inference. However, it relies on qualified types, an alternative source of complexity. In Links, we prefer unlimited function types $\tau \to^\bullet \tau'$ to linear function types $\tau \to^\circ \tau'$ when inferring the types of functions. The programmer is always free to override this choice by explicitly providing types. This approach preserves the simplicity of the language and of type reconstruction, but at the cost of some completeness.

12.6 Conclusion and Future Work

We have presented an account of lightweight functional session types, extending a core session-typed linear λ-calculus [13] with: the row typing of the core language for Links [11], the subkinding for linearity of Mazurak et al.'s lightweight linear types [17], and the asynchrony and access points of Gay and Vasconcelos's linear type theory for asynchronous session types [7].

There is a significant gap between variants of FST with and without access points. We would like to investigate abstractions that add some of the expressive power of access points, but are better behaved. In particular, it would be interesting to explore richer type systems for enforcing deadlock and race freedom, while allowing some amount of stateful concurrency. More immediately, it would also be natural to exploit the existing effect type system of Links to statically enforce desirable properties, for instance, by associating the use of access points with a particular effect type.

References

[1] R. Atkey. Parameterised notions of computation. *J. Funct. Program.*, 19(3–4):335–376, 2009.

[2] G. Bernardi and M. Hennessy. Using higher-order contracts to model session types. *CoRR*, abs/1310.6176v4, 2015.

[3] V. Bono and L. Padovani. Typing copyless message passing. *Logical Methods in Computer Science*, 8(1), 2012.

[4] L. Caires and F. Pfenning. Session types as intuitionistic linear propositions. In *CONCUR*. Springer, 2010.

[5] J. Carette, O. Kiselyov, and C. Shan. Finally tagless, partially evaluated: Tagless staged interpreters for simpler typed languages. *J. Funct. Program.*, 19(5):509–543, 2009.

[6] J. Garrigue, G. Keller, and E. Sumii, editors. *ICFP*. ACM, 2016.

[7] S. J. Gay and V. T. Vasconcelos. Linear type theory for asynchronous session types. *J. Funct. Program.*, 20(01):19–50, 2010.

[8] N. Kobayashi. Type systems for concurrent programs. In *10th Anniversary Colloquium of UNU/IIST*. Springer, 2002.

[9] P. J. Landin. The mechanical evaluation of expressions. *Computer Journal*, 6(4):308–320, 1964.

[10] J. Lévy and L. Maranget. Explicit substitutions and programming languages. In *FSTTCS*, volume 1738 of *LNCS*, pages 181–200. Springer, 1999.

[11] S. Lindley and J. Cheney. Row-based effect types for database integration. In B. C. Pierce, editor, *TLDI*. ACM, 2012.

[12] S. Lindley and J. G. Morris. Sessions as propositions. In *PLACES*, 2014.

[13] S. Lindley and J. G. Morris. A semantics for propositions as sessions. In J. Vitek, editor, *ESOP*, volume 9032 of *Lecture Notes in Computer Science*, pages 560–584. Springer, 2015.

[14] S. Lindley and J. G. Morris. Embedding session types in haskell. In G. Mainland, editor, *Haskell*, pages 133–145. ACM, 2016.

[15] S. Lindley and J. G. Morris. Lightweight functional session types (extended version). http://homepages.inf.ed.ac.uk/slindley/papers/fst-extended.pdf, 2016.

[16] S. Lindley and J. G. Morris. Talking bananas: structural recursion for session types. In Garrigue et al. [6], pages 434–447.

[17] K. Mazurak, J. Zhao, and S. Zdancewic. Lightweight linear types in System F°. In A. Kennedy and N. Benton, editors, *TLDI*. ACM, 2010.

[18] J. G. Morris. The best of both worlds: linear functional programming without compromise. In Garrigue et al. [6], pages 448–461.

[19] R. Pucella and J. A. Tov. Haskell session types with (almost) no class. In A. Gill, editor, *Haskell*. ACM, 2008.

[20] B. Toninho, L. Caires, and F. Pfenning.   Higher-order processes, functions, and sessions: A monadic integration.   In *ESOP*. Springer, 2013.

[21] J. A. Tov and R. Pucella. Practical affine types. In T. Ball and M. Sagiv, editors, *POPL*, pages 447–458. ACM, 2011.

[22] D. Vytiniotis, S. Weirich, and S. L. Peyton Jones.   FPH: first-class polymorphism for Haskell. In J. Hook and P. Thiemann, editors, *ICFP*. ACM, 2008.

[23] P. Wadler. Propositions as sessions. *J. Funct. Program.*, 24(2–3):384–418, 2014.

[24] D. Walker.   Substructural Type Systems.   In B. C. Pierce, editor, *Advanced Topics in Types and Programming Languages*, chapter 1. MIT Press, 2005

13

Distributed Programming Using Java APIs Generated from Session Types

Raymond Hu

Imperial College London, UK

Abstract

This is a tutorial on using Scribble [9], a toolchain based on multiparty session types [1, 4], for distributed programming in Java. The methodology is based on the generation of *protocol-specific Endpoint APIs* from Scribble specifications [6]. We start with a brief recap of TCP network programming using standard Java APIs, and their limitations with regards to safety assurances. The main tutorial content is an overview of the key stages of the Scribble toolchain, from global protocol specification, through Endpoint API generation, to Java endpoint implementation, with examples. We discuss the hybrid form of session safety promoted by the Endpoint API generation approach. We then consider Scribble specifications and implementations of HTTP as a real-world use case. Finally, we demonstrate some further Scribble features that leverage Endpoint API generation to safely support more advanced communication patterns.

13.1 Background: Distributed Programming in Java

The two core facilities for TCP-based network programming in Java (and other languages) are the *socket APIs* and Java *Remote Method Invocation* (RMI). The socket APIs allow the programmer to work directly with TCP connections, and are the basis over which many higher-level networking facilities are built. Java RMI is the Java adaptation of remote procedure call (RPC) functionality; with regards to this discussion, RMI is representative of the corresponding facilities in other languages or platform-independent frameworks, *e.g.*, RESTful Web services.

Running example: Math Service. As a running Hello World example, we specify and implement a two-party network service for basic arithmetic operations. For the purposes of this tutorial, we are not concerned with the most realistic development of such a service, but rather that this simple example features core constructs of protocol design, such as message sequencing, alternative cases and repeated sequences.

Figure 13.1 depicts the Math Service protocol as a UML sequence diagram [8, §17]. For some number of repetitions (loop), the client C sends to the server S a Val message with an Integer payload; C then selects between the two alternatives (alt), to send an Add or a Mult message carrying a second Integer. S respectively replies with a Sum or a Prod message carrying the result. Finally, C sends a Bye message, with no payload, ending the session.

13.1.1 TCP Sockets

Sockets are supported in the standard Java API by the java.net and java.nio.channels packages. Figure 13.2 gives a client implementation using Math Service for a factorial calculation via the java.net.Socket API. For simplicity, we assume serializable Java classes for each of the message types (*e.g.*, Add), with a field val for the Integer payload, and use standard Java object serialization via java.io.ObjectOutput/InputStream.

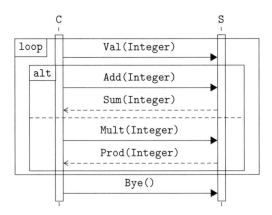

Figure 13.1   Sequence diagram for the Math Service protocol.

```
1  try (Socket s = new Socket("localhost", 8888);
2      ObjectOutputStream os = new ObjectOutputStream(s.getOutputStream());
3      ObjectInputStream is = new ObjectInputStream(s.getInputStream())) {
4    int i = 5, res = i; // Calculate 5!
5    while (i > 1) {
6      os.writeObject(new Val(i)); os.writeObject(new Add(-1)); os.flush();
7      i = ((Sum) is.readObject()).val;
8      os.writeObject(new Val(res)); os.writeObject(new Mult(i)); os.flush();
9      res = ((Prod) is.readObject()).val;
10   }
11   os.writeObject(new Bye()); os.flush();
12 }
```

Figure 13.2 A factorial calculation using Math Service via the `java.net.Socket` API.

From a networking perspective, TCP sockets offer a high-level abstraction in the sense of reliable and ordered message delivery. From an application perspective, however, the raw communication interface of a TCP channel is simply a pair of unidirectional bit streams—essentially a communications machine code, in contrast to the support for high-level data types in "local" computations.

Working directly with standard socket APIs thus affords almost no safety assurances with regards to the correctness of protocol implementations. The key kinds of application-level protocol errors are:

Communication mismatches when the sent message is not one of those expected by the receiver (also called a *reception error*). E.g., if c were to commence a session with an Add message: assuming an implementation of s in the style of Figure 13.2, this would likely manifest as a ClassCastException on the object returned by the readObject in s. Note, the dual error of the receiver applying an incorrect cast are equally possible.

Deadlock in situations where some set of participants are all blocked on mutually dependent input actions. E.g., if c were to call readObject after sending Val, but before sending Add or Mult; while s is (correctly) waiting for one of the latter.

Orphan messages if the receiver terminates without reading an incoming message. In practice, an orphan error often manifests as, *e.g.*, an EOFException, since TCP uses a termination handshake. E.g., if s skips the receive of Bye before c has sent it, leading c to attempt the write on a closed connection.

13.1.2 Java RMI

RMI is a natural approach towards addressing the mismatch between high-level, typed Java programming and low-level networking interfaces. Distributed computations can be (partially) abstracted away as regular method invocations, while benefiting from static typing of each call and its communicated arguments and return value.

The Math Service protocol may be fitted to a remote interface as in Figure 13.4(a), essentially by decomposing the protocol into separate call-return fragments; and Figure 13.3 re-implements the factorial calculation as a client of this interface. Individual remote calls are now statically typed with respect to their arguments and return. Unfortunately, RMI programs in general remain subject to the same potential protocol errors illustrated for the previous sockets example (although their concrete manifestations may differ). The typed RMI interface does not prevent, for example, a bad client from calling Add before Val.

Disadvantages of call-return based protocol decomposition are further illustrated by the (minimal) implementation of the remote interface in Figure 13.4(b), which suffices to serve a single client but is completely inadequate in the presence of concurrent clients. Basic RMI programs lose the notion of an explicit *session*-oriented abstraction in the code (cf., the threading of session control flow in Figure 13.2 wrt. the socket/stream variable usages), which complicates the correlation and management of application-level session flows across the separate methods.

13.2 Scribble Endpoint API Generation: Toolchain Overview

Using the Math Service running example, we demonstrate the stages of the Scribble toolchain, from global protocol specification, through Endpoint

```
RMIMath mathS = (RMIMath) registry.lookup("MathService");
int i = 5, res = i;
while (i > 1) { mathS.Val(i);   i = mathS.Add(i - 1);
                mathS.Val(res); res = mathS.Mult(i);   }
mathS.bye();
```

Figure 13.3 Factorial calculation as a client of the remote interface in Figure 13.4(a).

```
interface RMIMath              class RMIMathS implements RMIMath {
    extends Remote {             private int x;
  void Val(Integer x) .. ;       public void Val(Integer x) ...
  void Bye() throws .. ;           { this.x = x; }
  Integer Add(Integer y) .. ;    public void Bye() throws .. { }
  Integer Mult(Integer y) .. ;   public Integer Add(Integer y) ...
}                                  { return this.x + y; }
                                 public Integer Mult(Integer y) ...
                                   { return this.x * y; }
                                 ...
```

Figure 13.4   A remote Math Service: (a) interface, and (b) implementation.

API generation, to Java endpoint implementation. The source code of the toolchain [10] and tutorial examples [5] are available online.

13.2.1 Global Protocol Specification

The tool takes as its primary input a textual description of the source protocol or choreography from a global perspective. Figure 13.5 is a Scribble *global protocol* for the Math Service running example.

A *payload format type* declaration (line 1) gives an alias (Int) to data type definitions from external languages (java.lang.Integer) used for message formatting. The *protocol signature* (line 2) declares the name of the global protocol (MathSvc) and the abstraction of each participant as a named role (C and S).

Message passing is written, *e.g.*, Val(Int) from C to S. A *message signature* (Val(Int)) declares an *operator* name (Val) as an abstract message

```
1   type <java> "java.lang.Integer" from "rt.jar" as Int;
2   global protocol MathSvc(role C, role S) {
3     choice at C { Val(Int) from C to S;
4                   choice at C { Add(Int) from C to S;
5                                 Sum(Int) from S to C; }
6                   or { Mult(Int) from C to S;
7                        Prod(Int) from S to C; }
8                   do MathSvc(C, S); }
9             or { Bye() from C to S; }
10  }
```

Figure 13.5   Scribble global protocol for Math Service in Figure 13.1.

identifier (which may be, *e.g.*, a header field value in the concrete message format), and some number of payload types (a single `Int`). Message passing is output-asynchronous: dispatching the message is non-blocking for the sender (C), but the message input is blocking for the receiver (S). A *located choice* (*e.g.*, `choice at C`) states the subject role (C) for which selecting one of the cases (the `or`-separated blocks) to follow is a mutually exclusive *internal* choice. This decision is an *external* choice to all other roles involved in each block, and must be appropriately coordinated by explicit messages. A `do` statament enacts the specified (sub)protocol, including recursive definitions (*e.g.*, line 8).

The body of the `MathSvc` protocol may be equivalently written in a similar syntax to standard recursive session types:

```
rec X { choice at C { Val(int) from C to S; ... continue X; }
            or { Bye() from C to S; } }
```

Protocol validation. The tool validates the well-formedness of global protocols. We do not discuss the details of this topic in this tutorial, but summarise a few elements. Firstly, the source protocol is subject to a range of syntactic checks. Besides basic details, such as bound role names and recursion variables, the key conditions are *role enabling, consistent external choice subjects* and *reachability*. Role enabling is a consistency check on the (transitive) propagation of choice messages originating from a choice subject to the other roles involved in the choice. The following is a very simple example of bad role enabling:

```
choice at C { Val(Int) from S to C; ... } or { Bye() from C to S; }
```

Since the choice is at C, S should not perform any output before it is *enabled*, i.e, by receiving a message that directs it into the correct choice case. The second of the above conditions requires that every message of an external choice be communicated *from* the same role.

Reachability of protocol states is imposed on a per-role basis, *i.e*, on projections; Scribble protocols are also checked to be tail recursive per role. These rule out some basic syntactic inconsistencies (*e.g.*, sequential composition after a non-terminating recursion), and support the later FSM translation step (see below).

Together, the syntactic conditions support the main validation step based on explicit checking of safety errors (and progress violations), such as reception errors and deadlocks (outlined in § 13.1, on a bounded model of the protocol. For example, (wrongly) replacing `Bye` by `Val` will be found by the explicit error checking to be invalid.

```
choice at C { Val(Int) from C to S; ... } or { Val(Int) from C to S; }
```

The ambiguous (non-deterministic) receipt of the decision message by s from c (*i.e.*, a message identified by `Val` in both cases—Scribble does not introduce any implicit meta data or communications) may lead to various deadlock and orphan message errors, depending on the different permutations of c and s proceeding in the two cases. *E.g.*, if c proceeds in the right case and s in the left, then s will be stuck (in the "...") waiting for an `Add`/`Mult` (or will encounter a broken connection error).

Endpoint FSM generation. The next key step is the generation of an *Endpoint Finite State Machine* (EFSM) for each role in the protocol. We use the term EFSM for the particular class of multiparty communicating FSMs given by Scribble's syntax and validation. The construction is based on and extends the syntactic projection of *global types* to *local types* [4], followed by a translation to an EFSM, building on a correspondence between local types and communicating FSMs [2, 7]. The nodes of an EFSM represent the states in the localised view of the protocol for the target role, and the transitions are the communication actions performed by the role between states. The EFSM for every role of a valid global protocol is defined.

Figure 13.6 depicts the (dual) EFSMs for c and s in `MathSvc`. The initial states are numbered 1. The notation, *e.g.*, `S!Val(Int)` means output of message `Val(Int)` to s; ? dually denotes input. The recursive definition of this protocol manifests as the cycles returning to state 1.

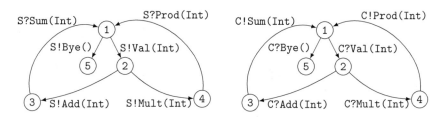

Figure 13.6 Endpoint FSMs for C and S in `MathSvc` (Figure 13.5).

13.2.2 Endpoint API Generation

For a given role of a valid global protocol, the toolchain generates an *Endpoint API* for implementing that role based on its EFSM. The current implementation generates Java APIs, but the principle may be applied or adapted to many statically-typed languages.

There are two main components of a generated Endpoint API, the *Session API* and the *State Channel API*. The generated APIs make use of a few protocol-independent base classes that are part of the Scribble runtime library: Role, Op, Session, MPSTEndpoint and Buf; the first three are abstract classes. These shall be explained below.

Session API. The frontend class of the Session API, which we refer to as the *Session Class*, is a generated final subclass of the base Session class with the same name as the source protocol, *e.g.*, MathSvc. It has two main purposes. One is to house the family of protocol-specific constants for type-directed session programming in Java, generated as follows.

A session type based protocol specification features various kinds of names, such as role names and message labels. A session type system typically requires these names to be present in the processes to drive the type checking (*e.g.*, [4, 1]). For the present Java setting, the Session API is generated to reify these abstract names as *singleton types* following a basic (eagerly initialised) singleton pattern. For each role or message operator name *n* in the source protocol, we generate:

- A final Java class named *n* that extends the appropriate base class (Role or Op). The *n* class has a single private constructor, and a public static final field of type *n* and with name *n*, initialised to a singleton instance of this class.
- In the Session Class, a public static final field of type *n* and with name *n*, initialised to the constant declared in the corresponding *n* class.

For example, for role C of MathSvc, the subclass C of Role is generated to declare the singleton constant public static final C C = new C();. The MathSvc class is generated to collect these various constants together, including the field public static final C C = C.C;.

The Session API comprises the Session Class with the singleton type classes. The other main purpose of the Session Class is for session initiation in endpoint implementations, as explained below.

An implementation of c **via Endpoint API generation.** At this point, we give, in Figure 13.7, a first version of the factorial calculation using the Endpoint API generated by the Scribble tool for c in MathSvc.

The code can be read similarly to the socket code in § 13.1.1; *e.g.*, s1 is a session channel variable. A difference from the earlier socket code is that the Scribble API is generated as a *fluent* interface, allowing consecutive I/O operations to be chained (*e.g.*, line 11). We refer to and explain this code through the following subsections.

Session initiation. Lines 3–6 in Figure 13.7 is a typical preamble for a (client) endpoint implementation using a Scribble-generated API. We start by creating a new MathSvc session by instantiating the Session Class. The session object, sess, is used to create a new session *endpoint* object of type MPSTEndpoint<MathSvc, C>, parameterised on the type of the session and the endpoint role. The c parameter in this type is the singleton type in the Session API; and the c argument in the constructor call is the single value of this type.

The third argument required by the MPSTEndpoint constructor is an implementation of the ScribMessageFormatter interface, that is responsible for the underlying serialization and deserialization of individual messages in this session. For this example, we use the default ObjectStreamFormatter provided by the Scribble runtime library based on the standard Java serialization protocol (messages communicated by this formatter must implement the Serializable interface).

```
1   int facto(int n) throws Exception { // Pre: n >= 1
2     Buf<Integer> i = new Buf<>(n), res = new Buf<>(i.val);
3     MathSvc sess = new MathSvc();
4     try (MPSTEndpoint<MathSvc, C> ep =
5           new MPSTEndpoint<>(sess, C, new ObjectStreamFormatter())) {
6       ep.connect(S, SocketChannelEndpoint::new, "localhost", 8888);
7       MathSvc_C_1 s1 = new MathSvc_C_1(ep);
8       while (i.val > 1)
9         s1 = sub1(s1, i) // sub1 on line 15
10          // State transitions: 1 -> 2 -> 4 -> 1 (see Fig.1.6)
11          .send(S, Val, res.val).send(S, Mult, i.val).receive(S, Prod, res);
12      s1.send(S, Bye); // 1 -> EndSocket
13      return res.val;
14    } }
15  MathSvc_C_1 sub1(MathSvc_C_1 s1, Buf<Integer> b) throws ... {
16    // State transitions: 1 -> 2 -> 3 -> 1 (see Fig.1.6)
17    return s1.send(S, Val, b.val).send(S, Add, -1).receive(S, Sum, b);
18  }
```

Figure 13.7 Factorial calculation using the Endpoint API generated for C.

Before proceeding to the main body of a protocol implementation, the MPSTEndpoint object is used to set up the session topology via connection establishment actions with the appropriate peer endpoints. On line 6, the MPSTEndpoint is used to perform the client-side connect to S. The second argument is a reference to the constructor of SocketChannelEndpoint in the Scribble runtime library, which embodies a standard TCP socket; alternatives include HTTP and shared memory endpoints. The connection setup phase is concluded when the MPSTEndpoint is passed as a constructor argument to the initial state channel constructor, MathSvc_C_1, expained next.

The MPSTEndpoint implements the Java AutoCloseable interface and should be handled using a try-with-resources, as on line 4; the encapsulated network resources are implicitly closed when control flow exits the try statement.

State Channel API. The State Channel API is generated to capture the protocol-specific behaviour of a role in the source global protocol, as represented by its EFSM, via the static typing facilities of Java.

- Each state in the EFSM is reified as a Java class for a *state-specific* session channel, thus conferring a distinct Java type to channels at each state in a protocol. We refer to instances of these generated channel classes as *state channels*.
- The I/O operations (methods) supported by a channel class are the transitions permitted by the corresponding EFSM state.
- The return type of each generated I/O operation is the channel type for the successor state following the corresponding transition from the current state. Performing an I/O operation on a state channel returns a new instance of the successor channel type.

By default, the API generation uses a simple state enumeration (*e.g.*, Figure 13.6) for the generated channel class names; *e.g.*, MathSvc_C_1 for the initial state channel. More meaningful names for states may be specified by the user as annotations in the Scribble source. The terminal state of an EFSM, if any, is generated as an EndSocket class that supports no further I/O operations. The channel class for the initial state is the only class with a public constructor, taking an MPSTEndpoint parameterised on the appropriate Session Class and role types; all other state channels are instantiated internally by the generated API operations.

Figure 13.8 summarises the generated channel classes and their main I/O operations for C in MathSvc. *E.g.*, a state channel of type MathSvc_C_1 supports methods for sending Val and Bye to S; these send methods are overloaded via

Gen. class	Session operation methods
MathSvc_C_1	MathSvc_C_2 send(S role, Val op, Integer pay1)
	EndSocket send(S role, Bye op)
MathSvc_C_2	MathSvc_C_3 send(S role, Add op, Integer pay1)
	MathSvc_C_4 send(S role, Mult op, Integer pay1)
MathSvc_C_3	MathSvc_C_1 receive(S role, Sum op, Buf<? super Integer> pay1)
MathSvc_C_4	MathSvc_C_1 receive(S role, Prod op, Buf<? super Integer> pay1)

Figure 13.8   State Channel API generated for C in MathSvc (Figure 13.5).

the parameters for the destination role and message operator (the singleton types of the Session API), as well as the message payloads. Sending a Val returns a MathSvc_C_2 channel, *i.e.*, the state of sending an Add or Mult; whereas a sending a Bye returns an EndSocket.

For unary input states, *i.e.*, an EFSM state with a single input transition, the generated receive method, *e.g.*, for Sum in MathSvc_C_3, takes Buf arguments parameterised according to the expected payload type(s), if any. Buf<T> is a simple generic one-slot buffer provided by the Scribble runtime library, whose value is held in a public val field. The receive method is generated to write the payload(s) of the received message to the respective Buf arguments.

In Figure 13.7, lines 7–14 use the State Channel API for C to perform the factorial calculation. Starting from the instance of MathSvc_C_1, assigned to s1, the implementation proceeds by performing *one* I/O operation on each current state channel to obtain the next. The fluent API permits convenient chaining of I/O operations, *e.g.*, line 17 in sub1 starts from state 1, and proceeds through states 2 and 3 (by sending Val and Add messages), before returning to 1 (by receiving the Sum). The endpoint implementation is complete upon reaching EndSocket.

Attempting any I/O action that is not permitted by the current protocol state, as designated by the target state channel, will be caught by Java type checking. For example (from the Eclipse IDE):

```
s1 = sub1(s1, i)//.send(S, Val, b.val) -- did not send the Val
    .send(S, Mult, i.val).receive(S, Prod, b);
```
> ⓧ The method send(S, Val, Integer) in the type MathService_C_1 is not applicable for the arguments (S, Mult, Integer)

13.2.3 Hybrid Session Verification

As demonstrated above, Scribble-generated Endpoint APIs leverage standard, static Java type checking to verify protocol conformance, *provided every state channel returned by an API operation is used exactly once up to the end of the session.* This is the implicit usage contract of a Scribble-generated API, to respect EFSM semantics in terms of following state transitions linearly up to the terminal state.

Much research has been conducted towards static analyses for such resource usage properties: to this end, it may be possible to combine these with API generation tools to recover fully static safety guarantees in certain contexts. However, designing such analyses for mainstream engineering languages, such as Java and C#, in full generality is a significant challenge, and often based on additional language extensions or imposing various conservative restrictions.

As a practical compromise, the Endpoint API generation of Scribble promotes a *hybrid* approach to session verification. The idea is simply to complement the static type checking of session I/O on state channels with *run-time* checks that each state channel is indeed used exactly once in a session execution.

Run-time checking of linear state channel usage. The checks on linear state channel usage are inlined into the State Channel API operations by the API generation. There are two cases by which state channel linearity may be violated.

Repeat use. Every state channel instance maintains a boolean state value indicating whether it has been *used, i.e.,* a session I/O operation has been performed on the channel. The API generation guards each I/O operation with a run-time check on this boolean. If the channel has already been used, a `LinearityException` is raised.

Unused. All state channels of a session instance share a boolean state value indicating whether the session is *complete* for the local endpoint. The API is generated to set this flag when a *terminal operation, i.e.* an I/O action leading to the terminal EFSM state, is performed. If control flow leaves the enclosing `try` statement of the associated `MPSTEndpoint`, the Scribble runtime checks this flag via the implicit `close` method of the `AutoCloseable` interface. If the session is incomplete, an exception is raised.

It is not possible for the completion flag to be set if any state channel remains unused on leaving the `try` statement of an `MPSTEndpoint`. IDEs (*e.g.*, Eclipse) support compile-time warnings in certain situations where AutoClose-able resources are not appropriately handled by a try.

Hybrid session safety. Together, a statically typed Endpoint API with run-time state channel linearity checking offers the following properties.

1. If a session endpoint implementation respects state channel linearity, then the generated API statically ensures freedom from the application errors outlined in § 13.1 (*i.e.*, *communication safety, e.g.*, [4, error-freedom]) when composed with conformant endpoints for the other roles in the protocol.
2. Regardless of state channel linearity, any statically well-typed endpoint implementation will never perform a message passing action that does not conform to the protocol.

These properties follow from the fact that the only way to violate the EFSM of the API, generated from a validated protocol, is to violate state channel linearity, in which case the API raises an exception *without* actually performing the offending I/O action. This hybrid form of session verification thus guarantees the absence of protocol violation errors during session execution up to premature termination, which is always a possibility in practice due to program errors outside of the immediate session code, or other failures, such as broken connections.

When following the endpoint implementation pattern promoted by a generated API, by associating session code to the MPSTEndpoint-try, the Java IOException of, *e.g.*, a broken connection will direct control flow out of the try, safely (w.r.t. session typing) avoiding further I/O actions in the failed session. Finer-grained treatment of session failures is a direction of ongoing development for Scribble (and MPST).

13.2.4 Additional Math Service Endpoint Examples

A first implementation of s. Figure 13.9 summarises the State Channel API generated for s in MathSvc. Unlike c, the EFSM for s features non-unary input states, which correspond at the process implementation level to the *branch* primitive of formal session calculi (*e.g.*, [1]). Java does not directly support a corresponding language construct, but API generation enables some different options.

One option, demonstrated here, is designed for standard Java switch patterns. For each branch state, a *branch-specific* enum is generated to enumerate

Generated class	Session operation methods
MathSvc_S_1	MathSvc_S_1_Cases branch(C role)
MathSvc_S_1_Cases	MathSvc_S_2 receive(Val op, Buf<? super Integer> pay1)
	EndSocket receive(Bye op)
MathSvc_S_2	MathSvc_S_2_Cases branch(C role)
MathSvc_S_2_Cases	MathSvc_S_3 receive(Add op, Buf<? super Integer> pay1)
	MathSvc_S_4 receive(Mult op, Buf<? super Integer> pay1)
MathSvc_S_3	MathSvc_S_1 send(C role, Sum op, Integer pay1)
MathSvc_S_4	MathSvc_S_1 send(C role, Prod op, Integer pay1)

```
1   try (ScribServerSocket ss = new SocketChannelServer(8888)) { // TCP
2     while (true) { // Persistent server
3       MathSvc sess = new MathSvc();
4       try (MPSTEndpoint<MathSvc, S> ep
5           = new MPSTEndpoint<>(sess, S, new ObjectStreamFormatter())) {
6         ep.accept(ss, C);
7         Buf<Integer> b1 = new Buf<>(), b2 = new Buf<>();
8         MathSvc_S_1 s1 = new MathSvc_S_1(ep);
9         Loop: while (true) {
10          MathSvc_S_1_Cases c1 = s1.branch(C);
11          switch (c1.op) {
12            case Bye: c1.receive(Bye); break Loop;
13            case Val:
14              MathSvc_S_2_Cases c2 = c1.receive(Val, b1).branch(C);
15              switch (c2.op) {
16                case Add: s1 = c2.receive(Add, b2)
17                            .send(C, Sum, b1.val + b2.val); break;
18                case Mult: s1 = c2.receive(Mult, b2)
19                            .send(C, Prod, b1.val * b2.val); break;
20  } } } } } }
```

Figure 13.9   State Channel API generated for S in MathSvc; and an implementation of S using the generated API.

the cases of the choice according to the source protocol. *E.g.*, for the initial state of S: enum MathSvc_S_1_Enum { *Val, Bye* }.

The channel class itself (Figure 13.9), MathSvc_S_1, is generated with a single branch operation. This method blocks until a message is received, returning a new instance of the generated MathSvc_S_1_Cases class, which holds the enum value corresponding to the received message in a final op field. Unfortunately, since the *static* type of the Cases object reflects the *range* of possible cases, the API requires the user to manually call the corresponding receive method of the Cases object, essentially as a form of cast to obtain the appropriately typed state channel.

Lines 11–20 in Figure 13.9 implement a `switch` on the `op` enum of `MathSvc_S_1_Cases`. The Java compiler is able to statically determine whether all enum cases are exhaustively handled. In each of the two cases (`Bye` and `Val`), the corresponding `receive`-cast is called on the `Cases` object to obtain the successor state channel of that (input) transition. Leveraging the hybrid verification approach, the generated API includes an implicit run-time check that the correct cast method is used following a `branch`; calling an incorrect method raises an exception.

§ 13.4 discusses an alternative API generation that allows session branches to checked by Java typing *without* additional run-time checks.

Alternative `c` **factorial implementation.** Following is an implementation of a factorial calculation using the `c` endpoint of `MathSvc` in a recursive method, illustrating the use of the State Channel API in an alternative programming style.

```
MathSvc_C_1 facto(MathSvc_C_1 s1, Buf<Integer> b) throws ... {
  if (b.val == 1) return s1; // Pre: b.val >= 1
  Buf<Integer> tmp = new Buf<>(b.val);
  return facto(sub1(s1, tmp), tmp) // sub1 from Fig. 13.7
      .send(S, Val, b.val).send(S, Mult, tmp.val).receive(S, Prod, b);
}
```

Besides conformance to the protocol itself, the state channel parameter and return types help to ensure that the appropriate I/O transitions are performed through the protocol states in order to enact the recursive method call correctly.

13.3 Real-World Case Study: HTTP (GET)

In this section, we apply the Scribble API generation methodology to a real-world protocol, HTTP/1.1 [3]. For the purposes of this tutorial, we limit this case study to the GET method of HTTP, and treat a minimal number of message fields required for interoperability with existing real-world clients and servers. The following implementations have been tested against Apache (as currently deployed by the dept. of computing, Imperial College London) and Firefox 5.0.1.

A key point illustrated by this experiment on using session types in practice is the interplay between *data types* (message structure) and *session types* (interaction structure) in a complete protocol specification. In particular, that aspects of the former can be refactored into the latter, while fully preserving protocol interoperability, to take advantage of the safety properties offered by Scribble-generated APIs in endpoint implementations.

13.3.1 HTTP in Scribble: First Version

HTTP is well-known as a client-server request-response protocol, typically conducted over TCP. Despite its superficial simplicity, *i.e.*, a sequential exchange of just two messages between two parties, the standards documentation for HTTP spans several hundred pages, as is often the case for Internet applications and other real-world protocols.

Global protocol. As a first version, we simply express the high-level notion of an HTTP request-response as follows:

```
sig <java> "...client.Req" from ".../Req.java" as Req;
sig <java> "...server.Resp" from ".../Resp.java" as Resp;
global protocol Http(role C, role S) {
  Req from C to S;
  Resp from S to C;
}
```

A small difference from the Scribble examples seen so far are the `sig` declarations for custom message formatting. Unlike `type` declarations, which pertain specifically to payload types, `sig` is used to work with host language-specific (*e.g.*, `<java>`) routines for arbitrary message formatting; *e.g.*, `Req.java` contains Java routines, provided as part of this protocol specification, for performing the serialization and deserialization between Java `Req` objects and the actual ASCII strings that constitute concrete HTTP requests on the wire.

Client implementation. For such a simple specification, we omit the EFSMs and Endpoint APIs for each role, and directly give client code using the generated API (omitting the usual preamble):

```
Buf<Resp> b = new Buf<>();
Http_C_1 s1 = new Http_C_1(client); // client: MPSTEndpoint<Http, C>
s1.send(S, new Req("/index.html", "1.1", host)).receive(S, Resp, b);
```

The generated API prevents errors such as attempting to receive `Resp` before sending `Req` or sending multiple `Req`s. However, one may naturally wonder if this is "all there is" to a correct HTTP client implementation—where is the complexity that is carefully detailed in the RFC specification?

The answer lies in the message formatting code that we have conveniently abstracted as the Req and Res message classes. A basic HTTP session does exchange only two messages, but these messages are richly structured, involving branching, optional and recursive structures. In short, this first version *assumes* the correctness of the Req and Resp classes (written by the protocol author, or obtained using other parsing/formatting utilites) as part of the protocol specification.

13.3.2 HTTP in Scribble: Revised

As defined in RFC 7230 [3] (§ 3 onwards), the message grammar is:

```
HTTP-message  = start-line *( header-field CRLF ) CRLF [ message-body ]
start-line    = request-line / status-line
request-line  = method SP request-target SP HTTP-version CRLF
header-field  = field-name ":" OWS field-value OWS
... // CRLF=carriage return line feed; SP=space; OWS=optional white space
```

Intuitively, the act of sending a HTTP request may be equivalently understood as sending a request-line, followed by sending zero or more header-fields terminated by CRLF, and so on. Following this intuition, we can refactor much of this structure from the data side of the specification to the session types side, giving a Scribble description that captures the target protocol specification in more explicit detail than previously. Consequently, the generated API will promote the Java endpoint to respect this finer-grained protocol structure by static typing, as opposed to assuming the correctness of the supplied message classes.

We are able to refine the Scribble for HTTP/TCP in this way because any application-level notion of "message" identified in the specification is ultimately broken down and communicated via the TCP bit streams, in a manner that is transparent to the other party (client or server). This approach may thus be leveraged for any application protocol conducted over a transport with such characteristics.

Global protocol. Figure 13.10 is an extract of a revised Scribble specification of HTTP. The monolithic request and response messages have been decomposed into smaller constituents; *e.g.*, RequestL and Host respectively denote the request-line and host-field in a request. For the most part, the Java code for formatting each message fragment as an HTTP ASCII string is reduced to a simple print instruction with compliant white spacing built in

```
1   sig <java> "...client.RequestLine" from ".../RequestLine.java" as RequestL;
2   sig <java> "...client.Host" from ".../Host.java" as Host;
3   sig <java> "...Body" from ".../Body.java" as Body;
4   sig <java> "...client.UserAgent" from ".../UserAgent.java" as UserA;
5   sig <java> "...server.HttpVersion" from ".../HttpVersion.java" as HttpV;
6   sig <java> "...server._200" from ".../_200.java" as 200;
7   sig <java> "...server._404" from ".../_404.java" as 404;
8   ...
9   global protocol Http(role C, role S) {
10    do Request(C, S);
11    do Response(C, S);
12  }
13  aux global protocol Request(role C, role S) {
14    RequestL from C to S; // GET /index.html HTTP/1.1
15    rec X {
16      choice at C { Host from C to S; continue X; } // host: www.doc.ic.ac.uk
17            or { UserA from C to S; continue X; }// User-Agent: Mozilla
18            or ...
19            or { Body from C to S; }
20  } } // (aux bypasses validating these "subprotocols" as "root" protos)
21  aux global protocol Response(role C, role S) {
22    HttpV from S to C; // HTTP/1.1
23    choice at S { 200 from S to C; } // 200 OK
24            or { 404 from S to C; } // 404 Not Found
25            ...
26  }
```

Figure 13.10   Extract from the revised specification of HTTP in Scribble.

(e.g, CRLFs). The structure by which these constituents should be composed to reform whole messages is now expressed in the Request and Response subprotocols.

Client implementation. Taking the revised Scribble HTTP, Endpoint API generation proceeds as usual, generating the EFSMs for each role to give the structure of the State Channel API. Lines 2–3 in Figure 13.11 is an almost minimal implementation of a *correctly formatted* request according to the Request subprotocol. The typed API ensures the initial, mandatory RequestLine; then amongst the recursive choice cases we opt to send only the Host field, before concluding the request by an empty Body. A complete client implementation is given by: doResponse(doRequest(s1)).

Besides limiting to a subset of the protocol, this revision is by no means a complete specification of HTTP in terms of capturing the entire message grammar in full detail; the fidelity of the Scribble specification may be pushed

```
1  Http_C_3 doRequest(Http_C_1 s1) throws Exception {
2    return s1.send(S, new RequestLine("/index.html", "1.1"))
3             .send(S, new Host("www.host.com")).send(S, new Body(""));
4  }
5  EndSocket doResponse(Http_C_3 s3) throws Exception {
6    Http_C_4_Cases cases = s3.async(S, HttpV, new Buf<>()).branch(S);
7    switch (cases.op) {
8      case _200: ...
9      case _404: ...
10     ...
11   }
12   ...
```

Figure 13.11   Extract from an implementation of a HTTP client via API generation.

further, perhaps towards a "character-perfect" specification, via suitably fine-grained message decomposition.

13.4 Further Endpoint API Generation Features

Branch-specific callback interfaces. Scribble also generates a *callback-based API* for branch states, which does *not* require additional run-time checks (cf. § 13.2.4). For each branch state, a *handler* interface is generated with a callback variant of receive for each choice case; *e.g.*, MathSvc_S_1_Handler in Figure 13.12. Apart from the operator and payloads, each method takes the continuation *state channel* as a parameter; the return type is void. Java typing ensures that a (concrete) implementation of this interface implicitly covers all cases of the branch. Finally, a variant of branch is generated in the parent channel class (*e.g.*, MathSvc_S_1) that takes an instance of the corresponding handler interface, with return void. As before, this branch blocks until a message is received; the API then delegates the handling of the message to the appropriate callback method of the supplied handler object.

Figure 13.12 gives a class that implements the handler interfaces of *both* branch states for s. Assuming an MPSTEndpoint<MathSvc, S> serv, this handler class may be used in an event-driven implementation of s by: new MathSvc_S_1(serv).branch(C, new MathSHandler()).

State-specific futures for unary inputs. For unary input states, Scribble additionally generates *state-specific input futures* as an alternative mechanism

Generated class	Session operation methods (additional to Fig. 1.9)
MathSvc_S_1	void branch(C role, MathSvc_S_1_Handler h)
MathSvc_S_1_Handler	void receive(MathSvc_S_2 s, Val op, Buf<Integer> pay1)
	void receive(EndSocket s, Bye op)
MathSvc_S_2	void branch(C role, MathSvc_S_2_Handler h)
MathSvc_S_2_Handler	void receive(MathSvc_S_3 s, Add op, Buf<Integer> pay1)
	void receive(MathSvc_S_4 s, Mult op, Buf<Integer> pay1)

```
1   class MathSHandler implements MathSvc_S_1_Handler, MathSvc_S_2_Handler {
2     private Buf<Integer> b1;
3     public void receive(MathSvc_S_2 s2, Val op, Buf<Integer> b1) .. {
4       this.b1 = b1;
5       s2.branch(C, this);
6     }
7     public void receive(EndSocket end, Bye op) throws ... { }
8     public void receive(MathSvc_S_3 s3, Add op, Buf<Integer> b2) throws ..
9       { s3.send(C, Sum, this.b1.val + b2.val).branch(C, this); }
10    public void receive(MathSvc_S_4 s4, Mult op, Buf<Integer> b2) throws ..
11      { s4.send(C, Prod, this.b1.val * b2.val).branch(C, this); }
12  }
```

Figure 13.12 Additional branch callback interfaces generated for S in MathSvc; and a corresponding implementation of S.

to the basic receive. For example, in the revised Scribble specification of HTTP (Figure 13.11), the channel class Http_C_3 corresponds to the state where C should receive the HTTP version element (HttpV) of the response status-line (Line 14 in Figure 13.10). For this state, Scribble generates the class Http_C_3_Future. Its key elements are input-specific fields for the message type (msg) or payloads (*e.g.*, pay1) to be received, and a sync method to force the future. For the Http_C_3 channel class itself, the following variant of receive is generated:

```
        Http_C_4 async(S role, HttpV op, Buf<Http_C_3_Future> fut)
```

Unlike a basic receive, calling async returns immediately with a new instance of Http_C_3_Future in the supplied Buf.

Calling sync first implicitly forces all pending prior futures, in order, for the same peer role. It then blocks the caller until the expected message is received, and writes the values to the generated fields of the future. This safely preserves the FIFO messaging semantics between each pair of roles in a session, so that endpoint implementations using generated futures retain the same safety properties as using only blocking receives. Repeat forcing of an input future has no effect.

An example usage of async was given in Figure 13.11 (line 6). There, the async is used to safely affect a *non-blocking input* action (the client is not interested in blocking on awaiting just the HttpV portion of the response). Since the HttpV future is never explicitly forced – unlike state channels, input-futures are not linear objects – async also affects a *user*-level form of *affine* input action, in the sense that the user never reads this message. Finally, async enables postponing input actions until later in a session, for safe user-level *permutation* of session I/O actions.

References

[1] M. Coppo, M. Dezani-Ciancaglini, N. Yoshida, and L. Padovani. Global progress for dynamically interleaved multiparty sessions. *Mathematical Structures in Computer Science*, 760:1–65, 2015.

[2] P.-M. Deniélou and N. Yoshida. Multiparty session types meet communicating automata. In *ESOP '12*, volume 7211 of *LNCS*, pages 194–213. Springer, 2012.

[3] R. Fielding, Y. Lafon, M. Nottingham, and J. Reschke. IETF RFCs 7230–7235 Hypertext Transfer Protocol 1.1. https://tools.ietf.org/html/rfc7230

[4] K. Honda, N. Yoshida, and M. Carbone. Multiparty asynchronous session types. In *POPL '08*, pages 273–284. ACM, 2008.

[5] R. Hu. Demo files for this BETTY tutorial chapter. https://github.com/scribble/scribble-java/tree/master/modules/demos/scrib/bettybook

[6] R. Hu and N. Yoshida. Hybrid session verification through endpoint API generation. In *FASE '16*, volume 9633 of *LNCS*, pages 401–418. Springer, 2016.

[7] J. Lange, E. Tuosto, and N. Yoshida. From communicating machines to graphical choreographies. In *POPL '15*, pages 221–232. ACM Press, 2015.

[8] OMG UML 2.5 specification. http://www.omg.org/spec/UML/2.5

[9] Scribble homepage. http://www.scribble.org

[10] Scribble GitHub repository. https://github.com/scribble/scribble-java

14

Mungo and StMungo: Tools for Typechecking Protocols in Java

Ornela Dardha[1], Simon J. Gay[1], Dimitrios Kouzapas[1], Roly Perera[1,2], A. Laura Voinea[1] and Florian Weber[1]

[1]School of Computing Science, University of Glasgow, UK
[2]School of Informatics, University of Edinburgh, UK

Abstract

We present two tools that support static typechecking of communication protocols in Java. Mungo associates Java classes with typestate specifications, which are state machines defining permitted sequences of method calls. StMungo translates a communication protocol specified in the Scribble protocol description language into a typestate specification for each role in the protocol by following the message sequence. Role implementations can be typechecked by Mungo to ensure that they satisfy their protocols, and then compiled as usual with javac. We demonstrate the Scribble, StMungo and Mungo toolchain via a typechecked POP3 client that can communicate with a real-world POP3 server.

14.1 Introduction

Modern computing is dominated by communication, at every level from manycore architectures through multithreaded programs to large-scale distributed systems; this contrasts with the original emphasis on data processing. Early recognition of the importance of structured data meant that high-level programming languages have always incorporated data types and supported programmers through the techniques of static and dynamic typechecking. The foundational status of structured data was explicitly recognised in the title of Wirth's classic 1976 text *Algorithms + Data Structures = Programs*, but a more appropriate modern slogan would be *Programs + Communication*

309

Structures = Systems. The new reality of communication-based software development needs to be supported by programming tools based on structuring principles and high-level abstractions. Given the success of data types, it is natural to apply type-theoretic techniques to the specification and verification of communication-based code. During the last twenty years, this goal has been pursued by the expanding and increasingly active research community on session types [12, 13, 24]. A session type is a formal structured description of a communication protocol, specifying the type, sequence and direction of messages. By embedding this description in the type system of a programming language, adherence to the protocol can be verified by static typechecking; if desired, dynamic monitoring can be introduced into the runtime system.

Several researchers have worked towards making typechecked communication structures available for mainstream software development, by transferring session types from their original setting of pi-calculus to functional and object-oriented languages [3, 5–8, 15, 17, 19]. Gay *et al.* [9] proposed an integration of session types and object-oriented programming through the concept of typestates [22], in which methods are constrained to be called only in particular sequences. They defined a translation from the session type of a communication channel endpoint into a typestate specification that constrains the use of send and receive methods on an object representing the channel endpoint. Their notation for typestate specifications was inspired by the syntax of session types.

Dardha, Gay, Kouzapas and Perera extended that work and implemented it as Mungo [16], a front-end typechecking tool for Java. They also generalised the translation from session types to typestate specifications, so that it handles multiparty [11] instead of binary session types, and made it concrete by implementing StMungo [16], a translator from the Scribble [20, 25] protocol description language into Mungo specifications. The Scribble description of a protocol is translated into an API with which to program implementations of protocol roles; the typestate specification associated with the API permits static checking of the correctness of the implementation of a role. Typestate specifications do not represent the notion of duality of session types; compatibility between roles depends on the assumption that their typestate specifications are derived from a single global session type. The paper by Kouzapas *et al.* [16] illustrated the use of Mungo and StMungo with a substantial case study of an SMTP client [21], including the low-level implementation details necessary to enable communication with standard SMTP servers. This achieved the long-standing goal of using session types to specify and verify implementations of real internet protocols.

The present chapter describes Mungo and StMungo in relation to three examples. The first, in Section 14.2.1, illustrates Mungo by defining and checking a typestate specification for an iterator. The second, in Section 14.3, is a simple multiparty scenario based on a travel agency. Finally, in Section 14.4, we show how Mungo and StMungo can be used to typecheck a client for the POP3 protocol [18].

14.2 Mungo: Typestate Checking for Java

Mungo is a static analysis tool that checks typestate properties in Java programs. Mungo implements two main components. The first is a Java-like syntax to define typestate specifications for classes, and the second is a typechecker that checks whether objects that have typestate specifications are used correctly. Mungo typechecks standard Java code without syntactic extensions; typestate specifications are defined in separate files and are associated with Java classes by means of the Java annotation mechanism. After typechecking with Mungo, programs can be compiled and run using standard Java tools. The declaration of a typestate specification in a single file contrasts with other approaches that take the viewpoint of typestate as pre- and post-conditions on methods; we discuss this point in Section 14.5. If a class has a typestate specification, the Mungo typechecker analyses each variable of that class in the program and extracts the method call behaviour (sequences of method calls) through the variable's life. Finally, it checks the extracted information against the sequences of method calls allowed by the typestate specification.

Mungo is implemented in the JastAdd [10] framework, which is a Reference Attribute Grammar (RAG) meta-compiler suite compatible with Java. JastAdd provides a Java parser and typechecker, and was also used to implement a parser for the typestate specification language.

Mungo supports typechecking for a subset of Java. The programmer can define classes with typestate specifications and classes without them. The typechecking procedure tracks variables storing instances of classes with typestate specifications, through argument passing and return values. Moreover, the typechecking procedure for the fields of a class follows the typestate specification of the class to infer a typestate usage for the fields. For this reason fields that have typestate specifications must be defined in a class that also has a typestate specification.

Mungo first runs the Java typechecker provided by the JastAdd framework. If there are no errors then Mungo performs additional well-formedness checks before it runs the typestate checking procedure. First, the tool checks for well-formed typestate specifications: they must be deterministic and all states must be reachable from the initial state. Second, it checks that a class with a typestate specification implements all the methods required in the typestate. Third, arrays cannot store objects that have typestates, because array access, and thus inference for objects that are stored in an array, cannot be determined at statically. Finally, fields with typestate specifications must be private and non-static, to disallow external interference with their state.

Completing the coverage of Java will require further work. Some features we anticipate to be relatively straightforward extensions, such as synchronised statements, the conditional operator ?:, inner and anonymous classes, and static initialisers. Generics, inheritance and exceptions are non-trivial. Currently, generics are not supported, while inheritance is supported for classes without associated typestate behaviour. Exceptions are supported syntactically but are type-checked under the (unsound) assumption that no exceptions are thrown; a try{...} catch(Exception e) {...} statement is typechecked by typechecking the try block but not the catch block. If the program does not throw exceptions then there will be no violations of typestate specifications, but exception handlers may violate typestates.

14.2.1 Example: Iterator

We introduce some of the features of Mungo through an example that enforces correct usage of a Java Iterator. The example shows how a programmer can define an API and associate it with a typestate specification in order to constrain the order in which methods can be called. In the code below we define class StateIterator to wrap a Java Iterator. We use the Java annotation syntax @Typestate("StateIteratorProtocol") to associate the class StateIterator with the typestate specification StateIteratorProtocol. We often refer to a typestate specification as a protocol, following the established terminology of "object protocol" in the typestate literature.

```
1  package iterator;
2  import java.util.Iterator;
3
4  @Typestate("StateIteratorProtocol")
5  class StateIterator {
```

```
6     private Iterator iter;
7
8     public StateIterator(Iterator i) { iter = i; }
9     public Object next()              { return iter.next(); }
10    public void remove()             { iter.remove(); }
11    public Boolean hasNext() {
12      if(iter.hasNext() == true)
13        return Boolean.True;
14      return Boolean.False;
15    } }
```

We assume that the underlying implementation of the Java Iterator
includes the remove() method. The implementation of method hasNext() uses
the Iterator to discover whether the underlying collection has more elements.
It assumes the definition of the enumeration

```
1     enum Boolean { True, False }
```

which is provided as part of the Mungo framework. This enumeration is used
to specify dependency of the protocol on the result of a method.

Overall, the StateIteratorProtocol protocol ensures that the Java Iterator will be used in a way that throws no exceptions (method next() throws
NoSuchElementException when there are no more elements in the underlying
collection, and method remove() throws IllegalStateException when there
is no element to removed). The code below defines the typestate specification
StateIteratorProtocol.

```
1    package iterator;
2
3    typestate StateIteratorProtocol {
4      HasNext = { Boolean hasNext(): <True: Next, False: end> }
5      Next =    { Object next(): HasNextOrRemove }
6      HasNextOrRemove = {
7        void remove(): HasNext,
8        Boolean hasNext(): <True: NextOrRemove, False: end>
9      }
10     NextOrRemove = {
11       void remove(): Next,
12       Object next(): HasNextOrRemove
13     } }
```

A new iterator object is in state HasNext, because that is the first state in the definition. The only method available is hasNext(). If method next() were available then NoSuchElementException might be thrown in the case where there are no (more) elements in the underlying collection. Similarly, the availability of method remove() might result in IllegalStateException. A call of method hasNext() means that the continuation of the protocol depends on the return value of the method. In the case of False no further interaction with the iterator is possible, thus preventing possible exceptions. If the value True is returned then the state changes to Next, which forces the programmer to call the next() method and proceed to state HasNextOrRemove. Method remove() is not available because it should only be called after next() in order to remove the element returned by next(). Method hasNext() is not available because calling it would be redundant

The state HasNextOrRemove offers a choice between methods remove and hasNext(). In the former case the iterator removes the current object and proceeds to the HasNext state. Alternatively, calling hasNext() either proceeds to state NextOrRemove or ends the protocol otherwise. In state NextOrRemove there is still the possibility of removing the last returned object and proceeding to the Next state (this is because a poll has already been done), or getting the next element of the collection using method next() and proceeding to the HasNextOrRemove state.

To summarise, if we assume semantic correctness of the methods of iter (for example, that iter.hasNext() correctly reports the state of iter), then by using Mungo to typecheck code that uses a StateIterator, we can ensure that NoSuchElementException and IllegalStateException will not occur. Specifically, we guarantee: i) not calling the next() method on an empty collection; ii) not calling the remove() when there is no element to remove from the underlying collection; iii) additionally, not having redundant calls of the hasNext() method.

To avoid conflicting state changes, objects with typestates must not be aliased. Mungo uses linear typing to prevent aliasing.

The code below, which is well-typed according to Mungo, creates and uses a StateIterator object. It creates a HashSet containing the positive integers smaller than 32, and then removes the even numbers.

```
1  Collection c = new HashSet();
2  Integer i = 0; while(i < 32) c.add(i++);
3  StateIterator iter = new StateIterator(c.iterator());
4  iterate:
```

```
5  do {
6    switch(iter.hasNext()) {
7      case True:
8        System.out.println(i = (Integer) iter.next());
9        if(i%2 == 0) iter.remove();
10       continue iterate;
11     case False:
12       break iterate;
13   }
14 } while(true);
```

The HashSet's iterator is wrapped in a StateIterator object, which is subsequently used according to its protocol. The loop structure in the protocol is matched by the pattern label: do { ... } while(true); together with the continue label; and break label; statements. The switch statement handles the possible results of hasNext(), controlling the continuation or termination of the loop. The code on line 9 chooses whether or not to call remove(); the state here is HasNextOrRemove.

14.3 StMungo: Typestates from Communication Protocols

StMungo (Scribble to Mungo) is a transpiler from Scribble to Java, which also generates Mungo typestate specifications. It is based on the integration of session types and typestates [9] which consists of a formal translation of session types for communication channels into typestate specifications for channel objects. The latter define the order in which the methods of the channel objects can be called. This specification of the permitted sequences of method calls is naturally viewed as a channel protocol. We take a step further: we extend this formal translation from binary to multiparty session types [11] and implement it as StMungo, which translates Scribble local protocols into typestate specifications and prototype implementation code based on TCP/IP sockets. After refinement, the implementation is typechecked using Mungo.

A Scribble local protocol describes the communication between one role and all the other participants in a multiparty scenario, including the way in which messages sent to different participants are interleaved. StMungo is based on the principle that each role in the multiparty communication can be abstracted as a Java class following the typestate corresponding to the role's local protocol. The typestate specification generated by StMungo, together with the Mungo typechecker, guide the programmer in the design and implementation of distributed multiparty communication-based programs with guarantees of communication safety and soundness. StMungo is the first tool to provide a practical embedding of multiparty session type protocols into object-oriented languages with typestate specifications.

The diagram shows how the toolchain consisting of Scribble, StMungo and Mungo is used to generate a Java program from a Scribble protocol.

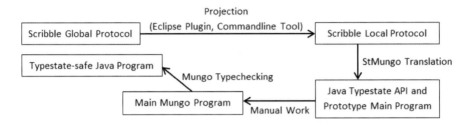

We start with a global protocol written in Scribble, which is then validated and projected into local protocols, one for each role specified in the global protocol. At this point we run StMungo on the local projections for which we want to generate a typestate. The tool generates a typestate specification, a Java API and a prototype main program. After completing the main program, typechecking with Mungo verifies that it correctly implements the protocol.

14.3.1 Example: Travel Agency

We now illustrate the toolchain of Scribble, StMungo and Mungo by means of a travel agency example, which models the process of booking a flight through a university travel agent.

Three participants are involved: Researcher (abbreviated R), who intends to travel; Agent (A), who is able to make travel reservations; and Finance (F), who approves expenditure from the budget. In the Scribble [25] language, we first define the global protocol among three *roles*, which are abstract representations of the participants. The protocol consists of sequences of interactions. Every message (e.g. request) can be associated with a payload type (e.g. Travel), a sender, and one or more receivers. Typically payload types are structured data types defined separately from the protocol specification.

In the global protocol, after the check message requesting authorisation for a trip, F can choose to approve or refuse the request.

```
1  global protocol BuyTicket(role R, role A, role F) {
2    request(Travel) from R to A;
3    quote(Price) from A to R;
```

```
4    check(Price) from R to F;
5    choice at F {
6      approve(Code) from F to R,A;
7      ticket(String) from A to R;
8      invoice(Code) from A to F;
9      payment(Price) from F to A;
10   } or {
11     refuse(String) from F to R,A;
12   } }
```

The Scribble tools can be used to validate the protocol definition and to derive a *local* version of the protocol for each role, according to the theory of multiparty session types [11]. This is known as *endpoint projection*. Here is the projection for R, which describes only the messages involving that role. The self keyword indicates that R is the local endpoint.

```
1   local protocol BuyTicket_R(self R, role A, role F) {
2     request(Travel) to A;
3     quote(Price) from A;
4     check(Price) to F;
5     choice at F {
6       approve(Code) from F;
7       ticket(String) from A;
8     } or {
9       refuse(String) from F;
10    } }
```

Notice that the exchange of invoice and payment between A and F is not included. Similarly, the local projection for A omits the check message; we omit its local projection. Finally, the local projection for F omits the request, quote and ticket messages.

```
1   local protocol BuyTicket_F(role R, role A, self F) {
2     check(Price) from R;
3     choice at F {
4       approve(Code) to R,A;
5       invoice(Code) from A;
6       payment(Price) to A;
7     } or {
8       refuse(String) from F to R,A;
9     } }
```

The common theme between protocols and typestate specifications is the requirement to do operations in particular orders. Our methodology for implementing the roles in a Scribble protocol is to define a Java class that encapsulates socket connections to provide the necessary communication, and provides methods that send and receive the messages in the protocol. This class constitutes an API for role programming. To ensure that communication methods are called in the order required by the protocol, we associate a typestate specification with the API, so that Mungo can check the correctness of code that uses the API. StMungo generates a Java API and a Mungo specification. If we are implementing all of the endpoints in a system, then the generated APIs are immediately interoperable with each other. However, interoperability with pre-existing endpoints such as a POP3 server (Section 14.4) typically requires an extra layer in order to translate between the abstract message labels defined in Scribble and the detailed textual message formats required by the protocol.

For the R role, StMungo converts the BuyTicket_R local projection into the following Mungo definitions:

1. RProtocol, a typestate specification capturing the interactions local to the R role.
2. RRole, a Java class that implements RProtocol by communication over Java sockets. This is an API that can be used to implement the R endpoint.
3. RMain, a prototype Java implementation of the R endpoint. This runs as a Java process, and provides a main() method which uses RRole to communicate with the other parties in the session. For testing purposes it provides a command-line interface to choose and display message parameters.

To complete the ticket buying example, we now describe the result of translating the local protocol for R. For each choice there is an enumerated type, named according to the numerical position of the choice in the sequence of choices within the local protocol. The values of the enumerated type are the names of the first message in each branch of the choice. For the choice in BuyTicket_R we have the following definition.

```
1  enum Choice1 { APPROVE, REFUSE; }
```

Every role involved in the choice will have an enumerated type with the same set of values, but the names of the types are not necessarily the same for every role.

The typestate specification RProtocol defines the allowed sequences of method calls. As it includes method headers, it also provides similar documentation to an interface. The initial state is the first one defined.

```
1  typestate RProtocol {
2    State0 = { void send_requestTravelToA(Travel): State1 }
3    State1 = { Price receive_quotePriceFromA(): State2 }
4    State2 = { void send_checkPriceToF(Price): State3 }
5    State3 = { Choice1 receive_Choice1LabelFromF():
6                      <APPROVE: State4, REFUSE: State6>  }
7    State4 = { Code receive_approveCodeFromF(): State5 }
8    State5 = { String receive_ticketStringFromA(): end }
9    State6 = { String receive_refuseTravelFromF(): end } }
```

The API is defined by the class RRole, which is also generated. When instantiated, it establishes socket connections to the other role objects in the session (ARole and FRole); we omit the details here.

```
1  @Typestate("RProtocol") public class RRole {
2    public RRole(){
3      ... // Bind the sockets and accept a client connection
4      try { // Create the read and write streams
5        socketAIn = new BufferedReader(..);
6        socketAOut = new PrintWriter(..);
7      } catch (IOException e) {
8        System.out.println("Read failed");  System.exit(-1);
9      } }
10   public void send_requestTravelToA(Travel payload) {
11     this.socketAOut.println(payload);  }
12   public Price receive_quotePriceFromA() {
13     String line = "";
14     try {  line = this.socketAIn.readLine();
15     } catch (IOException e) {
16       System.out.println("Input/Output error.");  System.exit(-1);
17     }
18     // Parse line to the appropriate type and then return it
19     return Price.parsePrice(line);  }
20     ... // Define all other methods in RProtocol }
```

The RMain class provides a prototype implementation of the R endpoint, using the RRole class to communicate with the other roles in the system.

Mungo statically checks the correctness of an R implementation (either based on the prototype or written separately), by checking that methods are called in allowed sequences and that all possible responses are handled. For example, main below is correct.

```
1   public static void main(String[] args) {
2     RRole r = new RRole();
3     Travel t = // input travel;
4     r.send_requestTravelToA(t);
5     Price p = r.receive_quotePriceFromA();
6     r.send_checkPriceToF(p);
7     switch(r.receive_Choice1LabelFromF().getEnum()) {
8       case APPROVE:
9         Code c = r.receive_approveCodeFromF();
10        println(r.receive_ticketStringFromA());
11        break;
12      case REFUSE:
13        println(r.receive_refuseStringFromF());
14        break;
15   } }
```

This code is checked by computing the sequences of method calls that are made on an RRole object, inferring the minimal typestate specification that allows those sequences, and then comparing this specification with the declared specification RProtocol. The comparison is based on a simulation relation. Typically the programmer would modify the prototype implementation by defining extra business logic, but she is also free to rewrite it completely. Mungo statically checks RMain, or any client of the RRole class, to ensure that methods of the protocol are called in a valid sequence and that all possible responses are handled.

14.4 POP3: Typechecking an Internet Protocol Client

As a more substantial example, we use a standard internet protocol, POP3 [18] (Post Office Protocol Version 3), to show the applicability of session types in the real world and the use of session type tools to typecheck protocols. The protocol allows an email client to retrieve messages from a server. The diagram below is based on RFC 1939 [18], the official specification of the protocol. The labels "+OK" and "-ERR" are part of the textual message format. For simplicity, several transitions from state TRANSACTION have been omitted.

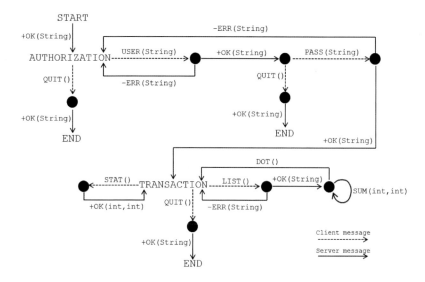

The protocol starts with the client connecting to the server and the server authenticating the connection. The client then has the choice to either submit a username to log into a mailbox, or to end the authorization. Upon receiving the username, the server has the choice to accept the username or to send an error message to the client, for example if the username does not exist. After the username has been accepted, the client is then required to send a password or to end the authorization. If the password is accepted, the transaction stage begins. In the transaction stage, the client has a choice of various commands: the diagram shows just STAT (status) and LIST (summary list). Some of these requests involve a choice at the server side to either fulfil the request or to send an error message.

Alternatively the client can choose to QUIT. The specification of the messages and state transitions of POP3 can be converted into a Scribble global protocol, as shown below.

```
1  global protocol POP3(role S, role C) {
2    OKN(String) from S to C;
3    rec authentication_username {
4      choice at C {
5        USER(String) from C to S;
6        choice at S {
7          OK(String) from S to C;
8          rec authentication_password {
9            choice at C   {
```

```
10              PASS(String) from C to S;
11              choice at S {
12                OK(String) from S to C;
13                rec transaction {
14                  choice at C {
15                    STAT() from C to S;
16                    OKN(int, int) from S to C;
17                    continue transaction;
18                  } or {
19                    LIST() from C to S;
20                  choice at S {
21                    OK(String) from S to C;
22                    rec summary_choice_retrieve {
23                      choice at S {
24                        DOT() from S to C;
25                        continue transaction;
26                      } or {
27                        SUM(int, int) from S to C;
28                        continue summary_choice_retrieve; } }
29                  } or {
30                    ERR(String) from S to C;
31                    continue transaction; }
32                } or {
33                  QUIT() from C to S;
34                  OKN(String)from S to C; } }
35                } or {
36                  ERR(String) from S to C;
37                  continue authentication_password; }
38              } or {
39                QUIT() from C to S;
40                OKN(String) from S to C; } }
41          } or {
42            ERR(String) from S to C;
43            continue authentication_username; }
44        } or {
45          QUIT() from C to S;
46          OKN(String) from S to C; } } }
```

Projection using the Scribble tools produces local protocols for the client and the server. For the rest of this section we focus on the client protocol. For brevity we omit the authentication phase.

```
1  local protocol POP3 (role S,self C) {
2    OKN(String) from S;
3    ...
4                  rec transaction {
5                    choice at C {
6                      STAT() to S;
7                      OKN(int,int) from S;
8                      continue transaction;
9                    } or {
10                     LIST() to S;
11                     choice at S {
12                       OK(String) from S;
13                       rec summary_choice_retrieve {
14                         choice at S {
15                           DOT() from S;
16                           continue transaction;
17                         } or {
18                           SUM(int,int) from S;
19                           continue summary_choice_retrieve; } }
20                     } or {
21                       ERR(String) from S;
22                       continue transaction; }
23                   } or {
24                     QUIT() to S;
25                     OKN(String) from S; } }
26               ...
27         QUIT() to S;
28         OKN(String) from S; } } }
```

We use StMungo to translate the Scribble local protocol into a typestate specification CProtocol, which defines the order in which the communication methods are called.

```
1  typestate CProtocol {
2  State0 = {String receive_OKStringFromS(): State1}
3  ...
4  State9 = {void send_STATToS(): State10,
```

```
5                   void send_RETR_NToS(): State12,
6                   void send_QUITToS(): State19}
7    State10 = {void send_STATToS(): State11}
8    State11 = {IntInt receive_OKNIntIntFromS(): State9}
9    State12 = {void send_LISTToS(): State13}
10   State13 = {Choice1 receive_Choice1LabelFromS():
11      <OK: State14, ERR: State18>}
12   State14 = {String receive_OKStringFromS(): State15}
13   State15 = {Choice2 receive_Choice2LabelFromS():
14      <DOT: State16, SUM: State17>}
15   State16 = {void receive_DOTFromS(): State9}
16   State17 = {String receive_SUMIntIntFromS(): State15}...}
```

14.4.1 Challenges of Using Mungo and StMungo in the Real World

Programming with loops A POP3 server responds to the LIST command by sending any number of lines, terminated by the DOT message.

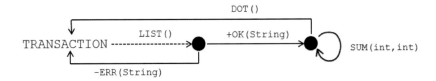

The Scribble description of the state reached by +OK() uses explicit recursion in which continue jumps to a named state.

```
1    rec summary_choice_list {
2      choice at S    {
3        DOT() from S to C
4        continue transaction;
5      } or {
6        SUM(int, int) from S to C;
7        continue summary_choice_list; } }
```

The Java code generated by StMungo is a direct translation, using labelled statements and continue. Given that we are generating imperative code rather than recursive functions, this seems to be the only systematic way to handle the arbitrary structure of Scribble's rec. Although continue is a goto, its use is controlled and checked by Mungo: it is only allowed when the recursion point in the protocol has been reached. The SJ language [15] introduces `sendWhile` and `receiveWhile` loops to match particular protocol patterns, but we have chosen not to extend Java with new loop constructs.

```
1  _summary_choice_list: do {
2    switch(currentC.receive_Choice2LabelFromS().getEnum()){
3      case Choice2.DOT:
4        Void payload10 = currentC.receive_DOTVoidFromS();
5        System.out.println("Received from S: ." + payload10);
6        continue _transaction;
7      case Choice2.SUM:
8        SUMIntInt payload11 = currentC.receive_SUMIntIntFromS();
9        System.out.println("Received from S: " + payload11);
10       continue _summary_choice_list; } }
11 while(true);
```

Abstract vs. concrete messages When designing a complete system and implementing all the roles, StMungo can generate concrete textual messages in a uniform way; alternatively, we could use a structured message format such as JSON. However, in POP3 and other standard protocols, the client has to work with the textual message formats defined by the protocol. For example, the Scribble message OK(int, int) from S to C; corresponds to a line of text such as +OK 2 200. In the current implementation of the POP3 example, conversion between abstract and concrete messages is done by hand-written code, but we are working on a tool to generate message converters from a specification.

Naming StMungo converts Scribble message names into Java method names. The method definition depends on whether or not the message appears at the beginning of a Scribble choice, and this cause naming conflicts if the same name is used for messages in both kinds of position. For example, OK and OKN in POP3 would more naturally both be OK.

Non-standard implementations Real-world servers do not always follow the RFC exactly. The specification of POP3 states that if the client sends an unknown username, it is rejected and the username must be sent again. However, the server used for this case study, namely GMX.co.uk, accepts an unknown username and expects the client to send the password again. Consequently, even after completing the prototype client generated by StMungo and checking it with Mungo, it is necessary to test the client thoroughly with existing servers if we want to ensure correct operation in all cases. When deviations from the RFC are discovered, the Scribble definition of the protocol can be generalised accordingly. This problem could be reduced by promoting the use of formal protocol descriptions within RFCs.

14.5 Related Work

Session types. The main pieces of related work on session types and Java are the Session Java (SJ) language [15] and the API generation approach [14], both by Hu *et al*. The API generation approach has been used to to analyse an SMTP client in Java. The API for SMTP implements multiparty session types using a pattern in which each communication method returns the receiver object with a new type that determines which communication methods are available at the next step. Standard Java typechecking can verify the correctness of communication when the pattern is used properly, with runtime monitoring being used to ensure linearity constraints are fulfilled. In contrast with this approach, Mungo's approach is completely static.

SJ [15] builds on earlier work [4, 5, 7] to add binary session type channels to Java. SJ implements a library for binary sessions that have a pre-defined interface. The syntax of Java is extended with communication statements to allow typechecking. The scope of a session is restricted to the body of a single method. Mungo removes this restriction by allowing the abstraction of multiparty session types as user-defined objects that can be passed and used throughout different program scopes.

Typestates. There have been many projects that add typestates to practical languages, since the introduction of the concept by Strom and Yemini [22]. Plural [2] is a noteworthy example. It is based on Java and has been used to study access control systems. Plural implements typestates by using annotations to define pre- and post-conditions on methods, referring to abstract states and predicates on instance variables. By contrast, Mungo explicitly defines the possible sequences of method calls. Plural and Mungo both allow the typestate after a method call to depend on the return value.

Plaid [1, 23] introduces typestate-oriented programming as a paradigm. Instead of class definitions, a program consists of state definitions containing methods that cause transitions to other states. Transitions are specified in a similar way to Plural's pre- and post-conditions. Similarly to classes, states can be structured into an inheritance hierarchy. As opposed to Plaid, Mungo focuses on the object-oriented paradigm in order to be applicable to Java.

Our previous paper [16] discusses related work in more detail.

Acknowledgements This research was supported by the UK EPSRC project "From Data Types to Session Types: A Basis for Concurrency and Distribution" (EP/K034413/1) and by COST Action IC1201 "Behavioural Types for Reliable Large-Scale Software Systems". We thank the reviewers for their detailed comments.

References

[1] J. Aldrich, J. Sunshine, D. Saini, and Z. Sparks. Typestate-oriented programming. In *OOPSLA '09*, pages 1015–1022. ACM Press, 2009.

[2] K. Bierhoff, N. E. Beckman, and J. Aldrich. Practical API protocol checking with access permissions. In *ECOOP '09*, volume 5653 of *Springer LNCS*, pages 195–219, 2009.

[3] S. Capecchi, M. Coppo, M. Dezani-Ciancaglini, S. Drossopoulou, and E. Giachino. Amalgamating sessions and methods in object-oriented languages with generics. *Theoret. Comp. Sci.*, 410:142–167, 2009.

[4] M. Dezani-Ciancaglini, S. Drossopoulou, D. Mostrous, and N. Yoshida. Objects and session types. *Information and Computation*, 207(5): 595–641, 2009.

[5] M. Dezani-Ciancaglini, E. Giachino, S. Drossopoulou, and N. Yoshida. Bounded session types for object oriented languages. In *FMCO '06*, volume 4709 of *Springer LNCS*, pages 207–245, 2006.

[6] M. Dezani-Ciancaglini, D. Mostrous, N. Yoshida, and S. Drossopoulou. Session types for object-oriented languages. In *ECOOP '06*, volume 4067 of *Springer LNCS*, pages 328–352, 2006.

[7] M. Dezani-Ciancaglini, N. Yoshida, A. Ahern, and S. Drossopoulou. A distributed object-oriented language with session types. In *TGC '05*, volume 3705 of *Springer LNCS*, pages 299–318, 2005.

[8] S. J. Gay and V. T. Vasconcelos. Linear type theory for asynchronous session types. *Journal of Functional Programming*, 20(1):19–50, 2010.

[9] S. J. Gay, V. T. Vasconcelos, A. Ravara, N. Gesbert, and A. Z. Caldeira. Modular session types for distributed object-oriented programming. In *POPL '10*, pages 299–312. ACM Press, 2010.

[10] G. Hedin. An introductory tutorial on JastAdd attribute grammars. In *Generative and Transformational Techniques in Software Engineering III*, volume 6491 of *Springer LNCS*, pages 166–200, 2011.

[11] K. Honda, N. Yoshida, and M. Carbone. Multiparty asynchronous session types. In *POPL '08*, pages 273–284. ACM Press, 2008.

[12] K. Honda. Types for dyadic interaction. In *CONCUR '93*, volume 715 of *Springer LNCS*, pages 509–523, 1993.

[13] K. Honda, V. Vasconcelos, and M. Kubo. Language primitives and type discipline for structured communication-based programming. In *ESOP '98*, volume 1381 of *Springer LNCS*, pages 122–138, 1998.

[14] R. Hu and N. Yoshida. Hybrid session verification through endpoint API generation. In *FASE 16*, volume 9633 of *Springer LNCS*, pages 401–418, 2016.

[15] R. Hu, N. Yoshida, and K. Honda. Session-based distributed programming in Java. In *ECOOP '08*, volume 5142 of *Springer LNCS*, pages 516–541, 2008.

[16] D. Kouzapas, O. Dardha, R. Perera, and S. J. Gay. Typechecking protocols with Mungo and StMungo. In *PPDP '16*, pages 146 159. ACM Press, 2016.

[17] M. Neubauer and P. Thiemann. An implementation of session types. In *PADL '04*, volume 3057 of *Springer LNCS*, pages 56–70, 2004.

[18] Post office protocol version 3, RFC 1939. https://www.ietf.org/rfc/rfc1939.

[19] R. Pucella and J. A. Tov. Haskell session types with (almost) no class. In *Proceedings of the 1st ACM SIGPLAN Symposium on Haskell*, pages 25–36. ACM Press, 2008.

[20] Scribble project homepage. www.scribble.org.

[21] Simple mail transfer protocol, RFC 821. https://tools.ietf.org/html/rfc821.

[22] R. E. Strom and S. Yemini. Typestate: A programming language concept for enhancing software reliability. *IEEE Trans. Softw. Eng.*, 12(1): 157–171, 1986.

[23] J. Sunshine, K. Naden, S. Stork, J. Aldrich, and É. Tanter. First-class state change in Plaid. In *OOPSLA '11*, pages 713–732. ACM Press, 2011.

[24] K. Takeuchi, K. Honda, and M. Kubo. An interaction-based language and its typing system. In *PARLE '94*, volume 817 of *Springer LNCS*, pages 398–413, 1994.

[25] N. Yoshida, R. Hu, R. Neykova, and N. Ng. The Scribble protocol language. In *TGC '13*, volume 8358 of *Springer LNCS*, pages 22–41, 2013.

15

Protocol-Driven MPI Program Generation

Nicholas Ng and Nobuko Yoshida

Imperial College London, UK

Abstract

This chapter presents Parameterised Scribble (Pabble), an extension of the Scribble language to capture scalable protocols, and a top-down, code generation framework of Message-Passing Interface (MPI) programs.

The code generation process begins with defining a Pabble protocol for the topology of the MPI application. An MPI parallel program skeleton is automatically generated from the protocol, which can then be merged with code kernels defining their behaviours. The merging process is fully automated through the use of an aspect-oriented compilation tool.

Pabble protocols are parameterised over the number of roles at runtime, and are grounded on theories of parameterised multiparty session types (MPST) where valid Pabble protocols can ensure safety and progress of communication in the generated MPI programs. Using the framework, programmers only need to supply the intended Pabble protocol and provide code kernels to obtain parallelised programs. Since the skeleton generation and the merging process are automatic, the framework not only simplifies the development of MPI programs, the output programs are efficient and scalable MPI applications, that are guaranteed, free from communication mismatch, type errors or deadlocks by construction, improving productivity of programmers.

15.1 Introduction

The Message Passing Interface (MPI) [8] is the de-facto standard for parallel programming on high-performance computing systems. Despite the advances in novel techniques and models such as the Partitioned Global Address Space

(PGAS) used by X10 [3, 10, 17] for simplifying parallel programming, MPI is still by far the most widely used parallel programming library in the scientific community. However, parallel programming with the MPI library is a well-documented difficult task, in which reasoning about interactions between distributed processes is difficult at scale, and communication mismatches are amongst the most common pitfalls by MPI users [6].

To apply behavioural types in safe, scalable parallel programming, this chapter presents a parallel programming workflow based on a protocol language Pabble, which we will explain in more details in Section 15.5. Figure 15.1 shows the overview of our approach, and this chapter explains the core use case of the approach highlighted in the figure. A Pabble protocol is an abstract representation of the communication topology, or *parallel communication patterns* of a parallel application. We consider every application a coupling between sequential, computation code that defines functional behaviours of processes in the application, and a communication topology that connects the processes together as a coherent application. Hence, to build a parallel application, we first define the communication protocol, written in Pabble. A valid Pabble protocol is guaranteed free of interactions and patterns that introduce communication errors and deadlocks. The Pabble protocol is used to generate an annotated MPI program backbone (Section 15.6), specifying the interactions between parallel processes. Based on the Pabble protocol, computation kernels are written in C language (C99), using queues to pass data locally between the kernels. The kernels are then merged with the MPI backbone by LARA [2], an aspect-oriented compilation tool, to transform the backbone and the kernels into a complete MPI application (Section 15.7).

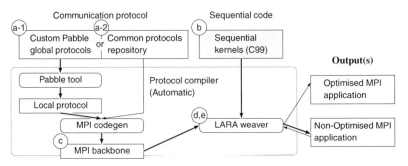

Figure 15.1   Pabble-based MPI program generation workflow (core flow highlighted).

In addition to the merge, LARA can also perform pragma directed optimisations on the source code to overlap communication and computation, improving the runtime performance. The details of the optimisations, rigorous evaluations of the approach, and a pre-generated repository of common protocols are omitted from this chapter, but can be found in the original paper [12].

15.2 Pabble: Parameterised Scribble

In this section we introduce *Parameterised Scribble* (Pabble) [14, 15], a developer friendly notation for specifying application level interaction protocol based on the theory of *parameterised* multiparty session types [5]. As the name suggests, Pabble is a parametric evolution of Scribble [16, 18], which itself is based on the theory of multiparty session types [1, 9]. We begin with an example Scribble protocol to explain the basic syntax of Pabble and the Scribble family of protocol languages, and why parameterisation is important for protocols describing scalable, parallel program topologies.

Scribble

```
1  module example;
2  global protocol Ring(role Worker1, role Worker2, role Worker3) {
3    rec LOOP {
4      Data(T) from Worker1 to Worker2;
5      Data(T) from Worker2 to Worker3;
6      DataLast(T) from Worker3 to Worker1;
7      continue LOOP; }
8  }
```

Listing 15.1 Ring protocol in Scribble.

This `Ring` protocol describes a series of communications in which the role `Worker1` passes a message of type `Data(T)` to `Worker3` by forwarding through `Worker2`, and receives back a `DataLast(T)` message from `Worker3` to complete the ring. It is easy to notice that explicitly describing all interactions among distinct roles is verbose and inflexible: for example, when extending the protocol with an additional role `Worker4`, we must rewrite the whole protocol. On the other hand, we observe that these worker roles have identical communication patterns that can be logically grouped together: $Worker_{i+1}$ receives a message from $Worker_i$ and the last `Worker` sends a message to $Worker_1$. In order to capture these replicable patterns, we introduce an extension of Scribble with dependent types, namely Pabble. In Pabble, multiple participants can be grouped in the same role and indexed.

This greatly enhances the expressive power and modularity of the protocols. Here 'parameterised' refers to the number of participants in a role that can be changed by parameters.

```
1  module example;
2  const N = 3;
3  global protocol Ring(role Worker[1..N]) {
4    rec LOOP {
5      Data(T) from Worker[i:1..N-1] to Worker[i+1];
6      DataLast(T) from Worker[N] to Worker[1];
7      continue LOOP; }
8  }
```

Listing 15.2   Parametrised Ring protocol in Pabble.

Our ring example is rewritten in the syntax of Pabble shown above. The role Worker[1..N] declares workers with indices 1 up to an arbitrary integer N. The Worker roles can be identified individually by their indices, for example, Worker[1] refers to the first and Worker[N] refers to the last. In the body of the protocol, the sender, Worker[i:1..N-1], declares multiple Workers, bound by the bound variable i, and iterates from 1 to N-1. The receivers, Worker[i+1], are calculated on their indices for each instance of the bound variable i. The second line is a message sent back from Worker[N] to Worker[1].

```
3  local protocol Ring at Worker[1..N](role Worker[1..N]){
4    rec LOOP {
5      if Worker[i:2..N]    Data(T) from Worker[i-1];
6      if Worker[i:1..N-1] Data(T) to Worker[i+1];
7      if Worker[1]        DataLast(T) from Worker[N];
8      if Worker[N]        DataLast(T) to Worker[1];
9      continue LOOP; }
10 }
```

The above code shows the *local protocol* of Ring, which is a localised version of Listing 15.2 at the Worker role. It represents the Worker[1..N] parameterised role, and corresponds to multiple endpoints in the same logical grouping. A Pabble local protocol is automatically generated from its global protocol following the projection algorithm in [14], and programmers only need to define the global protocol to use Pabble for MPI development.

Above servers as a primer on the Pabble language, sufficient for our introductory example; Later, the full syntax and explanations of the Pabble language will be given in Section 15.5.

15.3 MPI Backbone

A typical MPI program follows a Single Program, Multiple Data (SPMD) parallel programming model, where a single source code is executed by multiple parallel processors. This model shares a lot of similarities with the parameterised local protocols in Pabble which groups together similar roles in a single protocol, except that local protocols can be generated from global protocols which are easier to express overall communication or topologies. As a running example, we use the Pabble protocol presented earlier to demonstrate the framework and implement a ring accumulator that calculates a sum of values from each Worker and distribute to all.

C/MPI Backbone

```
 1  int main(int argc, char *argv[])
 2  { MPI_Init(&argc, &argv);
 3    MPI_Comm_rank(MPI_COMM_WORLD, &meta.pid);
 4    MPI_Comm_size(MPI_COMM_WORLD, &meta.nprocs);
 5  #pragma pabble type T
 6    typedef void T; ⇒ typedef double T;
 7    MPI_Datatype MPI_T; ⇒ MPI_Datatype MPI_T = MPI_DOUBLE;
 8
 9    T *bufData_r, *bufData_s;
10    /** Other buffer declarations **/
11    /** Definitions of cond0, cond1, ... **/
12  #pragma pabble kernel Init ⇒ init(Init, "input.txt")
13  #pragma pabble predicate Ring
14    while (1) { ⇒ while(iter())
15      if (cond0) { /*if Worker[i:2..N]*/
16        bufData_r = (T *)calloc(meta.buflen(Data), sizeof(T));
17        MPI_Irecv(bufData_r, meta.buflen(Data), MPI_T, /*Worker[i-1]*/...);
18        MPI_Wait(&req[0], &stat[0]);
19        pabble_recvq_enqueue(Data, bufData_r);
20  #pragma pabble kernel Data ⇒ accumulate(Data);
21      }
22      if (cond1) { /*if Worker[i:1..(N-1)]*/
23  #pragma pabble kernel Data ⇒ accumulate(Data);
24        bufData = pabble_sendq_dequeue();
25        MPI_Isend(bufData, meta.buflen(Data), MPI_T, /*Worker[i+1]*/...);
26        MPI_Wait(&req[1], &stat[1]);
27        free(bufData);
28      }
29      // Similarly for DataLast between Worker[1] and Worker[N]
30      MPI_Finalize();
31    }
32    return EXIT_SUCCESS; }
```

Listing 15.3 MPI backbone generated from the Ring protocol.

15.3.1 MPI Backbone Generation from `Ring` Protocol

Based on the Pabble `Ring` protocol in the introduction, our code generation framework generates an *MPI backbone* code (e.g. Listing 15.3). First it automatically generates *local protocols* from a global protocol as an intermediate step to make MPI code generation more straightforward. The MPI backbone generation procedure is described in details later in Section 15.6, here we focus on the generated MPI backbone code output.

An MPI backbone is a C99 program with boilerplate code for initialising and finalising the MPI environment of a typical MPI application (lines 2–4 and 30 respectively), and MPI primitives for message passing (e.g. `MPI_Isend`/`MPI_Irecv`[1]). Therefore the MPI backbone realises the interaction between participants as specified in the Pabble protocol, without supporting any specific application functionality. The backbone has three kinds of #pragma annotations as placeholders for kernel functions, types and program logic. The annotations are explained in Section 15.7.1. The boxed code in Listing 15.3 represents how the backbone are converted to code that calls the kernel functions in the MPI program.

On lines 5 and 6, *generic type* T and `MPI_T` are defined datatypes for C and MPI respectively. T and `MPI_T` are refined later when an exact type (e.g. `int` or composite `struct` type) is known with the kernels.

Following the type declarations, other variable declarations including the buffers (line 9), and their allocation and deallocation are managed by the backbone. They are generated as guarded blocks of code, which come directly from the local protocol. lines 15–21 shows a guarded receive that correspond to if `Worker[i:2..N] Data(T)from Worker[i-1]` in the protocol and lines 22–28 for if `Worker[i:1..N-1] Data(T)to Worker[i+1]`.

Given the MPI backbone, we can then implement computation kernels for the MPI program.

15.4 Computation Kernels

Computation kernels are C functions that describe the algorithmic behaviour of the application. Conceptually, each message interaction defined in Pabble (e.g. `Label(T)from Sender to Receiver`), and – through the automatic MPI backbone generation – the MPI backbone, can be associated to a kernel by its label (e.g. `Label`).

---

[1] We use `MPI_Isend`/`MPI_Irecv` with `MPI_Wait` in place of the equivalent `MPI_Send`/`MPI_Recv` respectively. To simplify presentation we write `MPI_Send`/`MPI_Recv` in the rest of the chapter.

Figure 15.2   Global view of Label(T) from Sender to Receiver;.

Figure 15.2 shows how kernels are invoked in a message-passing state-ment between two processes named Sender and Receiver respectively. Since a message interaction statement involves two participants (e.g. Sender and Receiver), the kernel serves two purposes: (1) produce a message for sending and (2) consume a message after it has been received. The two parts of the kernel are defined in the same function, but runs on the sending process and the receiving process respectively. The kernels are top-level functions and do not send or receive messages directly through MPI calls. Instead, messages are passed between kernels and the MPI backbone (derived from the Pabble protocol) via a queue API: in order to send a message, the producer kernel (e.g. (1)) of the sending process enqueues the message to its send queue; and a received message can be accessed by a consumer kernel (e.g. (2)), dequeuing from its receive queue. This allows the decoupling between computation (as defined by the kernels) and communication (as described in the MPI backbone).

15.4.1 Writing a Kernel

We now explain how a user writes a kernel file, which contains the set of kernel functions related to a Pabble protocol for an application. As an example, we implement accumulator in a ring topology below.

A minimal kernel file must define a variable meta of meta_t type, which contains the process id (i.e. meta.pid), total number of spawned pro-cesses (i.e. meta.nprocs) and a callback function that takes one param-eter (message label) and returns the send/receive size of message payload (i.e. unsigned int meta.bufsize(int label)). The meta.buflen function returns the buffer size for the MPI primitives based on the label given, as a lookup table to manage the buffer sizes centrally. Process id and total number of spawned processes will be populated automatically by the backbone code generated. The kernel file includes the definitions of the kernel functions, annotated with pragmas, associating the kernels with message labels. The pragmas that are allowed are detailed in Section 15.7. The kernels can use file (i.e. static) scope variables for local data storage. Our ring accumulator kernel file starts with the following declarations for local data and meta:

Kernel file header

```
 1  typedef struct {
 2    double* values; int N;
 3  } local_data_t;
 4  static local_data_t *local;
 5
 6  unsigned int buflen(int label) { return 1; } // 1-size buffer for all.
 7  meta_t meta = {/*pid*/0, /*nprocs*/1, MPI_COMM_NULL, &buflen};
```

15.4.1.1 Initialisation

Most parallel applications require explicit partitioning of input data. In these cases, the programmer writes a kernel function for partitioning, such that each participant has a subset of the input data. Input data are usually partitioned with a layout similar to the layout of the participants. In our ring accumulator example, the processes are arranged linearly, and the input file contains an array of at least meta.nprocs elements, so meta.nprocs initial values are read into the local->values array. In our example initialisation function below, we also set the current accumulated value to be our initial value of local->values[meta.pid].

Kernel: Init

```
 9  #pragma pabble kernel Init
10  void init(int id, const char *filename)
11  { FILE *fp = fopen(filename, "r");
12    local = (local_data_t *)malloc(sizeof(local_data_t));
13    local->values = NULL; local->N = 0;
14    ... // allocate etc.
15    int nprocs = meta.nprocs; // Number of processes (known at runtime).
16    for (int i=0; i<nprocs; i++)
17      fscanf(fp, "%f", &local->values[i]); // Copy data to local
18    fclose(fp); local->N = nprocs;
19    local->accumulated = local->values[meta.pid]; /* initial value on proc */
20  }
```

15.4.1.2 Passing data between backbone and kernel through queues

The kernels are void functions with at least one parameter, which is the label of the kernel. Inside the kernel, no MPI primitive should be used to perform message passing. Data received from another participant or data that need to be sent to another participant can be accessed using a receive queue

Kernel: Data   User C Kernel

```
20  #pragma pabble kernel Data
21  void accumulate(int id)
22  { double *rcvd_val; // Ptr to received value (temp).
23    if (!pabble_recvq_isempty() && pabble_recvq_top_id() == id) {
24      rcvd_val = (double *)pabble_recvq_dequeue(); // allocated by backbone
25      local->accumulated += *rcvd_val;
26    } else { // Allocate and send value
27      accumulated_val = (double *)calloc(meta.buflen(id), sizeof(double));
28      *accumulated_val = local->accumulated
29      pabble_send_enqueue(id, accumulated_val);
30    }
31  }
```

and send queue. Consider the following kernel for the label Data in the ring accumulator example:

Each kernel has access to a send and receive queue local to the whole process, which holds pointers to the buffer to be sent and the buffer containing the received messages, respectively. The queues are the only mechanism for kernels to interface the MPI backbone. The simplest kernel is one that forwards incoming messages from the receive queue directly to the send queue. In the above function, when the kernel function is called, it either consumes a message from the receive queue if it is not empty (i.e. after a receive), or produce a message for the send queue (i.e. before a send).

Kernels can have extra parameters. For example, in the init function above, filename is a parameter that is not specified by the protocol (i.e. Init ()). When such functions are called, all extra parameters are supplied by command-line arguments in the final generated MPI application.

15.4.1.3 Predicates

A predicate kernel is similar to a normal void kernel, but with a function signature that returns an int (as a boolean), it is used as a conditional variable, where the value of the variable is determined by the body of the kernel. In the iter() predicate kernel, we use the number of processes to determine when the ring protocol has completed a cycle (i.e. executed meta.nproc times) and terminate the while-loop.

Kernel: Ring   User C Kernel

```
32  #pragma pabble predicate Ring
33  int iter() { static int i = 0; return i++ < meta.nprocs }
```

Aftere writing the computation kernels, we can then use the framework to merge the MPI backbones with the computation kernels, and we get a complete MPI program. The resulting MPI program is shown in Listing 15.3 (boxed code).

15.5 The Pabble Language

In this section, we present more details of the Pabble language, including its syntax, and the well-formedness conditions (i.e. syntactic restrictions to ensure protocol correctness) of the language.

15.5.1 Global Protocols Syntax

Figure 15.3 lists the core syntax of Pabble, which consists of two protocol declarations, *global* and *local*. A global protocol is declared with the protocol name (*str* denotes a string) with role and group parameters followed by the body G. Role R is a name with argument expressions. The argument expressions are ranges or arithmetic expressions h, and the number of arguments corresponds to the dimension of the array of roles: for example, Worker [1..4][1..2] denotes a 2-D array with size 4 and 2 in the two dimensions respectively, forming a 4-by-2 array of roles.

Declared roles can be grouped by specifying a named group using the keyword group, followed by the group name and the set of roles. For example,

```
group EvenWorker={Worker[2][2], Worker[4][2]}
```

creates a group which consists of two Workers. A special built-in group, All, is defined as *all processes in a session*. We can encode collective operators such as many-to-many and many-to-one communication with All, which will be explained later.

Apart from specifying ranges by constants, ranges can also be specified using expressions. Expression e consists of operators for numbers, logarithm, left and right logical shifts (<<, >>), numbers, variables (i, j, k), and constants (M, N). Constants are either *bound* outside the protocol declaration or are left *free* (unbound) to represent an arbitrary number. As in [11], when the constants are bound, they are declared by numbers outside the protocol, e.g. const N = 10 or lower and upper bounds, e.g. const N = 1..10. We also allow leaving the declaration *free* (unbound), e.g. const N, as a shorthand to represent an arbitrary constant with lower and upper bounds 0 and max respectively, i.e. const N = 0..max, where max is a special

Global Pabble

global protocol $str(para)$ { G }

Parameter

$para$	::=	role R_d, ...,	Role declaration
		group str = {R_d, ...}, ...	Group declaration

Global protocol body

G	::=	$l(T)$ from R to R;	Interaction
	|	choice at R { G_1 } or ... or { G_N }	Choice
	|	foreach (b) { G }	Foreach
	|	allreduce $op_c(T)$;	Reduction
	|	rec l { G }	Recursion
	|	continue l;	Continue
	|	G G	Sequential composition

Payload type

T	::=	int | float | ...	Data types

Expression

e	::=	e op e	Binary expressions
	|	num	Integers
	|	$i, j, k, ...$ | N	Variables, constants
op	::=	op_c | - | / | % | << | >> | log | ...	Binary operations
op_c	::=	+ | * | ...	Commutative operations

Role

R_d	::=	str	Role declaration
	|	$str[e..e]...[e..e]$	Param. role declaration
R	::=	str	Roles
	|	$str[h]...[h]$	Param. roles
	|	All	*All* group role
h	::=	b | e	Role parameter
b	::=	$i : e..e$	Binding range

Local Pabble

local protocol str at $R_d(para)$ { L }

Local protocol body

L	::=	[if R] $l(T)$ from R;	(Conditional) Receive
	|	[if R] $l(T)$ to R;	(Conditional) Send
	|	choice at R { L_1 } or ... or { L_N }	Choice
	|	foreach (b) { L }	Foreach
	|	allreduce $op_c(T)$;	Reduction
	|	rec l { L }	Recursion
	|	continue l;	Continue
	|	L L	Sequential composition

Figure 15.3 Pabble syntax.

value representing the maximum possible value or practically unbounded. Binding range expression b takes the form of $i : e_1..e_n$ which means i is ranged from e_1 to e_n. Binding variables always bind to a range expression and not individual values. Indices in a Pabble protocol must be bound with the binding range expression, the details are omitted here, please see *indices well-formed conditions* in [14].

In a global protocol G, $l(T)$ from R_1 to R_2 is called an *interaction statement*, which represents passing a message with label l and type T from one role R_1 to another role R_2. R_1 is a *sender role* and R_2 is a *receiver role*. choice at R { G_1 } or ... or { G_N } means the role R will select one of the global types $G_1,\ldots,G_N$. rec l { G } is recursion with the label l which declares a label for continue l statement. foreach (b) {G} denotes a for-loop whose iteration is specified by b. For example, foreach (i:1..n){ G } represents the iteration from 1 to n of G where G is parameterised by i.

Finally, allreduce $op_c(T)$ means all processes perform a distributed reduction of value with type T with the operator op_c (like MPI_Allreduce in MPI), and sends the resulting value from the reduction to all processes. It takes a mandatory predefined operator op_c where op_c must be a commutative and associative arithmetic operation so they can correspond to MPI reduction operations which have the same requirements. Pabble currently supports sum and product.

We allow using simple expressions (e.g. Worker[i:0..2*N-1]) to parameterise ranges. In addition, indices can also be calculated by expressions on bound variables (e.g. Worker[i+1]) to refer to relative positions of roles.

There are restrictions on the indices on such as relative indices calculations and index bounds presented below. The restrictions ensure termination of the projection algorithm and safety of the communication topology at runtime.

15.5.1.1 Restriction on constants

In Pabble protocols, constants can be defined by

(1) A single numeric value (const N = 3); or
(2) Lower and upper bound constraints not involving the max keyword; or
(3) A range defined with the max keyword.

(1) sets a fixed value to a constant, as exemplified in Listing 15.2. (2) gives runtime constants a lower bound and an upper bound, e.g. the number of processes spawned in a scalable protocol, which is unknown at design time and will be defined and immutable once the execution begins. To ensure Pabble protocols are communication-safe in all possible values of constants, we must ensure that all parameterised role indices stay within their declared range.

Such conditions prevent sending or receiving from an invalid (non-existent) role which will lead to communication mismatch at runtime.

The following explains how to determine whether the protocol will be valid for all combinations of constants:

```
1   const M = 1..3;
2   const N = 2..5;
3   global protocol P(role R[1..N]) {
4     T from R[i:1..M] to R[i+1];
5   }
```

The basic constraints from the constants are:

$$1 \leq M, M \leq 3, 2 \leq N \text{ and } N \leq 5$$

We then calculate the range of R[i+1] as R[2..M+1]. Since the objective is to ensure that the role parameters in the protocol body (i.e. 1..M and 2..M+1) stay within the bounds of 1..N, we define a constraint set to be:

$$1 \leq 1 \ \& \ M \leq N \text{ and } 1 \leq 2 \ \& \ M + 1 \leq N$$

which are lower and upper bound inequalities of the two ranges. From them, we obtain this inequality as a result:

$$M + 1 \leq N$$

By comparing this against the basic constraints on the constants, we can check that not all outcomes belong to the regions and thus this is not a communication-safe protocol (an example of a unsafe case is $M = 3$ and $N = 2$). On the other hand, if we alter line 4 to T from R[i:1..N-1] to R [i+1];, the constraints are unconditionally true and so we can guarantee all combinations of constants M and N will not cause communication errors.

(3) is a special case of (2), where the upper bound of a constant is set to the max keyword. We write const N = 0..max to represent a range without upper bound, here it means the constant N can be any integer value larger than 1. Since it is not possible to enumerate all values of N, we apply a more restrictive constraint on the expressions, allowing only range calculation that uses addition or subtractions on integers (e.g. i+1).

15.5.2 Local Protocols

As mentioned in Section 15.2, *local protocols* are localised versions of the global protocols at each role, and are used directly for skeleton generation. They are generated from a global protocol by a projection algorithm detailed in [14]. Local protocol L consists of the same syntax of the global type except the input from R (receive) and the output to R (send). The main declaration local protocol *str* at R_e (...) { L } means the protocol is located at role R_e. We call R_e *the endpoint role*. In Pabble, multiple local protocol instances can reside in the same parameterised local protocol. This is because each local protocol is a local specification for a participant of the interaction. When there are multiple participants with a similar interaction structure that fulfil the same *role* in the protocol, such as the Worker role from our Ring example from the introduction, the participants are grouped together as a single parameterised role. The local protocol for a collection of participants can be specified in a single parameterised local protocol, using *conditional statements* on the role indices to capture corner cases. For example, in a general case of a pipeline interaction, all participants receive from one neighbour and send to another neighbour, except the first participant which initiates the pipeline and is only a sender and the last participant which ends the pipeline and does not send. In these cases we use conditional statements to guard the input or output statements. To express conditional statements in local protocols, if R may be prepended to an input or output statement. if R input/output statement will be ignored if the local role does not match R. More complicated matches can be performed with a parameterised role, where the role parameter range of the condition is matched against the parameter of the local role. For example, if Worker[1..3] will match Worker[2] but not Worker[4]. It is also possible to bind a variable to the range in the condition, e.g. if Worker[i:1..3], and i can be used in the same statement.

15.6 MPI Backbone Generation

Below we explain how Pabble statements are translated into MPI blocks.

15.6.1 Interaction

An interaction statement in a Pabble protocol is projected in the local protocol as two parts: receive and send. The correspondence is shown in Figure 15.4.

The first line of the local protocol shows a receive statement, written in Pabble as if P[dstId] from P[srcId]. The statement is translated to a block of MPI code in 3 parts. First, memory is dynamically allocated for the receive buffer (line 2), the buffer is of Type and its size fetched from the function meta.bufsize(Label). The function is defined in the kernels and

Global Protocol

Projected Local Protocol

$Label(Type)$ from P$[srcIdx]$ $\rightarrow$ if P$[dstIdx]$ $Label(Type)$ from P$[srcIdx]$;
                to P$[dstIdx]$;   if P$[srcIdx]$ $Label(Type)$ to P$[dstIdx]$;

Interaction

Output C/MPI Backbone

```
 1  if (meta.pid == role_P(dstIdx)) {
 2   buf = (Type *)calloc(meta.bufsize(Label), sizeof(Type));
 3   MPI_Recv(buf, meta.bufsize(Label), MPI_Type, role_P(srcIdx), Label, ...);
 4   pabble_recvq_enqueue(Label, buf);
 5  #pragma pabble kernel Label
 6  }
 7  if (meta.pid == role_P(srcIdx)) {
 8  #pragma pabble kernel Label
 9   buf = pabble_recvq_dequeue();
10   MPI_Send(buf, meta.bufsize(Label), MPI_Type, dstIdx, Label, ...);
11   free(buf);
12  }
```

Figure 15.4   Pabble interaction statement and its MPI backbone.

returns the size of message for the given message label. Next, the program calls MPI_Recv to receive a message (line 3) from participant P[srcRole] in Pabble. role_P(srcIdx) is a lookup macro from the generated backbone to return the process id of the sender. Finally, the received message, stored in the receive buffer buf, is enqueued into a global receive queue with pabble_recvq_enqueue() (line 4), followed by the pragma indicating a kernel of label Label should be inserted. The block of receive code is guarded by an if-condition, which executes the above block of MPI code only if the current process id matches the receiver process id.

The next line in the local protocol is a send statement, converse of the receive statement, written as if P[srcIdx] Label(Type)to P[dstIdx]. The MPI code begins with the pragma annotation, then dequeuing the global send queue with pabble_sendq_dequeue() and sends the dequeued buffer with MPI_Send. After this, the send buffer, which is no longer needed, is deallocated. The block of send code is similarly guarded by an if-condition to ensure it is only executed by the sender. By allocating memory before receive and deallocating memory after send, the backbone manages memory for the user systematically. Since the protocol and the backbone makes no assumption about memory management on user's computation kernel, this mechanism helps the separation of concern between the protocol (i.e. the generated backbone) and the user kernels, and leaves open the possibility of optimal memory management during merge without breaking existing kernels.

15.6.2 Parallel Interaction

A Pabble parallel interaction statement is written as Label(Type)from P[i:1..N-1] to P[i+1], meaning all processes with indices from 1 to N-1 send a message to its next neighbour. P[1] initiates sending to P[2], and P[2] receives from P[1] then sends a message to P[3], and so on. As shown in Figure 15.5, the local protocol encapsulates the behaviour of all P[1..N] processes, and the statement is realised in the local as conditional receive followed by a conditional send, similar to ordinary interaction. The difference is the use of a range of process ids in the condition, and *relative* indices in the sender/receiver indices. The generated MPI code makes use of expression with meta.pid (current process id) to calculate the relative index.

15.6.3 Internal Interaction

When role with name __self is used in a protocol, it means that both the sending and receiving endpoints are internal to the processes, and there is no interaction with external processes. This statement applies to all processes, and is not to be confused with self-messaging, e.g. Label()from P[1] to P[1], which would lead to deadlock. The statement does not use any MPI primitives. The purpose of using this special role is to create optional insertion point for the MPI backbone, which may be used for optional kernels such as initialisation or finalisation, hence it generates a pragma in the MPI backbone.

Global Protocol		Projected Local Protocol
Label(*Type*) from P[i:1..N-1] to P[i+1];	$\rightarrow$	if P[i:2..N] *Label*(*Type*) from P[i-1]; if P[i:1..N-1] *Label*(*Type*) to P[i+1];

Parallel Interaction   Output C/MPI Backbone

```
1   if (role_P(2)<=meta.pid&&meta.pid<=role_P(N)) {
2     buf = (Type *)calloc(meta.bufsize(Label), sizeof(Type));
3     MPI_Recv(..., /*prevRank:*/ meta.pid-1, Label, ...);
4     pabble_recvq_enqueue(Label, buf);
5   #pragma pabble kernel Label
6   }
7   if (role_P(1)<=meta.pid&&meta.pid<=role_P(N-1)) {
8   #pragma pabble kernel Label
9     buf = pabble_sendq_dequeue();
10    MPI_Send(..., /*nextRank:*/ meta.pid+1, Label, ...); free(buf);
11  }
```

Figure 15.5   Pabble parallel interaction statement and its MPI backbone.

Global/Local Protocol   Output C/MPI Backbone

```
Internal() from __self to __self;      1  #pragma pabble Internal
```

Figure 15.6   Pabble internal interaction statement and its MPI backbone.

15.6.4 Control-flow: Iteration and For-loop

rec and foreach are iteration statements. Specifically rec/continue is recursion, where the iteration conditions are not specified explicitly in the protocol, and translates to while-loops. The loop condition is the same in all processes, otherwise be known as *collective loops*. The loop generated by rec has a #pragma pabble predicate annotation, so that the loop condition can be later replaced by a kernel (see Section 15.7.1).

The foreach construct, on the other hand, specifies a counting loop, iterating over the integer values in the range specified in the protocol from the lower bound (e.g. 0) to the upper bound value (e.g. N-1). This construct can be naturally translated into a C for-loop.

15.6.5 Control-flow: Choice

Conditional branching in Pabble is performed by label branching and selection. We use the example given in Figure 15.8 to explain. The deciding process, e.g. P[master], makes a choice and executes the statements in the selected branch. Each branch starts by sending a unique label, e.g. Branch0, to the decision receiver, e.g. P[worker]. Hence for a well-formed Pabble protocol, the first line of each branch is from the deciding process to the same process but using a different label.

Note that the decision is only known between the two processes in the first statement, and other processes should be explicitly notified or use broadcast to propagate the decision. The MPI backbone is generated with a different structure as the local protocol. First, the MPI backbone contains an outer

Global/Local Protocol   Global/Local Protocol

```
rec LoopName { ... continue LoopName; }   foreach (i:0..N-1) { ... }
```

Iteration   Output C/MPI Backbone   **Foreach**   Output C/MPI Backbone

```
1  #pragma pabble predicate LoopName     1  for (int i=0; i<=N-1; i++) {
2  while (1) {                           2    ...
3    ... }                               3  }
```

Figure 15.7   Control-flow: Pabble iteration statements and their corresponding MPI backbones.

Global Protocol   Projected Local Protocol

```
choice at P[master] {              choice at P[master] {
  Branch0(Type) from P[master]       if P[worker] Branch0(Type) from P[master];
                to P[worker];  →     if P[master] Branch0(Type) to P[worker];
  ...                                ...
} or { ... }                       } or { ... }
```

Choice   Output C/MPI Backbone

```
1   if (rank==role_P(master)) { // Choice sender
2   #pragma pabble predicate Branch0
3     if (1) {
4       // Block of send.
5       MPI_Send(..., MPI_Type, role_P(worker), Branch0, ...);
6     } else
7   #pragma pabble predicate Branch1
8     if (1) { ... }
9   } else { // Choice receiver
10    MPI_Probe(role_P(master), MPI_ANY_TAG, comm, &status);
11    switch (status.MPI_TAG) {
12      case Branch0:
13      // Ordinary block of recv.
14      if (rank==role_P(worker)) {
15        MPI_Recv(..., MPI_Type, role_P(master), Branch0, ...);
16        pabble_recvq_enqueue(Branch0, buf); }
17      ... break;
18  #pragma pabble Branch1
19      case Branch1: ...
20    }
21  }
```

Figure 15.8   Control-flow: Pabble choice and its corresponding MPI backbone.

if-then-else, splitting the deciding process (lines 1–9) and the decision receiver (lines 9–21). In the deciding process, a block of if-then-else-if code is generated to perform a send with different label (called MPI tag), e.g. line 5. This statement is generated with all the queue and memory management code as described above for ordinary interaction statements. Each of the if-condition is annotated with #pragma pabble predicate BranchLabel, so that the conditions can be replaced by predicate kernels (see Section 15.7.1). For the decision receiver, MPI_Probe is used to peek the received label, then the switch statement is used to perform the correct receive (for different branches).

15.6.6 Collective Operations: Scatter, Gather and All-to-all

Collective operations are written in Pabble as multicast or multi-receive message interactions. While it is possible to convert these interactions into multiple blocks of MPI code following the rules in Figure 15.7 (e.g. loop through receivers for scatter), we take advantage of the efficient and expressive collective primitives in MPI. Figure 15.9 shows the conversion of Pabble statements into MPI collective operations. We describe only the most generic collective operations, i.e. MPI_Scatter, MPI_Gather and MPI_Alltoall.

Translating collective operations from Pabble to MPI uses both global Pabble protocol statements and local protocol. If a statement involves the All role as sender, receiver or both, it is a collective operation. Figure 15.9 shows that translated blocks of MPI code do not use if-statements to distinguish between sending and receiving processes. This is because collective

Global Protocol

```
Label(Type) from P[rootRole] to All; // One-to-Many: (a) Scatter
Label(Type) from All to P[rootRole]; // Many-to-One: (b) Gather
Label(Type) from All to All;         // Many-to-Many: (c) All-to-All
```

Collective operation: (a) Scatter   Output C/MPI Backbone

```
1  rbuf = (Type *)calloc(meta.buflen(Label), sizeof(Type));
2  #pragma pabble kernel Label
3  sbuf = pabble_sendq_dequeue();
4  MPI_Scatter(sbuf, meta.buflen(Label), MPI_Type,
5             rbuf, meta.buflen(Label), MPI_Type, role_P(rootRole), ...);
6  pabble_recvq_enqueue(Label, rbuf);
7  #pragma pabble kernel Label
8  free(sbuf);
```

Collective operation: (b) Gather   Output C/MPI Backbone

```
1  rbuf = (Type *)calloc(meta.buflen(Label)*meta.nprocs, sizeof(Type));
2  #pragma pabble kernel Label
3  sbuf = pabble_sendq_dequeue();
4  MPI_Gather(sbuf, meta.buflen(Label), MPI_Type,
5            rbuf, meta.buflen(Label), MPI_Type, role_P(rootRole), ...);
6  pabble_recvq_enqueue(Label, rbuf);
7  #pragma pabble kernel Label
8  free(sbuf);
```

Collective operation: (c) All-to-All   Output C/MPI Backbone

```
1  rbuf = (Type *)calloc(meta.buflen(Label)*meta.nprocs, sizeof(Type));
2  #pragma pabble kernel Label
3  sbuf = pabble_sendq_dequeue();
4  MPI_Alltoall(sbuf, meta.buflen(Label), MPI_Type,
5              rbuf, meta.buflen(Label), MPI_Type, ...);
6  pabble_recvq_enqueue(Label, rbuf);
7  #pragma pabble kernel Label
8  free(sbuf);
```

Figure 15.9 Collective operations: Pabble collectives and their corresponding MPI backbones.

primitives in MPI are executed by *both* the senders and the receivers, and the runtime decides whether it is a sender or a receiver by inspecting the rootRole parameter (which is a process rank) in the MPI_Scatter or MPI_Gather call. Otherwise the conversion is similar to their point-to-point counterparts in Figure 15.4.

15.6.7 Process Scaling

In addition to the translation of Pabble statements into MPI code, we also define the process mapping between a Pabble protocol and a Pabble-generated MPI program. Typical usage of MPI programs can be parameterised on the number of spawned processes at runtime via program arguments. Hence, given a Pabble protocol with *scalable* roles, we describe the rules below to map (parameterised) roles into MPI processes.

A Pabble protocol for MPI code generation can contain any number of constant values (e.g. const M = 10), which are converted in the backbone as C constants (e.g. #define M 10), but it can use at most one *scalable constant* [13], and will scale with the total number of spawned processes. A scalable constant, defined in Section 15.5.1.1 as constant type (3), is written:

```
const N = 1..max;
```

The constant can then be used for defining parameterised roles, and used in indices of parameterised message interaction statements. For example, to declare an $N \times N$ role P, we write in the protocol:

```
global protocol P (role P[1..N][1..N])
```

which results in a total of N^2 participants in the protocol, but N is not known until execution time. MPI backbone code generated based on this Pabble protocol uses N throughout. Since the only parameter in a scalable MPI program is its size (i.e. number of spawned processes), the following code is generated in the backbone to calculate, from size, the value of C local variable N:

```
MPI_Comm_size(MPI_COMM_WORLD, &meta.nprocs); // # of processes
int N = (int)pow(meta.nprocs, 1/2); // N = sqrt(meta.nprocs)
```

15.7 Merging MPI Backbone and Kernels

15.7.1 Annotation-Guided Merging Process

To combine the MPI backbone with the kernels, our aspect-oriented design-flow inserts kernel function calls into the MPI backbone code. The insertion points are realised as #pragmas in the MPI backbone code, generated from the input protocol as placeholders where functional code is inserted. There are multiple types of annotations whose syntax is given as:

```
#pragma pabble [<entry point type>] <entry point id> [(param0, ...)]
```

where *entry point type* is one of kernel, type or predicate, and *entry point id* is an alphanumeric identifier.

15.7.2 Kernel Function

#pragma pabble kernel Label defines the insertion point of kernel functions in the MPI backbone code. Label is the label of the interaction statement, e.g. Label(T) from Sender to Receiver, and the annotation is replaced by the kernel function associated to the label Label. Programmers must use the same pragma to manually annotate the implementation of the kernel function. The first row in Table 15.1 shows an example.

15.7.3 Datatypes

#pragma pabble type TypeName annotates a generic type name in the backbone, and also annotates the concrete definition of the datatype in the kernels. In the second row of Table 15.1, the C datatype T is defined to be void since the protocol does not have any information to realise the type. The kernel defines T to be a concrete type of double, and hence our tool transforms the typedef in the backbone into double and infers the corresponding MPI_Datatype (MPI derived datatypes) to the built-in MPI integer primitive type, i.e. MPI_Datatype MPI_T = MPI_DOUBLE. From the

Table 15.1 Annotations in backbone and kernel

		Generated MPI backbone	User supplied kernel	Merged code

Kernel Function		`#pragma pabble kernel Label`	`#pragma pabble kernel Label`
			`void kernel_func(int label)`
			`{ ... }`
				`kernel_func(Label);`

Datatypes		`#pragma pabble type T`
		`typedef void T;`
		`MPI_Datatype MPI_T;`
			`#pragma pabble type T`
			`typedef double T;`
				`typedef double T;`
				`MPI_Datatype MPI_T`
				`   = MPI_DOUBLE;`

Conditionals		`#pragma pabble predicate Cond`
		`while (1)`
		`{ ... }`
			`#pragma pabble predicate Cond`
			`int condition()`
			`{ ... return bool; }`
				`while (condition())`
				`{ ... }`

given type we can also generate MPI datatypes for structures of primitive types, e.g. struct { int x, int y, double m } is transformed to its MPI-equivalent datatype.

15.7.4 Conditionals

#pragma pabble predicate Label annotates predicates, e.g. loop conditions or if-conditions, in the backbone. Since a Pabble communication protocol (and transitively, the MPI backbone) does not specify a loop condition, the default loop condition is 1, i.e. always true. This annotation introduces a way to insert a conditional expression defined as a kernel function. It precedes the while-loop, as shown in the third row of Table 15.1, to label the loop with the name Label. The kernel function that defines expressions must use the same annotation as the backbone, e.g. #pragma pabble predicate Label. After the merge, this kernel function is called when the loop condition is evaluated.

15.8 Related Work

The general approach of describing parallel patterns and reusing them with different computation modules can date back to [4] by Darlington et al., where parallel patterns are described as higher order *skeleton* functions, written in a functional language. Parallel applications are implemented as functions that combine with the skeletons and transformed. Their system targets specialised parallel machines, and our approach targets MPI, a standard for parallel programming in a range of hardware configurations. The approach, also known as *algorithmic skeleton frameworks* for parallel programming, is surveyed in [7]. Some of these tools also target MPI for high-level structured parallel programming, and only works with a limited set of parallel patterns. Our code generation workflow based on Pabble supports generic patterns written in Pabble and guarantees communication safety in the generated MPI code.

15.9 Conclusion

In this chapter we presented a protocol-based workflow for constructing safe and efficient parallel applications. The framework consists of two parts, a safe-by-construction parallel interaction backbone, generated from the Pabble protocol language, and an aspect-oriented compilation workflow to mechanically insert computation code into the backbone. Our approach simplifies parallel programming by making use of parallel communication patterns, described with our Pabble protocol description language, and building independent kernel code around the patterns as sequential C code. This approach is flexible, where multiple sets of kernels can share a common parallel communication pattern, since the computation and the communication are maintained separately.

Acknowledgements This work is supported EPSRC projects EP/K034413/1, EP/K011715/1, EP/L00058X/1 and EP/N027833/1; and by EU FP7 612985 (UpScale).

References

[1] L. Bettini, M. Coppo, L. DAntoni, M. D. Luca, M. Dezani-Ciancaglini, and N. Yoshida. Global Progress in Dynamically Interleaved Multiparty Sessions. In *CONCUR 2008*, volume 5201 of *LNCS*. Springer, 2008.

[2] J. a. M. Cardoso, T. Carvalho, J. G. Coutinho, W. Luk, R. Nobre, P. Diniz, and Z. Petrov. LARA: an aspect-oriented programming language for embedded systems. In *AOSD '12*. ACM Press, 2012.

[3] P. Charles, C. Grothoff, V. Saraswat, C. Donawa, A. Kielstra, K. Ebcioglu, C. von Praun, and V. Sarkar. X10: an object-oriented approach to non-uniform cluster computing. In *OOPSLA '05*. ACM Press, 2005.

[4] J. Darlington, A. Field, P. Harrison, P. H. J. Kelly, D. W. N. Sharp, and Q. Wu. Parallel programming using skeleton functions. In *PARLE'93*, 1993.

[5] P.-M. Denielou, N. Yoshida, A. Bejleri, and R. Hu. Parameterised Multiparty Session Types. *Logical Methods in Computer Science*, 8(4):1–46, October 2012.

[6] J. DeSouza, B. Kuhn, B. R. de Supinski, V. Samofalov, S. Zheltov, and S. Bratanov. Automated, scalable debugging of MPI programs with Intel Message Checker. In *SE-HPCS '05*. ACM Press, 2005.

[7] H. González-Vélez and M. Leyton. A Survey of Algorithmic Skeleton Frameworks: High-level Structured Parallel Programming Enablers. *Softw. Pract. Exper.*, 40(12):1135–1160, 2010.

[8] W. Gropp, E. Lusk, and A. Skjellum. *Using MPI: Portable Parallel Programming with the Message-Passing Interface*. MIT Press, 1999.

[9] K. Honda, N. Yoshida, and M. Carbone. Multiparty Asynchronous Session Types. *JACM*, 63(1):9:1–9:67, 2016.

[10] J. K. Lee and J. Palsberg. Featherweight X10. In *PPoPP '10*. ACM Press, 2010.

[11] J. Magee and J. Kramer. *Concurrency – state models and Java programs (2. ed.)*. Wiley, 2006.

[12] N. Ng, J. G. Coutinho, and N. Yoshida. Protocols by Default: Safe MPI Code Generation based on Session Types. volume 9031 of *LNCS*. Springer, 2015.

[13] N. Ng and N. Yoshida. Pabble: Parameterised Scribble for Parallel Programming. In *PDP 2014*, pages 707–714, 2014.

[14] N. Ng and N. Yoshida. Pabble: parameterised Scribble. *SOCA*, 9(3–4), 2015.

[15] Pabble project on GitHub. https://github.com/pabble-lang

[16] Scribble homepage. http://scribble.org/

[17] X10 homepage. http://x10-lang.org

[18] N. Yoshida, R. Hu, R. Neykova, and N. Ng. The Scribble Protocol Language. In *TGC 2013*, volume 8358 of *LNCS*. Springer, 2013.

16

Deductive Verification of MPI Protocols

Vasco T. Vasconcelos[1], Francisco Martins[1], Eduardo R. B. Marques[2],
Nobuko Yoshida[3] and Nicholas Ng[3]

[1]LaSIGE, Faculty of Sciences, University of Lisbon, PT
[2]CRACS/INESC-TEC, Faculty of Sciences, University of Porto, PT
[3]Imperial College London, UK

Abstract

This chapter presents the PARTYPES framework to statically verify C programs that use the Message Passing Interface, the widely used standard for message-based parallel applications. Programs are checked against a protocol specification that captures the interaction in an MPI program. The protocol language is based on a dependent type system that is able to express various MPI communication primitives, including point-to-point and collective operations. The verification uses VCC, a mechanical verifier for concurrent C programs. It takes the program protocol written in VCC format, an annotated version of the MPI library, and the program to verify, and checks whether the program complies with the protocol.

16.1 Introduction

Message Passing Interface (MPI) [3] is a portable message-passing API for programming parallel computers running on distributed memory systems. To these days, MPI remains the dominant framework for developing high performance parallel applications.

Usually written in C or Fortran, MPI programs call library functions to perform point-to-point send/receive operations, collective and synchronisation operations (such as broadcast and barrier), and combination of partial results of computations (gather and reduce operations). Developing MPI applications is an error-prone endeavour. For instance, it is quite easy to write programs that cause processes to wait indefinitely for a message, or that exchange data of unexpected sorts or lengths.

Verifying that MPI programs are exempt from communication errors is far from trivial. The state-of-the-art verification tools for MPI programs use advanced techniques such as runtime verification [7, 15, 16, 21] and model checking [5–7, 15, 17, 20]. These approaches frequently stumble upon the problem of scalability since the search space grows exponentially with the number of processes. It is often the case that the verification of real applications limits the number of processes to less than a dozen [18].

We approach the problem of verifying C+MPI code using a type theory for parallel programs. In our framework—PARTYPES—types describe the communication behaviour programs, that is, protocols. Programs that conform to one such type are guaranteed to follow the protocol and not to run into deadlocks. The verification is scalable, as it does not depend on the number of processes or other input parameters.

A previous work introduces the type theory underlying protocol specification, shows the soundness of the methodology by designing a core language for parallel programming and proving a progress result for well-typed programs, and provides a comparative evaluation of PARTYPES against other state-of-the-art tools [10].

This chapter takes a pragmatic approach to the verification of C+MPI code, by explaining the procedure from the point of view of someone interested in verifying actual code, omitting theoretic technical details altogether. Protocols are written in a dependent type language that includes specific constructors for some of the most common communication primitives found in MPI programs. The conformance of a program against a protocol is checked using VCC, a software verifier for the C programming language [1]. In a nutshell, one checks C+MPI source code against a protocol as follows:

1. Write a protocol for the program, that can be translated mechanically to VCC format;
2. Introduce special, concise marks in the C+MPI source code to guide the automatic generation of VCC annotations required for verification;
3. Use the VCC tool to check conformance of the source code against the protocol.

If VCC runs successfully, then the program is guaranteed to follow the protocol and to be exempt from deadlocks, regardless of the number of processes, problem dimension, number of iterations, or any other parameters. The verification process is guided by two tools—the Protocol Compiler and the Annotation Generator—and by the PARTYPES MPI library. All these can be found at the PARTYPES website [14]. The tools and the library almost completely insulate the user from working with the VCC language.

The rest of this chapter is organised as follows. The next section introduces a running example and discusses typical faults found in MPI programs. Then Section 16.3 describes the protocol language and Sections 16.4 and 16.5 provide an overview of the verification process. Section 16.6 discusses related work and Section 16.7 concludes the chapter.

16.2 The Finite Differences Algorithm and Common Coding Faults

This section introduces a running example and discusses common pitfalls encountered when developing MPI programs.

The *finite differences* algorithm computes an approximation of derivatives by the finite difference method. Given an initial vector X_0, the algorithm calculates successive approximations to the solution $X_1, X_2, \ldots$, until a pre-defined maximum number of iterations has been reached. A distinguished process, say the one with rank 0, disseminates the problem size (that is, the length of array X) through a broadcast operation. The same process then divides the input array among all processes. Each participant is responsible for computing its local part of the solution. When the pre-defined number of iterations is reached, process rank 0 obtains the global error through a reduce operation and collects the partial arrays in order to build a solution to the problem (Figure 16.1, left). In order to compute its part of the solution, each process exchanges boundary values with its left and right neighbours on every iteration (Figure 16.1, right).

Figure 16.2 shows C+MPI source code that implements the finite differences algorithm, adapted from a textbook [4]. The main function describes the

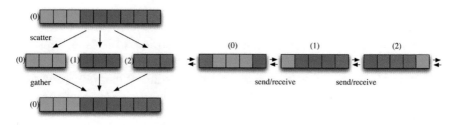

Figure 16.1   Communication pattern for the finite differences algorithm.

```
 1  int main(int argc, char** argv) {
 2    int rank, procs, n; // process rank; number of processes; problem size
 3    ...
 4    MPI_Init(&argc, &argv);
 5    MPI_Comm_rank(MPI_COMM_WORLD, &rank);
 6    MPI_Comm_size(MPI_COMM_WORLD, &procs);
 7    if (rank == 0) {
 8      n = read_problem_size(procs);
 9      read_vector(work, n);
10    }
11    MPI_Bcast(&n, 1, MPI_INT, 0, MPI_COMM_WORLD);
12    int local_n = n / procs;
13    MPI_Scatter(work, local_n, MPI_FLOAT, &local[1], local_n, MPI_FLOAT, ...);
14    int left  = rank == 0 ? procs - 1 : rank - 1; // left neighbour
15    int right = rank == procs - 1 ? 0 : rank + 1; // right neighbour
16    for (iter = 1; iter <= NUM_ITER; iter++) {
17      ...
18      if (rank == 0) {
19        MPI_Send(&local[1],         1, MPI_FLOAT, left,  ...);
20        MPI_Send(&local[local_n],   1, MPI_FLOAT, right, ...);
21        MPI_Recv(&local[local_n+1], 1, MPI_FLOAT, right, ...);
22        MPI_Recv(&local[0],         1, MPI_FLOAT, left,  ...);
23      } else if (rank == procs - 1) {
24        MPI_Recv(&local[local_n+1], 1, MPI_FLOAT, right, ...);
25        MPI_Recv(&local[0],         1, MPI_FLOAT, left,  ...);
26        MPI_Send(&local[1],         1, MPI_FLOAT, left,  ...);
27        MPI_Send(&local[local_n],   1, MPI_FLOAT, right, ...);
28      } else {
29        MPI_Recv(&local[0],         1, MPI_FLOAT, left,  ...);
30        MPI_Send(&local[1],         1, MPI_FLOAT, left,  ...);
31        MPI_Send(&local[local_n],   1, MPI_FLOAT, right, ...);
32        MPI_Recv(&local[local_n+1], 1, MPI_FLOAT, right, ...);
33      }
34      ...
35    }
36    MPI_Reduce(&localerr, &globalerr, 1, MPI_FLOAT, MPI_MAX, 0, ...);
37    MPI_Gather(&local[1], local_n, MPI_FLOAT, work, local_n, MPI_FLOAT, 0, ...);
38    ...
39    MPI_Finalize();
40    return 0;
41  }
```

Figure 16.2   Excerpt of an MPI program for the finite differences problem.

behaviour of all processes together; the behaviour of each individual process may diverge based on its process number, designated by *rank*, and set on line 5 using the MPI_Comm_rank primitive. The number of processes (procs in the figure) is obtained through primitive MPI_Comm_size on line 6. Rank 0 starts by reading the problem size and the corresponding input vector X_0 (lines 8–9, variables n and work). The same participant then broadcasts the problem size (line 11, call to MPI_Bcast) and distributes the input vector to all other participants (line 13, call to MPI_Scatter).

Each participant is then responsible for computing its part of the solution. The program enters a loop (lines 16–35), specifying point-to-point message exchanges (MPI_Send, MPI_Recv) between each process and its left and `right` neighbours, based on a ring topology. The various message exchanges distribute boundary (`local[0]` and `local[local_n+1]`) values necessary to local calculations. Different send/receive orders for different ranks (lines 19–22, lines 24–27, and lines 29–32) aim at avoiding deadlock situations (`MPI_Send` and `MPI_Recv` are blocking, synchronous, unbuffered operations). The loop ends when a pre-defined number of iterations is attained. Once the loop is over, rank 0 computes the global error through a reduction operation (`MPI_reduce`, line 36) and gathers the solution obtaining from each process (including itself) a part of the vector (`MPI_Gather`, line 37).

For space reasons we have omitted a few actual parameters in some calls to MPI operations: the ellipsis in Figure 16.2 denote parameters 0 (the message tag number) and `MPI_COMM_WORLD` (the communicator) in all operations, except in `MPI_Recv` where they denote, in addition, parameter `&status`.

The code in Figure 16.2 is extremely sensitive to variations in the structure of MPI operations. We distinguish five kinds of situations that are further discussed below:

1. Type mismatches in messages,
2. Array length mismatches in messages,
3. Missing send or receive operations,
4. Wrong send-receive order in messages, and
5. Incompatible MPI operations for the different processes.

The first two situations are related to how MPI primitives describe data transmitted in messages: usually in the form of a pointer to a buffer, the length of the buffer, and the type of elements in the buffer. A *type mismatch* in a message exchange occurs when, for example, one replaces `MPI_FLOAT` by `MPI_DOUBLE` in line 19. Then process rank 0 sends a value of type `double`, while process rank `procs-1` expects a `float`. An *array length mismatch* happens, for example, if one replaces 1 with 2 as the second parameter on line 19. Then process rank 0 sends two floating point numbers, while process rank `procs-1` expects exactly one (line 24). It should be emphasised that these mismatches are caught at runtime, if caught at all.

The last three cases all lead to *deadlocks*. In general, MPI programs enter deadlocked situations when a communication operation is not matched by all the processes involved. For example, the *omission of the send operation* on line 19 will leave process rank `procs-1` eternally waiting for a message to come on line 24. For another example, *exchanging the two receive operations* in lines 21 and 22 leads to a deadlock where ranks 0 and 1 will be forever waiting for one another.

Incompatible MPI operations for the different processes come in different flavours. For example, replacing the receive operation by an MPI_Bcast on line 24 leads to a situation where process rank 0 tries to send a message, while rank procs-1 tries to broadcast. For another example, replace the root process of the reduce operation at line 36 from 0 to rank. We are left with a situation where each process executes a different reduce operation, each trying to collect the maximum of the values provided by all processes. For a last example, enclose the gather operation on line 37 by a conditional of the form if(rank == 0). In this case process rank 0 will be forever waiting for the remaining processes to provide their parts of the array.

16.3 The Protocol Language

This section introduces the protocol language, following a step-by-step construction of the protocol for our running example.

In the beginning, process rank 0 broadcasts the problem size, a natural number. We write this as

```
broadcast 0 natural
```

That process rank 0 divides X_0 (an array of floating pointing numbers) among all processes is described by a scatter operation.

```
scatter 0 float[]
```

Now, each process loops for a given number of iterations, nIterations. We write this as follows.

```
foreach iteration: 1..nIterations
```

Variable nIterations must be somehow introduced in the protocol. It denotes a value that must be known to all processes. Typically, there are two ways for processes to get to know this value:

- The value is exchanged, resorting to a collective communication operation, in such a way that *all* processes get to know it, or
- The value is known to all processes before computation starts, for example because it is hardwired in the source code or is read from the command line.

In the former case we could add another broadcast operation in the first lines of the protocol. In the latter case, the protocol language relies on the val constructor, allowing a value to be introduced in the program:

```
val nIterations: positive
```

Either solution would solve the problem. If a broadcast is used then processes must engage in a broadcast operation; if val is chosen then no value exchange is needed, but the programmer must identify the value in the source code that will replace variable nIterations.

We may now continue analysing the loop body (Figure 16.2, lines 17–34). In each iteration, each process sends a message to its left neighbour and another message to its right neighbour. Such an operation is again described as a foreach construct that iterates over all processes. The first process is 0; the last is size-1, where size is a distinguished variable that represents the number of processes. The inner loop is then written as follows.

```
foreach i: 0..size-1
```

When i is the rank of a process, a conditional expression of the form i=size-1 ? 0 : i+1 denotes the process' right neighbour. Similarly, the left neighbour is i=0 ? size-1 : i-1.

To send a message from process rank r1 to process rank r2 containing a value of a datatype D, we write message r1 r2 D. In this way, to send a message containing a floating point number to the left process, followed by a message to the right process, we write.

```
message i (i=0 ? size-1 : i-1) float
message i (i=size-1 ? 0 : i+1) float
```

So, now we can assemble the loops.

```
foreach iteration: 1..nIterations
  foreach i: 0..size-1 {
    message i (i=0 ? size-1 : i-1) float
    message i (i=size-1 ? 0 : i+1) float
  }
```

Once the loop is completed, process rank 0 obtains the global error. Towards this end, each process proposes a floating point number representing the local error. Rank 0 then reads the maximum of all these values. We write all this as follows:

```
reduce 0 max float
```

Finally, process rank 0 collects the partial arrays and builds a solution X_n to the problem. This calls for a gather operation.

```
gather 0 float[]
```

Before we put all the operations together in a protocol, we need to discuss the nature of the arrays distributed and collected in the scatter and gather operations. In brief, the scatter operation distributes X_0, dividing it in small pieces, while gather collects the subarrays to build X_n. So, we instead write:

```
scatter 0 float[n]
...
gather 0 float[n]
```

Variable n, describing the length of the global array, must be introduced in the protocol. This is typically achieved by means of a val or a broadcast operation. In this case n stands for the problem size that was broadcast before. So we name the value that rank 0 provides as follows.

```
broadcast 0 n:natural
```

But n cannot be an arbitrary non-negative number. It must evenly divide X_0. In this way, each process gets a part of X_0 of equal length, namely length(X0)/size, and we do not risk accessing out-of-bound positions when manipulating the subarrays. So we would like to make sure that the length of X_0 equal divides the number of processes. For this we use a *refinement* datatype. Rather that saying that n is a natural number we say that it is of datatype {x: natural | x % size = 0}. The complete protocol is in Figure 16.3.

As an aside, natural can be expressed as {x: integer | x >= 0}. Similarly, positive abbreviates {x: integer | x > 0}, and float[n] abbreviates a refinement type of the form {x: float[] | length(x) = n}.

Further examples of protocols can be found in a previous work [10] and at the PARTYPES web site [14]. The current version protocol language supports:

```
 1  protocol FiniteDifferences {
 2    val nIterations: positive
 3    broadcast 0 n: {x: natural | x % size = 0}
 4    scatter 0 float[n]
 5    foreach iteration: 1 .. nIterations
 6      foreach i: 0 .. size-1 {
 7        message i (i = 0 ? size-1 : i-1) float
 8        message i (i = size-1 ? 0 : i+1) float
 9      }
10    reduce 0 max float
11    gather 0 float[n]
12  }
```

Figure 16.3   Protocol for the finite differences algorithm.

- Different MPI communication primitives such as `message`, `broadcast`, `reduce`, `allreduce`, `scatter`, `gather`, and `allgather`;
- Control flow primitives, including sequential composition (`;`), primitive recursion (`foreach`), conditional (`if-then-else`), and `skip` (that is, the empty block of operations).

Protocols are subject to certain formation rules [10], including:

- Variables must be properly introduced with `val`, `broadcast`, `allreduce`;
- Ranks must lie between 0 and `size-1`;
- The two ranks in a `message` must be different;
- The length of arrays in `scatter` and `gather` must equally divide `size`.

The PROTOCOLCOMPILER checks protocol formation and, in addition, generates a C header file containing the VCC code that describes the protocol. The tool comes as an Eclipse plugin; it may alternatively be used on a web browser from the PARTYPES web page [14]. Figure 16.4 shows a screenshot of Eclipse when the compiler did not manage to prove that the value of expression `i=size ? 0 : i+1` lies between 0 and `size-1`.

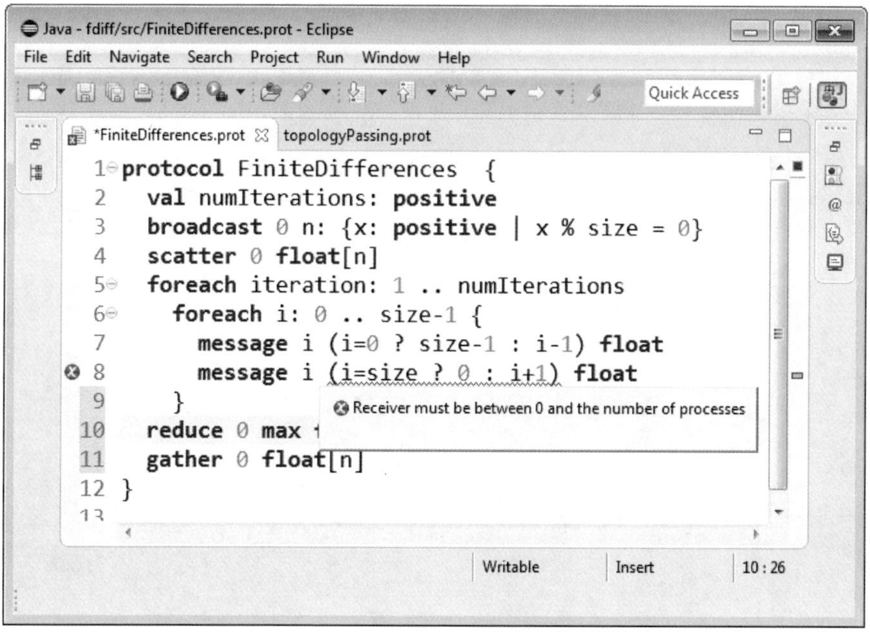

Figure 16.4 Protocol compiler running under the Eclipse IDE.

16.4 Overview of the Verification Procedure

This section and the next present the PARTYPES methodology. Figure 16.5 illustrates the workflow of the verification procedure. Two inputs are required:

- The C+MPI source code (example in Figure 16.2);
- The protocol for the program (example in Figure 16.3).

First, the C+MPI source code must be adapted for verification, the reason being that VCC accepts only a subset of the C programming language. Then, special *marks* are inserted in the C source code. One of our tools, the ANNO-TATIONGENERATOR (AG in the figure), expands the marks. The output is C source code with VCC annotations, which we denote by C+MPI+VCC. The VCC annotations allow the verification of the C code against the protocol.

A second tool, the PROTOCOLCOMPILER (PC in the figure), checks protocol formation and generates a C header file containing the protocol in VCC format. At this point two C header files need to be included in the C source code: the PARTYPES MPI library, and the protocol in VCC format. The PARTYPES MPI library, mpi.h, is a surrogate C header file containing the type theory (as described in a previous work [10]) in VCC format and available at PARTYPES web page [14].

The C code is now ready to be submitted to VCC. The outcome is one of three situations:

- VCC signals success. We know that the C+MPI code, as is, conforms to the protocol, hence is exempt from all the problems discussed in Section 16.2;

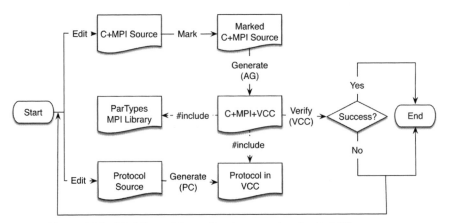

Figure 16.5   Workflow of the verification procedure for C+MPI programs.

- VCC complains presenting the list of failed assertions. In this case, the source of the problem may lie at three different places:

  - the protocol does not capture the communication pattern of the program and needs to be rectified;
  - the C+MPI program is not well annotated, either because it needs additional marks or because some existing marks are misplaced;
  - the C+MPI program itself has a fault that needs to be fixed. In our example, the problem size (stored in variable n) must be a multiple of the number of processes (stored in variable procs), so that the source code may conform to the protocol. Since the problem size is the value of function read_problem_size (line 8, Figure 16.2), we may add an explicit contract to the function:

```
int read_problem_size(int procs)
  _(ensures \result>=0 && \result%procs==0);
{
  . . .
}
```

  In such cases PARTYPES users must make use of the VCC specification language.

- VCC times out. This situation typically happens when the underlying SMT solver fails to check some refinement condition. The PARTYPES user should revise protocol refinements and possibly rewrite them. For instance, to describe that the process with rank i sends a floating point value to its right neighbour in a ring topology, we could have written

```
message i (i+1)%size float
```

It is well-known that non-linear integer arithmetics is undecidable in general and inefficiently supported by SMT solvers. Expressions such as (i+1)%size may complicate the verification procedure, possibly leading to timeouts. Instead, we include in our protocol (Figure 16.3) an equivalent proposition that is more amenable for the solver, namely, i=size-1 ? 0 : i+1.

The rest of this section describes the source code adaptation required to run VCC. In general, the original C+MPI source code requires routine adjustments in order to be accepted by VCC. Adjustments comprise the deletion or the replacement of code that is not supported by VCC. In particular we:

- delete functions with a variable number of arguments (such as printf and scanf);
- suppress all floating point arithmetic;
- replace multidimensional by single dimensional arrays and adjust the code accordingly.

VCC is a verifier for concurrent C. Even though C+MPI code is generally single-threaded, VCC tries to verify that the source code is thread-safe in any concurrent environment that respects the contracts on its functions and data structures. This complicates the verification process and demands additional VCC annotations that are not directly related to the verification of the adherence of code to protocols. In particular, the PARTYPES user needs to guarantee that memory accesses do not introduce data races. He does so by proving that memory locations are not concurrently written (i.e., \thread_local in VCC terms) upon reading, and not concurrently written or read upon writing (\mutable or \writable).

In our running example, and in order to facilitate the explanation and to concentrate on the adherence to the protocol, we inlined all subsidiary functions in the main function, made all arrays local to main, and omitted the code concerned with the actual computation of the finite differences. This greatly simplifies the annotation process as we must only deal with local memory, and do not have to cope with other verification demands such as maintaining loop invariants or proving that integer arithmetics does not overflow. Such adjustments must be exerted with great care so as not to alter the interactive behaviour of the original program.

16.5 The Marking Process

This section completes the PARTYPES methodology for checking C+MPI code by addressing the marking step.

In general, simple protocols require no marks. Imagine the protocol

```
reduce 0 sum integer
```

describing a simple algorithm where each process computes its part of the solution and process rank 0 collects the solution by adding the parts. Because the protocol uses a simple communication primitive no source code marking is required.

We require no marking for the MPI primitives supported by PARTYPES since their usage is taken care of by the contracts provided in the PAR-TYPES MPI library (mpi.h). The PARTYPES user must aid verification through appropriate marks when more advanced protocol features come into play, such as dependent functions (val), primitive recursion (foreach), and conditionals (if-then-else).

We start with val. We have seen in Section 16.3 that this primitive introduces a constant in the protocol:

```
val nIterations: positive
```

Users must provide the actual program value for nIterations. Analysing the code in Figure 16.2, one realises that the protocol variable nIterations corresponds to the program constant NUM_ITER. We then add the mark

```
@apply(MAX_ITER)
```

after the three MPI initialisation primitives (MPI_Init, MPI_Comm_rank, and MPI_Comm_size), that is, after line 6.

Next, we address foreach. Again, we seek the assistance of the user in pointing out the portion of the source code that matches each occurrence of this primitive. In the protocol of Figure 16.3, loop

```
foreach iteration: 1 .. nIterations
```

is meant to be matched against the for loop in Figure 16.2 starting at line 16. We then introduce the mark

```
@foreach(iter, 1, NUM_ITER)
```

just before the body of the for loop, thus associating the protocol loop variable and its bounds with those in the C code.

For the inner loop in the protocol (that is, lines 6–9 in Figure 16.3) we could proceed similarly would the source code be perfectly aligned with the protocol, as in the excerpt below meant to replace lines 18–33 in Figure 16.2:

```
for (i = 0; i < procs; i++) {
  if (rank == i)
    MPI_Send(&local[1],        1, MPI_FLOAT, left,  ...);
  else if (rank == left)
    MPI_Recv(&local[0],        1, MPI_FLOAT, i,     ...);
  if (rank == i)
    MPI_Send(&local[local_n],  1, MPI_FLOAT, right, ...);
  else if (rank == right)
    MPI_Recv(&local[local_n+1], 1, MPI_FLOAT, i,    ...);
}
```

However, efficient implementations do not exhibit loops to implement this kind of `foreach` protocols. The loop in the protocol states that each process (0, ..., `size-1`) must send a message to its left and to its right neighbour. This means that each process will be involved in *exactly four* message passing operations: send left, send right, receive from left, receive from right. Therefore the above `for` loop can be *completely unrolled* into a series of conditional instructions, each featuring two message send and two message receive operations, as witnessed by the code in Figure 16.2, lines 18–33.

How do we check `foreach` protocols against conditional instructions in source code? A possible approach would be to let the verifier unroll the protocol loop. This may work when `size` is known to be a small natural number. In general, however, protocols do not fix the number of processes. That is the case with our running example which must run on any number of processes (starting at 2, for processes cannot send messages to themselves). In such cases VCC takes `size` to be a 64 bits non-negative integer. This poses significant difficulties to the unrolling process both in terms of memory and verification time.

In the running example, the apparent mismatch between the protocol and the program is that there are three different behaviours in the program depending on the rank (Figure 16.2, lines 18–33), while the protocol specifies a single behaviour, namely:

```
message i (i = 0 ? size-1 : i-1) float
message i (i = size-1 ? 0 : i+1) float
```

At first sight, it may seem as if the protocol does not specify the required diversity of behaviours, but in fact it does. To see why, let us unroll the inner `foreach` loop. This is what we get when we omit the type of the message (`float`):

```
message 0 size-1; message 0 1;        // when i = 0
message 1 0; message 1 2;             // when i = 1
...
message size-2 1; message size-2 size-1;// i = size-2
message size-1 size-2; message size-1 0 // i = size-1
```

From the unrolled protocol we conclude that the behaviour of process rank 0 is the following:

1. send a message to its left neighbour (`size-1`);
2. send a message to its right neighbour (`1`);

3. receive a message from its right neighbour; and, finally,
4. receive a message from its left neighbour.

The behaviour is straightforward to obtain: just identify the messages that mention rank 0, and use a send when 0 is the source of the `message` or a receive otherwise. This exactly coincides with the four send/receive operations in the C code for rank 0, lines 19–22.

For the last rank (that is, `size-1`) the relevant send/receive operations are the following:

1. receive a message from its right neighbour (`0`);
2. receive a message from its left neighbour (`size-2`),
3. send a message to its left neighbour; and, finally,
4. send a message to its right neighbour.

This pattern coincides with the source code, lines 24–27. All other behaviours (when rank is between 1 and `size-2`) are similarly obtained and are left as an exercise for the interested reader. The pattern thus obtained should match the code, lines 29–32. Notice that the order of the messages is important, and that we have identified as many behaviours as there are conditional branches in the source code (lines 18–33).

Based on this analysis, and in order to guide the verification process we seek the help of the user by selecting the relevant `foreach` steps (iterations) in each branch of the program. A *relevant step* for rank k corresponds to one `foreach` iteration where either the source or the target of a `message` appearing in the loop body is k. A step that does not mention k (as source or target) is equivalent to `skip`, the empty protocol, and hence irrelevant for verification purposes. In order to check that all non-relevant steps are `skip`, we must provide the loop bounds (`0` and `procs-1` in this case), in addition to the relevant steps.

For example, when rank is 0 the relevant steps are when i is equal to rank, right, and left, in this order. So we insert the mark

```
@foreach_steps(rank, right, left, 0, procs-1)
```

just before the code block in lines 19–22. For rank equal to `size-1` the relevant steps are the `right`, the `left`, and the `rank`, again in this order. The required mark at line 23 is

```
@foreach_steps(right, left, rank, 0, procs-1)
```

and the annotation to include in line 28 is

```
@foreach_steps(left, rank, right, 0, procs-1).
```

Figure 16.6 presents the marked version of the program in full.

```
 1   int main(int argc, char** argv) {
 2     int rank, procs, n; // process rank; number of processes; problem size
 3     ...
 4     MPI_Init(&argc, &argv);
 5     MPI_Comm_rank(MPI_COMM_WORLD, &rank);
 6     MPI_Comm_size(MPI_COMM_WORLD, &procs);
 7     @apply(NUM_ITER)
 8     if (rank == 0) {
 9       n = read_problem_size(procs);
10       read_vector(work, n);
11     }
12     MPI_Bcast(&n, 1, MPI_INT, 0, MPI_COMM_WORLD);
13     int local_n = n / procs;
14     MPI_Scatter(work, local_n, MPI_FLOAT, &local[1], local_n, MPI_FLOAT, ...);
15     int left  = rank == 0 ? procs - 1 : rank - 1; // left neighbour
16     int right = rank == procs - 1 ? 0 : rank + 1; // right neighbour
17     for (iter=1; iter<=NUM_ITER; iter++) @foreach(iter, 1, NUM_ITER) {
18       ...
19       if (rank == 0) @foreach_steps(rank, right, left, 0, procs-1) {
20         MPI_Send(&local[1],         1, MPI_FLOAT, left, ...);
21         MPI_Send(&local[local_n],   1, MPI_FLOAT, right, ...);
22         MPI_Recv(&local[local_n+1], 1, MPI_FLOAT, right, ...);
23         MPI_Recv(&local[0],         1, MPI_FLOAT, left, );
24       }else if (rank == procs-1) @foreach_steps(right, left, rank, 0, procs-1)
25       {
26         MPI_Recv(&local[local_n+1], 1, MPI_FLOAT, right, ...);
27         MPI_Recv(&local[0],         1, MPI_FLOAT, left, ...);
28         MPI_Send(&local[1],         1, MPI_FLOAT, left, ...);
29         MPI_Send(&local[local_n],   1, MPI_FLOAT, right, ...);
30       } else @foreach_steps(left, rank, right, 0, procs-1) {
31         MPI_Recv(&local[0],         1, MPI_FLOAT, left, ...);
32         MPI_Send(&local[1],         1, MPI_FLOAT, left, ...);
33         MPI_Send(&local[local_n],   1, MPI_FLOAT, right, ...);
34         MPI_Recv(&local[local_n+1], 1, MPI_FLOAT, right, ...);
35       }
36       ...
37     }
38     MPI_Reduce(&localerr, &globalerr, 1, MPI_FLOAT, MPI_MAX, 0, ...);
39     MPI_Gather(&local[1], local_n, MPI_FLOAT, work, local_n, MPI_FLOAT,0,...);
40     ...
41     MPI_Finalize();
42     return 0;
43   }
```

Figure 16.6   The code of Figure 16.2 with verification marks inserted.

16.6 Related Work

There are different aims and different methodologies for the verification of MPI programs [6]. The verification of interaction-based properties typically seeks to establish the absence of deadlocks and otherwise ill-formed communications among processes (e.g., compatible arguments at both ends in a point-to-point communication, in close relation to type checking safe communication). Several tools exist with this purpose, either for static or runtime verification, usually employing techniques from the realm of model checking and/or symbolic execution. All these tools are hindered by the inherent scalability and state-explosion problems. Notable examples include CIVL [19], DAMPI [21], ISP [15], MOPPER [2], MUST [7], and TASS [20].

In contrast to these tools, PARTYPES follows a deductive verification approach with the explicit aim of attaining scalable verification. A previous work [10] conducts a comparative evaluation by benchmarking PARTYPES against three state-of-the-art tools: ISP [15], a runtime verifier that employs dynamic partial order reduction to identify and exercise significant process interleavings in an MPI program; MUST [7], also a runtime verifier, but that employs a graph-based deadlock detection approach; and TASS [20], a static analysis tool based on symbolic execution. For the tools and the programs considered, PARTYPES runs in a constant time (the tool is insensitive to the number of processes, problem size, and other parameters), in clear contrast to the running time of all the other tools, which exhibited exponential growth in a significant number of cases.

In addition to PARTYPES, the theory of multi-party session types [9] inspired other works in the realm of message-passing programs and MPI in particular. Scribble [8, 22] is a language to describe global protocols for a finite set of participants in message-passing programs using point-to-point communication. Through a notion of projection, a local protocol can be derived for each participant from a global Scribble protocol. Programs based on the local protocols can be implemented using standard message-passing libraries, as in Multiparty Session C [13]. Pabble [12], an extension of Scribble, is able to express interaction patterns of MPI programs where the number of participants in a protocol is decided at runtime, rather than fixed a priori, and was used to generate safe-by-construction MPI programs [11].

In comparison to these works, PARTYPES is specifically aimed at protocols for MPI programs and the verification of the compliance of arbitrary programs against a given protocol. In conceptual terms, we address collective communication primitives in addition to plain point-to-point communication, and require no explicit notion of protocol projection.

16.7 Conclusion

This chapter presents PARTYPES, a type-based methodology to statically verify message-passing parallel programs. By checking that a program follows a given protocol, one guarantees a series of important safety properties, in particular that the program does not run into deadlocks. In contrast to other state-of-the-art approaches that suffer from scalability issues, our approach is insensitive to parameters such as the number of processes, problem size, or the number of iterations of a program.

The limitations of PARTYPES can be discussed along two dimensions:

- Even though PARTYPES addresses the core features of MPI, it leaves important primitives uncovered. These include non-blocking operations and wildcard receive (the ability to receive from any source), among many others.
- Our methodology is sound (in the sense that it does not yield false positives) but too intentional at times. For instance, it requires protocol loops and source code loops to be perfectly aligned, while the type theory [10] allows more flexibility, loop unrolling in particular.

Acknowledgements This work was supported by FCT (projects Advanced Type Systems for Multicore Programming PTDC/EIA–CCO/12254, HYRAX CMUP-ERI/FIA/0048/2013, and the LaSIGE Research Unit UID/CEC/00408/2013), the NORTE 2020 program (project SMILES, NORTE–01–0145–FEDER–000020), EPSRC (projects EP/K034413/1, EP/K011715/1, EP/L00058X/1, and EP/N027833/1), and EU (UPSCALE FP7 612985, and Betty COST Action IC1201).

References

[1] E. Cohen, M. Dahlweid, M. Hillebrand, D. Leinenbach, M. Moskal, T. Santen, W. Schulte, and S. Tobies. VCC: A practical system for verifying concurrent C. In *TPHOLs*, volume 5674 of *LNCS*, pages 23–42. Springer, 2009.

[2] V. Forejt, D. Kroening, G. Narayanswamy, and S. Sharma. Precise predictive analysis for discovering communication deadlocks in MPI programs. In *FM*, volume 8442 of *LNCS*, pages 263–278. Springer, 2014.

[3] MPI Forum. *MPI: A Message-Passing Interface Standard—Version 3.0*. High-Performance Computing Center Stuttgart, 2012.

[4] I. Foster. *Designing and building parallel programs*. Addison-Wesley, 1995.

[5] X. Fu, Z. Chen, H. Yu, C. Huang, W. Dong, and J. Wang. Symbolic execution of mpi programs. In *ICSE*, pages 809–810. IEEE Press, 2015.

[6] G. Gopalakrishnan, R. M. Kirby, S. F. Siegel, R. Thakur, W. Gropp, E. Lusk, B. R. De Supinski, M. Schulz., and G. Bronevetsky. Formal analysis of MPI-based parallel programs. *CACM*, 54(12):82–91, 2011.

[7] T. Hilbrich, J. Protze, M. Schulz, B. R. de Supinski, and M. S. Müller. MPI runtime error detection with MUST: advances in deadlock detection. In *SC*, pages 30:1–30:11. IEEE/ACM, 2012.

[8] K. Honda, R. Hu, R. Neykova, T. C. Chen, R. Demangeon, P. Denielou, and N. Yoshida. Structuring communication with session types. In *COB*, volume 8665 of *LNCS*, pages 105–127. Springer, 2014.

[9] K. Honda, N. Yoshida, and M. Carbone. Multiparty asynchronous session types. In *POPL*, pages 273–284. ACM, 2008.

[10] H. A. López, E. R. B. Marques, F. Martins, N. Ng, C. Santos, V. T. Vasconcelos, and N. Yoshida. Protocol-based verification of message-passing parallel programs. In *OOPSLA*, pages 280–298. ACM, 2015.

[11] N. Ng, J. G. F. Coutinho, and N. Yoshida. Protocols by default: Safe MPI code generation based on session types. In *CC*, volume 9031 of *LNCS*, pages 212–232. Springer, 2015.

[12] N. Ng and N. Yoshida. Pabble: parameterised scribble. *Service Oriented Computing and Applications*, 9(3–4):269–284, 2015.

[13] N. Ng, N. Yoshida, and K. Honda. Multiparty Session C: Safe parallel programming with message optimisation. In *TOOLS Europe*, volume 7304 of *LNCS*, pages 202–218. Springer, 2012.

[14] Partypes homepage. http://gloss.di.fc.ul.pt/ParTypes.

[15] S. Pervez, G. Gopalakrishnan, R. M. Kirby, R. Palmer, R. Thakur, and W. Gropp. Practical model-checking method for verifying correctness of MPI programs. In *PVM/MPI*, volume 4757 of *LNCS*, pages 344–353. Springer, 2007.

[16] M. Schulz and B. R. de Supinski. A flexible and dynamic infrastructure for MPI tool interoperability. In *ICPP*, pages 193–202. IEEE, 2006.

[17] S. F. Siegel and G. Gopalakrishnan. Formal analysis of message passing. In *VMCAI*, volume 6538 of *LNCS*, pages 2–18. Springer, 2011.

[18] S. F. Siegel and L.F. Rossi. Analyzing BlobFlow: A case study using model checking to verify parallel scientific software. In *EuroPVM/MPI*, volume 5205 of *LNCS*, pages 274–282. Springer, 2008.

[19] S. F. Siegel, M. Zheng, Z. Luo, T. K. Zirkel, A. V. Marianiello, J. G. Edenhofner, M. B. Dwyer, and M. S. Rogers. CIVL: The concurrency intermediate verification language. In *SC*, pages 61:1–61:12. IEEE Press, 2015.

[20] S. F. Siegel and T. K. Zirkel. Loop invariant symbolic execution for parallel programs. In *VMCAI*, volume 7148 of *LNCS*, pages 412–427. Springer, 2012.

[21] A. Vo, S. Aananthakrishnan, G. Gopalakrishnan, B. R. de Supinski, M. Schulz, and G. Bronevetsky. A scalable and distributed dynamic formal verifier for MPI programs. In *SC*, pages 1–10. IEEE, 2010.

[22] N. Yoshida, R. Hu, R. Neykova, and N. Ng. The Scribble protocol language. In *TGC*, volume 8358 of *LNCS*, pages 22–41. Springer, 2013.

Index

About the Editors

Simon Gay received his Ph.D. from Imperial College London and is now Professor of Computing Science at the University of Glasgow. He is Director of Research in the School of Computing Science, and leader of the Formal Analysis, Theory and Algorithms research section. From 2012 to 2016 he was Chair of COST Action IC1201: Behavioural Types for Reliable Large-Scale Software Systems.

António Ravara received his Ph.D. from the Technical University of Lisbon and is now Assistant Professor of Informatics at the NOVA University of Lisbon. He is a founding member of the NOVA Laboratory for Computer Science and Informatics (NOVA LINCS), which is a leading Portuguese research unit in the area of Computer Science and Engineering. From 2012 to 2016 he was Vice-Chair of COST Action IC1201: Behavioural Types for Reliable Large-Scale Software Systems.